ANALYTICAL CHEMISTRY SYMPOSIA SERIES — volume 19

analytical spectroscopy

Proceedings of the 26th Conference on Analytical Chemistry in Energy Technology, Knoxville, TN, October 11—13, 1983

ANALYTICAL CHEMISTRY SYMPOSIA SERIES — volume 19

analytical spectroscopy

Proceedings of the 26th Conference on Analytical Chemistry in Energy Technology, Knoxville, TN, October 11–13, 1983

edited by
W.S. Lyon

Analytical Chemistry Division, Oak Ridge National Laboratory,
Oak Ridge, TN, U.S.A.

Sponsored by the Analytical Chemistry Division, Oak Ridge National Laboratory, operated by Union Carbide Corporation for the Department of Energy under Contract No. W-7405-eng-26.

ELSEVIER
Amsterdam — Oxford — New York — Tokyo   1984

ELSEVIER SCIENCE PUBLISHERS B.V.
Molenwerf 1
P.O. Box 211, 1000 AE Amsterdam, The Netherlands

Distributors for the United States and Canada:

ELSEVIER SCIENCE PUBLISHING COMPANY INC.
52, Vanderbilt Avenue
New York, NY 10017, U.S.A.

Library of Congress Cataloging in Publication Data

Conference on Analytical Chemistry in Energy Technology
    (26th : 1983 : Knoxville, Tenn.)
    Analytical spectroscopy.

    (Analytical chemistry symposia series ; v. 19)
    Includes index.
    1. Spectrum analysis--Congresses. I. Lyon, W. S.
(William S.) II. Title. III. Series.
QD95.C633 1983   543'.0858 -   84-3975
ISBN 0-444-42312-5

ISBN 0-444-42312-5 (Vol. 19)
ISBN 0-444-41786-9 (Series)

© Elsevier Science Publishers B.V., 1984
All rights reserved. No part of this publication may be reproduced, stored in a retrieval system or transmitted in any form or by any means, electronic, mechanical, photocopying, recording or otherwise, without the prior written permission of the publisher, Elsevier Science Publishers B.V., P.O. Box 330, 1000 AH Amsterdam, The Netherlands

Printed in The Netherlands

CONTENTS

Analytical Chemistry Symposia Series XI
Preface XIII
Conference Committee XIV

LASERS

Recent Applications of Lasers in Analytical Chemistry, *F.E. Lytle* 3

Determination of Uranium in Plutonium Solutions by Laser Induced
    Fluorescence, *J.E. Young, P.T. Deason* 7

Instrumentation for Remote Sensing Over Fiber Optics, *T. Hirschfeld,*
    G. Haugen, F. Milanovich 13

Fiber Optics Temperature and Pressure Probe, *T. Hirschfeld, F. Wang,*
    G. Haugen, G. Hieftje 19

On-Line Uranium Determination Using Remote Fiber Fluorimetry,
    R.A. Malstrom, T. Hirschfeld 25

Versatility of Laser Induced Fluorescence Line Narrowing in Chemical
    Analysis, *M.J. McGlade, J.M. Hayes, G.J. Small, V. Heisig,*
    A.M. Jeffrey 31

Matrix-Isolation Photoacoustic Spectroscopy, *R.W. Shaw, H.E. Howell*
    G. Mamantov, E.L. Wehry 37

Selective Fluorometric Determinations by Laser-Induced Matrix Isolation
    Fluorescence Spectrometry, *E.L. Wehry* 43

Metal Surface Enhanced Fluorescence, *T.D. Harris, A.M. Glass, D.H. Olson*   49

Prospects for Single-Molecule Detection in Liquids by Laser-Induced
    Fluorescence, *M. Trkula, N.J. Dovichi, J.C. Martin, J.H. Jett,*
    R.A. Keller 53

Kinetic Analysis of Laser Induced Phosphorescence in Uranyl Phosphate
    for Improved Analytical Measurements, *B.A. Bushaw* 57

Performance Appraisal Studies of Laser-Enhanced Ionization in Flames --
The Determination of Nickel in Petroleum Products, *G.C. Turk,
G.J. Havrilla, J.D. Webb, A.R. Forster* 63

Laser Spectroscopic Studies of Solvation Phenomena, *M.J. Wirth* 69

Laser Spectroscopic Studies of Vibrational Overtones. Influence of
Molecular Conformation on Overtone Energies, *H.L. Fang, R.L. Swofford* 75

Infrared Laser Spectroscopy Using a Long Pathlength Absorption Cell,
K.C. Kim, R.A. Briesmeister... 81

MASS SPECTROMETRY

Some Advances in Inorganic Analytical Mass Spectrometry, *H.J. Svec* 89

A State-of-the-Art Mass Spectrometer System for Determination of Uranium
and Plutonium Isotopic Distributions in Process Samples, *C.A. Polson* .. 103

Americium Determination Using an Isotope Dilution Mass Spectrometry
Technique, *D.W. Crawford, R.D. Peavy* 111

An In-Line Analyzer for Monitoring Gas Composition in Tritium
Purification Processes - Design, *P. Chastagner* 119

Application of Mass Spectrometry to Fuels and Materials Testing at FFTF,
C.E. Plucinski, M.W. Goheen, J.J. McCown 127

Improved Precision and Speed of Thermal Ionisation Mass Spectrometry
With a Multicollector, *P.J. Turner, J.E. Cantle, R.C. Haines* 133

Resonance Ionization Mass Spectrometry of Iron--Quantitative Aspects,
J.D. Fassett, L.J. Moore, J.C. Travis 137

Isotopic Measurements of Uranium and Plutonium by Resonance Ionization
Mass Spectrometry, *D.L. Donohue, J.P. Young, D.H. Smith* 143

Sputter Initiated Resonance Ionization Spectroscopy for Trace Element
Analysis, *J.E. Parks, H.W. Schmitt, G.S. Hurst, W.M. Fairbank, Jr.* 149

Resonance Ionization Mass Spectrometry at Los Alamos National Laboratory, *N.S. Nogar, S.W. Downey, R.A. Keller, C.M. Miller* 155

The Determination of Uranium in Urine by Isotope Dilution Mass Spectrometry Using Resin Bead Loading, *M.P. May, R.L. Walker, T.G. Scott, F.F. Dyer, J.R. Stokely* 161

SIMS Study of Compositional Changes Observed in a PuO_2 Heat Source Cladding Alloy, *W.H. Christie, D.H. Taylor, R.E. Eby, D. Pavone* 167

Laser Enhanced Ionization in Glow Discharge Mass Spectrometry, *W.W. Harrison, P.J. Savickas, R.K. Marcus, K.R. Hess* 173

Examination of Excited State Populations in Sputtering Using Multiphoton Resonance Ionization, *F.M. Kimock, J.P. Baxter, D.L. Pappas, P.H. Kobrin, N. Winograd* 179

PLASMA

A Preliminary Study on the Determination of Boron and Cadmium Impurities in Uranium Oxide and Uranium Metals by Inductively Coupled Plasma-Atomic Emission Spectroscopy, *B.R. Bear, M.C. Edelson, B. Gopalan, V.A. Fassel* 187

Analyses of Nuclear Fuel and High-Level Waste by Inductively Coupled Plasma-Emission Spectroscopy, *C.J. Coleman* 195

Determination of Trace Impurities in Uranium Matrices by Inductively Coupled Plasma-Atomic Emission Spectrometry, *A.A. Halouma, R.B. Farrar, E.A. Hester, R.W. Morrow* 201

Measurement of Rhodium and Palladium in Selected Hanford Defense Wastes, *M.H. West, C.S. Homi, C.H. Hunter* 203

Considerations for the Installation of an Inductively Coupled Plasma for the Analysis of Radioactive Samples, *C.T. Apel, D.L. Gallimore* 213

Evolution of Containment Facilities for Spectroscopic Analysis at Rockwell Hanford Operations, *J.M. Hiller* 219

A New Facility for High Resolution Inductively Coupled Plasma-Atomic
Multielement Emission Spectroscopy of Radioactive Materials,
M.C. Edelson, V.A. Fassel 227

Plans for Inductively Coupled Plasma Atomic Emission Spectroscopy
(ICP-AES) Analysis of Impurities in Plutonium Materials at Rocky Flats,
C.E. Michel, G.E. Brown 235

NUCLEAR

Utilization of the Intense Pulsed Neutron Source (IPNS) at Argonne
National Laboratory for Neutron Activation Analysis, *R.R. Heinrich,
L.R. Greenwood, R.J. Popek, A.W. Schulke, Jr.* 243

Heavy-Ion Induced X-Ray Satellite Emission as a Chemical Probe,
*T.M. Rosseel, J.M. Dale, L.D. Hulett, H.F. Krause, S. Raman, C.R. Vane,
J.P. Young* 249

Analytical Measurements of Actinide Migration in a Laboratory-Simulated
Basalt HLW Repository, *D.L. Bowers, T.J. Gerding, S.M. Fried,
G.F. Vandegrift, M.G. Seitz* 255

Analytical Research at the Institute of Radiochemistry, Nuclear Research
Center, Karlsruhe, FRG, *E. Gantner, E. Mainka, H. Ruf, H.J. Ache* 263

Comparison of Burnup Methods for $(Th,U)O_2$ Fuel, *L.W. Green, R.M. Cassidy,
W.J. Edwards, C.H. Knight* 273

Irradiation and Gamma-Ray Spectrometric Parameters for ^{129}I Neutron
Activation Analysis, *J.H. Kaye, R.S. Strebin, Jr., F.P. Brauer,
W.A. Mitzlaff* 279

Monitoring of Radionuclides and Metallic Impurities in the KNK-II Primary
Sodium, *H.H. Stamm, K.Ch. Stade* 285

Surface Characterization of Leadscrews Taken from the TMI-2 Reactor
Vessel, *K.J. Hofstetter, H. Lowenschuss, V.F. Baston* 293

An n-Type High Purity Germanium Detector-Based On-Line Rapid Coal Analysis
System, *F.E. LeVert, W.W. Managan* 299

Automated Gamma Spectrometry and Data Analysis on Radiometric Neutron
Dosimeters, *W.Y. Matsumoto* 309

Environmental Applications for an Intrinsic Germanium Well Detector,
P. Stegnar, J.S. Eldridge, N.A. Teasley, T.W. Oakes 313

OTHER SPECTROSCOPIC TECHNIQUES

Sampling and Characterization of Aerosols Formed in the Atmospheric
Hydrolysis of UF_6, *W.D. Bostick, W.H. McCulla, P.W. Pickrell,
D.A. Branam* 321

Applications of Laser Spectroscopy for Species Determination in Fission
Product Release Experiments, *W.H. McCulla, G.E. Nelson, R.A. Lorenz* ... 327

Spectroscopic Diagnostics of an Argon Plasma Jet Seeded with Uranium,
G.K. Anderson 333

General-Purpose Chemical Analyzer for Online Analyses of Radioactive
Solutions, *W.A. Spencer, J.W. Kronberg* 337

Influence of Uranyl Dibutylphosphate on the UV/VIS Spectrophotometric
Online Monitoring of Uranium in Tributylphosphate/Hydrocarbon Solvent,
E.T. Creech, A.C. Rutenberg, R.W. Smithwick, R.D. Seals 343

Controlled-Potential-Coulometric Determination of Uranium at a Platinum
Electrode, *N.M. Saponara, D.D. Jackson* 345

Atomic Absorption Spectrophotometric Methods for the Determination of
Phosphorous and Silicon in Steel, *J.P. McCarthy, E.B. Nunn, C. Kinard* 349

Analysis of Molecular Hydrogen Using Spontaneous Raman Spectroscopy,
K. Veirs, G.M. Rosenblatt 355

Diffuse Reflectance Infrared Fourier Transform Spectroscopy: A
Versatile and Promising Analytical Technique for Solids, *N.R. Smyrl,
E.L. Fuller, Jr., G.L. Powell* 357

Diode Laser Spectrometry for Monitoring Acetylene On-Line in a
Liquefaction Plant, *W.L. Maddox, C.M. Turner* 359

Optical Measurement of SO$_2$ in Combustion Environments, *D. Lucas, N.J. Brown* 361

Quantitative Determination of Hydrogen by Pulsed Nuclear Magnetic Resonance Spectroscopy, *A. Attalla, R.C. Bowman, Jr.* 363

Applications of Fourier Self-Deconvolution, *R.L. Howell* 371

Capillary Supercritical Fluid Chromatography - Fourier Transform Infrared Spectrometry, *S.V. Olesik, S.B. French, M. Novotny* 373

Supercritical Fluid Methods in Analytical Chemistry, *R.D. Smith, B.W. Wright, H.R. Udseth* 375

Air-Stable Reference Material for Measurement of the Oxygen-to-Metal (O/M) Ratio of Nuclear Fuel, *C.F. Hammond, R.E. Honnell, J.E. Rein* 381

Index 387

ANALYTICAL CHEMISTRY SYMPOSIA SERIES

Volume 1 Recent Developments in Chromatography and Electrophoresis. Proceedings of the 9th International Symposium on Chromatography and Electrophoresis, Riva del Garda, May 15–17, 1978
edited by A. Frigerio and L. Renoz

Volume 2 Electroanalysis in Hygiene, Environmental, Clinical and Pharmaceutical Chemistry. Proceedings of a Conference, organised by the Electroanalytical Group of the Chemical Society, London, held at Chelsea College, University of London, April 17–20, 1979
edited by W.F. Smyth

Volume 3 Recent Developments in Chromatography and Electrophoresis, 10. Proceedings of the 10th International Symposium on Chromatography and Electrophoresis, Venice, June 19–20, 1979
edited by A. Frigerio and M. McCamish

Volume 4 Recent Developments in Mass Spectrometry in Biochemistry and Medicine, 6. Proceedings of the 6th International Symposium on Mass Spectrometry in Biochemistry and Medicine, Venice, June 21–22, 1979
edited by A. Frigerio and M. McCamish

Volume 5 Biochemical and Biological Applications of Isotachophoresis. Proceedings of the First International Symposium, Baconfoy, May 4–5, 1979
edited by A. Adam and C. Schots

Volume 6 Analytical Isotachophoresis. Proceedings of the 2nd International Symposium on Isotachophoresis, Eindhoven, September 9–11, 1980
edited by F.M. Everaerts

Volume 7 Recent Developments in Mass Spectrometry in Biochemistry, Medicine and Environmental Research, 7. Proceedings of the 7th International Symposium on Mass Spectrometry in Biochemistry, Medicine and Environmental Research, Milan, June 16–18, 1980
edited by A. Frigerio

Volume 8 Ion-selective Electrodes, 3. Proceedings of the Third Symposium, Mátrafüred, Hungary, October 13–15, 1980
edited by E. Pungor

Volume 9 Affinity Chromatography and Related Techniques. Theoretical Aspects/Industrial and Biomedical Applications. Proceedings of the 4th International Symposium, Veldhoven, The Netherlands, June 22–26, 1981
edited by T.C.J. Gribnau, J. Visser and R.J.F. Nivard

Volume 10 Advances in Steroid Analysis. Proceedings of the Symposium on the Analysis of Steroids, Eger, Hungary, May 20–22, 1981
edited by S. Görög

Volume 11 Stable Isotopes. Proceedings of the 4th International Conference, Jülich, March 23–26, 1981
edited by H.-L. Schmidt, H. Förstel and K. Heinzinger

Volume 12   Recent Developments in Mass Spectrometry in Biochemistry, Medicine and Environmental Research, 8. Proceedings of the 8th International Symposium on Mass Spectrometry in Biochemistry, Medicine and Environmental Research, Venice, June 18–19, 1981
edited by A. Frigerio

Volume 13   Chromatography in Biochemistry, Medicine and Environmental Research, 1. Proceedings of the 1st International Symposium on Chromatography in Biochemistry, Medicine and Environmental Research, Venice, June 16–17, 1981
edited by A. Frigerio

Volume 14   Chromatography and Mass Spectrometry in Biomedical Sciences, 2. Proceedings of the International Conference on Chromatography and Mass Spectrometry in Biomedical Sciences, Bordighera, Italy, June 20–23, 1982
edited by A. Frigerio

Volume 15   Computer Applications in Chemistry. Proceedings of the 6th International Conference on Computers in Chemical Research and Education (ICCCRE), held in Washington, DC, July 11–16, 1982
edited by S.R. Heller and R. Potenzone, Jr.

Volume 16   New Approaches in Liquid Chromatography. Proceedings of the 2nd Annual American-Eastern European Symposium on Advances in Liquid Chromatography, Szeged, Hungary, June 16–18, 1982
edited by H. Kalász

Volume 17   Chemical Sensors. Proceedings of the International Meeting on Chemical Sensors, held in Fukuoka, Japan, September 19–22, 1983
edited by T. Seiyama, K. Fueki, J. Shiokawa and S. Suzuki

Volume 18   Modern Trends in Analytical Chemistry. Part A: Electrochemical Detection in Flow Analysis; Proceedings of the Scientific Symposium, Matrafüred, Hungary, October 17–20, 1982. Part B: Pattern Recognition in Analytical Chemistry; Proceedings of the Scientific Symposium, Matrafüred, Hungary, October 20–22, 1982
edited by E. Pungor, G.E. Veress and I. Buzás

Volume 19   Analytical Spectroscopy. Proceedings of the 26th Conference on Analytical Chemistry in Energy Technology, Knoxville, TN, October 11–13, 1983
edited by W.S. Lyon

Volume 20   Topics in Forensic and Analytical Toxicology. Proceedings of the Annual Meeting of the International Association of Forensic Toxicologists, Munich, August 21–25, 1983
edited by R.A.A. Maes

PREFACE

The 26th Conference focused its attention on two relatively hot fields of analytical chemistry, lasers and mass spectrometry, and two pertinent applied areas, plasma and nuclear spectroscopy. The large number of papers were well received by the over 200 attendees, and the reader will find ample reason herein to confirm the collective judgement of the committee and attendees that this was an unusually successful conference.

Credit for the success of the meeting must go primarily to the participants: authors who presented papers and audience who listened and responded. This year's format, different in several ways, was originally suggested by W. D. Shults, Director, Analytical Chemistry Division.

This proceedings volume is the joint effort of authors, committee and editor. All played a part in its production. But special recognition and thanks are due Gail Vineyard, the proceedings secretary, who handled the details of manuscript receipt, proofreading, error correction, and final preparation including table of contents and index typing. Her assistance and helpful suggestions are reflected in many ways in this volume.

W. S. Lyon

CONFERENCE COMMITTEE

A. L. Harrod, General Chairman

W. S. Lyon, Technical Program Chairman

M. L. Emery, Treasurer

S. Cates, P. Mullins, S. Ranney
Conference Secretaries

G. Vineyard
Proceedings Secretary

M. V. Buchanan
F. F. Dyer
D. R. Heine
W. R. Laing
J. M. Ramsey
W. D. Shults
D. H. Smith

LASERS

RECENT APPLICATIONS OF LASERS IN ANALYTICAL CHEMISTRY

F. E. LYTLE
Department of Chemistry, Purdue University, West Lafayette, IN 47907

ABSTRACT

This review primarily will update two recent articles [ref. 1,2] concerning the use of lasers in analytical fluorimetry. The discussion will focus on the use of instrumental techniques developed to improve the working detection limit via increases in selectivity. Examples will include chromatography, time resolution and line narrowing. In addition, the topic of multiphoton ionization/mass spectrometry will be covered.

INTRODUCTION

One of the major accomplishments of analytical laser spectroscopy in the last decade has been the development of multiphoton ionization and fluorescence strategies useful for the detection of single atoms and/or molecules. Presently, the successful application of such methodology involves situations where the analyte appears in a very well controlled environment. As such, the next logical task involves the development of analytical strategies possessing higher degrees of selectivity. This would permit the direct, trace level determination of one or more components in an arbitrarily complicated mixture.

CHEMICAL AND CHROMATOGRAPHIC SPECIFICITY

The primary difficulty with most optical methods of determining molecular identity or concentration via absorption is the general lack of high resolution spectral features. As such the technique is susceptible to interference by almost any other molecule with similar or greater numbers of conjugated double bonds. Additionally, as demands are placed on methodology to handle decreasing amounts of material, other much weaker optical processes begin to interfere as the concentration of the normally passive matrix becomes drastically larger than the analyte. For a numerical example of the first type of interference consider an instrument having a lower limit of detection for some arbitrary analyte of 10^{-12} for the molarity-quantum yield product. This means that a compound with the same molar absorptivity and a quantum yield of 10^{-4} (normally considered non-fluorescent) would provide an interference equal to the signal when its concentration is only 10^{-8} M! Additionally, for the example given, the concentration differential between the solvent and the analyte is about

10^{13} producing Rayleigh scatter $\sim 10^6$ times as large as the fluorescence and solvent Raman about 100 times as large.

As a result of the blank limited nature of the fluorescence measurement, all successful laser based trace analyses have employed additional forms of selectivity. Table 1 gives some typical examples utilizing either chemical or chromatographic approaches.

TABLE 1

Typical lower limits of detection for laser based methodology.

Compound	Laser	LLD (ng mL^{-1})	Reference
aflatoxins	N_2	0.065	3
anthracene	N_2/dye/SHG	<0.004	4
carprofen	N_2/dye/SHG	2.5	5
fluoranthene	Kr^+	0.002	6
glucose-6-phosphate	He:Cd	0.52	7
insulin	Ar^+	0.4	8
rhodamine 6G	Ar^+	6.7×10^{-5}	9
zearalenone	Ne:Cd	5 ppb corn	10

The results for anthracene and rhodamine 6G represent analyses involving pure compounds in pure solvents. In other words, the specificity is provided by prior information about the chemical system. The rhodamine case is of particular interest since it involved hydrodynamic focusing in an attempt to reduce the scatter background while simultaneously providing a small sample volume. The fluoranthene case involves mixtures of pure compounds injected onto an HPLC column, while the aflatoxins were airborne samples collected and separated with TLC. The glucose-6-phosphate determination had added specificity due to an enzyme, while the insulin determinations used an immunoassay approach. Finally, the carprofen and zearalenone examples involved extensive chemical work-ups, and in the latter case HPLC.

LINE NARROWING AND TIME RESOLUTION

Line narrowing techniques reduce the width of the normally broad molecular bands by removing thermal effects and/or site inhomogeneities. In fluid solution this goal is achieved by using matrix isolation or Shpol'skii matrices to improve the uniformity of the solvation sites, and by the use of photoselection of a subset of sites in a very low temperature vitreous solvent. The results of such methodology were adequately covered in two previous reviews [ref. 1,2].

In the gas phase, spectral simplification is achieved by supersonically expanding a molecular beam and cooling the sample sufficiently to populate only the lowest rotational levels. In 1981, Small and co-workers [ref. 11] proposed using the technique as a detector coupled to a gas chromatograph. More recently, Hayes and Small [ref. 12] have experimentally demonstrated this technique by the quantitation and unambiguous identification of naphthalene, 1-methyl-naphthalene and 2-methyl-naphthalene in a sample of Wilmington crude oil. Additional experiments, not involving the use of a chromatograph, have been performed by Jortner and co-workers [ref. 13]. They demonstrated that supersonic expansion was useful for the identification of large organic molecules, sensitive detection of minor amounts of impurities of organic molecules, and isotopically selective spectroscopic analysis.

As with line narrowing techniques, the major goal of time-resolved fluorimetry is an increase in specificity. This can be achieved in two ways. First, the temporal behavior of the analyte can be utilized to distinguish its emission from other components of the sample. Second, the signal can purposely be caused to appear in one region of time while the major interferences will hopefully appear either in some other region of time or spread evenly over all time. The results of such methodology were adequately covered in two previous reviews [ref. 1,2].

MULTIPHOTON IONIZATION/MASS SPECTROMETRY

Multiphoton ionization/mass spectrometric (MPI/MS) studies of inorganic species have been pursued primarily to remove the isobaric interferences normally observed in isotope dilution mass spectrometry. The concept of the experiment is that the laser ionization will provide elemental selectivity, but not isotopic selectivity; while the high-precision mass spectrometer will provide only mass number selectivity due to its low resolution. The combination provides a high precision instrument with an elemental selectivity greater than 50,000. Three example systems will suffice. The Los Alamos group [ref. 14] has studied the determination of Lu and Yb in binary mixtures, the Oak Ridge group [ref. 15] has studied Nd and Sm mixtures, while the National Bureau of Standards group [ref. 16] has studied the ion formation of Mo, Re, and V.

Although several groups are actively working in the area of MPI/MS of organics, the extent that the laser will increase the overall selectivity is not yet clear. However, two varients show promise and the reader is referred to these papers and their included references. Lubman and Kronick [ref. 17] have used supersonic expansion to provide specific ionization of similar molecules. And, Reilly and co-workers [ref. 18] have combined MPI with GC/MS to provide detection limits as low as 200 fg for naphthalene.

REFERENCES

1. T. D. Harris and F. E. Lytle, in D. S. Kliger (Ed.), Ultrasensitive Laser Spectroscopy, Academic Press, N.Y., 1983, pp. 369-433.
2. F. E. Lytle, J. Chem. Ed., 59 (1983) 915-920.
3. M.K.L. Bickling, R. N. Kniseley, and H. J. Svec, Anal. Chem., 55 (1983) 200-204.
4. J. H. Richardson and M. E. Ando, Anal. Chem., 49 (1977) 955-959.
5. N. Strojny and J. A. F. deSilva, Anal. Chem., 52 (1980) 1554-1559.
6. S. Folestad, L. Johnson, and B. Josefsson, Anal. Chem., 54 (1982) 925-929.
7. T. Imasaka and R. N. Zare, Anal. Chem., 51 (1979) 2082-2085.
8. S. D. Lidofsky, T. Imasaka, and R. N. Zare, Anal. Chem., 51 (1979) 1602-1605.
9. N. J. Dovichi, J. C. Martin, J. H. Jett, and R. A. Keller, Science, 219 (1983) 845-847.
10. G. J. Diebold, N. Karny, R. N. Zare, and L. M. Seitz, J. Assoc. Off. Anal. Chem., 62 (1979) 564-569.
11. J. C. Brown, J. M. Hayes, J. A. Warren, and G. J. Small, in G. M. Hieftje, J. C. Travis, and F. E. Lytle (Eds), Lasers in Chemical Analysis, Humana Press, Clifton, N.J., 1981, Chapter 12.
12. J. M. Hayes and G. J. Small, Anal. Chem., 54 (1982) 1202-1204.
13. A. Amirov, U. Even, and J. Jortner, Anal. Chem., 54 (1982) 1666-1673.
14. C. M. Miller, N. S. Nogar, A. J. Gancarz, and W. R. Shields, Anal. Chem., 54 (1982) 2377-2378.
15. J. P. Young and D. L. Donohue, Anal. Chem., 55 (1983) 88-91.
16. J. D. Fassett, J. D. Travis, L. J. Moore, and F. E. Lytle, Anal. Chem., 55 (1983) 765-770.
17. D. M. Lubman and M. N. Kronick, Anal. Chem., 54 (1982) 660-665.
18. G. Rhodes, R. B. Opsal, J. T. Meek, and J. P. Reilly, Anal. Chem., 55 (1983) 280-286.

DETERMINATION OF URANIUM IN PLUTONIUM SOLUTIONS BY LASER INDUCED FLUORESCENCE

John E. Young and Paul T. Deason

Savannah River Plant, E. I. du Pont de Nemours and Company, Aiken, SC, (USA)

ABSTRACT

A highly sensitive method for determining uranium in plutonium solutions has been developed at the Savannah River Plant. It uses a pulsed nitrogen laser as an excitation source for a time resolved fluorimetric measurement. Since the sample matrix contains large quantities of fluorescence quenchers, sample purification is necessary to achieve 1 part per billion sensitivity. Liquid-liquid solvent extraction is used to separate uranium from the sample matrix into a fluorescence enhancing solution. The method has been successfully applied to PUREX process streams with a precision of $\pm 13\%$ (95% C.I.) with volumetric transfers.

INTRODUCTION

A common method for the determination of trace uranium is the sodium fluoride sintered pellet (SP) method. In the SP method, the sample is applied to a sodium fluoride pellet, sintered, and compared in a fluorimeter to a set of similarly prepared standards.[1] Since many ions interfere, liquid-liquid solvent extraction procedures have been developed to purify uranium prior to fluorimetric analysis.[2] The use of the solvent extraction SP method has been used for many years as a selective method for uranium. Some of the problems encountered with the extraction SP method include lengthy analysis time, poor precision, and an inadequate limit of detection (80 ppb).

Recently, P. G. Whitkop described a technique for the determination of uranium in aqueous samples by laser induced fluorescence.[3] The method was developed as an alternative to the SP method, primarily to achieve higher sensitivity. First the uranium is separated from interfering ions by solvent extraction into tri n-butyl phosphate. Then the uranium is back-extracted into the fluorescence enhancing solution, molar ortho-phosphoric acid.

This report describes the development and implementation of the laser technique for the analysis of trace uranium in PUREX plutonium solutions.

EXPERIMENTAL

The instrument used in this development was a Scintrex UA-3 uranium analyzer, manufactured by Scintrex, Inc., Concord, Ontario, Canada. The excitation source is a pulsed nitrogen laser filtered for the 337 nm line output. The instrument has been described elsewhere in detail.[4] The most common application of this instrument is the determination of trace uranium in aqueous samples for geochemical exploration. For the work described below, the instrument was set up in a contained radiobench for the analysis of radioactive samples.

Reagents

Synthetic plutonium standards were prepared by adding uranyl nitrate to solutions that were similar in composition to PUREX process solutions. These standards contained approximately 2 g/l of plutonium and from 20-500 ppb of uranium. The 20 volume % TBP in isooctane was prepared by diluting 200 ml of TBP to 1 liter with isooctane, and then scrubbing with 5M sodium carbonate to remove mono- and di-butyl phosphoric acids. The purified extractant was then equilibrated with 3M HNO_3.

Procedure

To a 2-dram vial, 1 ml of sample, 2 ml of 4M HNO_3, 0.1M $Al(NO_3)_3$, 0.2 ml of ferrous sulfamate (F.S.). and 1 ml of 20% TBP/isooctane is added. The sample is stirred for 1 minute with a vortex stirrer. 0.5 ml of the organic phase is transferred to a 4-dram vial containing 5.5 ml of 1M H_3PO_4. The vial is shaken and the organic phase is removed with a medicine dropper. 5 ml of the H_3PO_4 is transferred to a 1 x 2 x 4 cm fluorescence cuvet and is stirred with a magnetic stirrer. The photomultiplier voltage is adjusted until the meter reads 20-40% of full scale. An extracted reagent blank is then inserted and adjusted to 0.00 with the bucking potential supply. The sample is then inserted and the fluorescence is recorded. The photomultiplier voltage is used to calculate the spike volume and concentration necessary for 60% scale deflection. The sample is spiked, stirred for 15 seconds, and the spike fluorescence is measured.

RESULTS AND DISCUSSION

Direct analysis of plutonium solutions by diluting into a fluorescence enhancing medium yields poor detection limits due to the strong quenching effect of plutonium. (See Figure 1.) Iron has exponential quenching effects similar to plutonium, whereas aluminum and nitric acid have linear quenching effects. (See Figure 2.) Mathematical models have been devised for these quenchers. The models can be used to determine optimum sample volume for a dilution method. For example, a 1 g/l Pu solution in 2M HNO_3 containing 100 ppb uranium would increase in fluorescence to a maximum, and then drop when the quenchers are more concentrated. (See Figure 3.) By using the minimum detectable fluorescence value and these quenching curves, a detection limit of 100 ppm U/Pu is calculated for the dilution method.

Although quenching effects can be compensated for by a standard addition technique, loss of sensitivity is incurred. However, with the extraction procedure, detection limits are preserved by separating the uranium from the quenching agents.

The solvent extraction reagents were selected for high uranium recovery, good separation from plutonium and iron, minimal quenching by nitric acid and the residual solvent, and rapid solvent phase separation. Ortho-phosphoric acid was chosen over pyrophosphate as the fluorescence enhancing medium for two reasons: 1) plutonium is more soluble in phosphoric acid, and 2) the phosphoric acid system is less sensitive to [HNO_3] variations. The TBP/isooctane extractant was chosen because it does not quench fluorescence, and it efficiently separates uranium from plutonium. Impurity mono-butyl and di-butyl phosphoric acids in TBP interfere with the method by two different mechanisms and must be removed. The alkyl phosphoric acids form exceptionally stable complexes with uranium which favor the organic phase when contacted with H_3PO_4.[5] This prevents uranium from being quantitatively transferred from the organic to the H_3PO_4 phase. The alkyl phosphoric acids also tend to emulsify the aqueous and organic phases during solvent extraction. The solvent is purified by contacting with an alkaline sodium carbonate solution, neutralizing and precipitating sodium alkyl phosphate salts.

The variable sensitivity technique, as described in the procedure, allows the analysis of solutions between 1 ppb and 10 ppm without the need for additional sample dilutions. This 10,000-fold operating range is particularly useful for samples of unknown uranium concentration. The standard addition technique will also compensate for changes in the high voltage power supply output, as well as correcting for quenching effects.

The precision and accuracy of the procedure over a 28-day period are illustrated in Figure 4. During the period, the relative standard deviation for U determination of 20-200 ppb U standards was 9% (N = 36). No bias or trends were observed during the measurement period.

This new technique is roughly 80 times more sensitive, and can be completed in less than half the time of the SP method. This has proven to be a valuable improvement in analytical measurement capabilities at the Savannah River Plant.

ACKNOWLEDGMENTS

The authors wish to thank P. G. Whitkop for development work, J. P. Clark for the preparation of standards and statistical evaluation of data, and H. P. Holcomb for technical assistance.

The information contained in this article was developed during the course of work under Contract No. DE-AC09-76SR00001 with the U.S. Department of Energy.

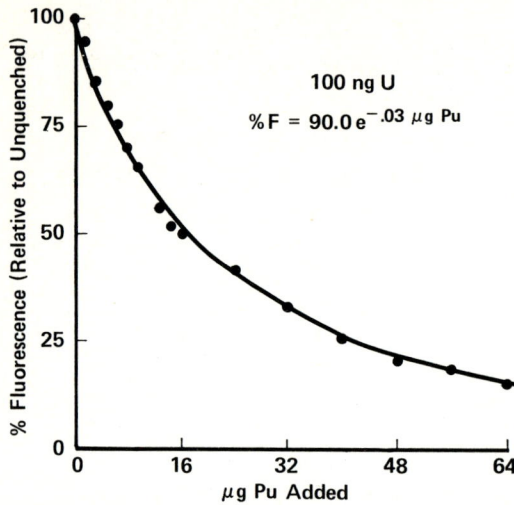

Fig. 1. Effect of Plutonium on the Fluorescence of Uranyl Phosphate.

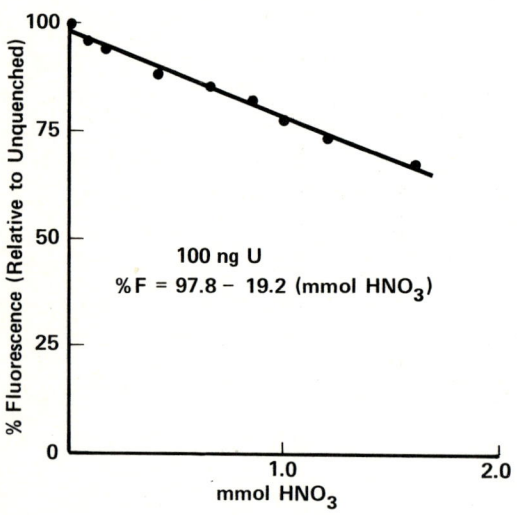

Fig. 2. Effect of Nitric Acid on the Fluorescence of Uranyl Phosphate.

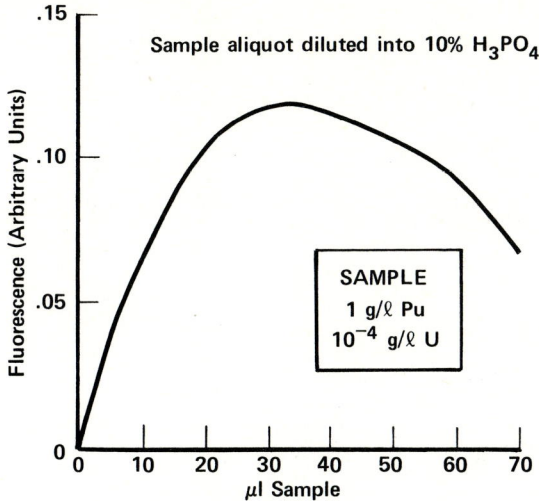

Fig. 3. Fluorescence as a Function of Sample Volume

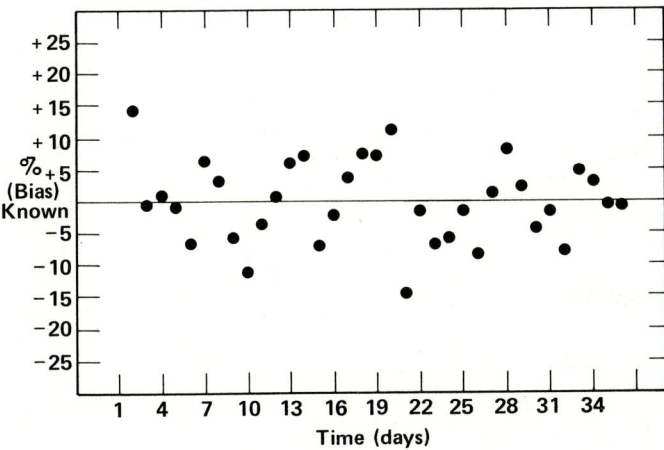

Fig. 4. Uranium in Plutonium Solutions by Laser Induced Fluorescence - Quality Control Data

LITERATURE CITED

1. C. J. Rodden, "Analysis of Essential Nuclear Reactor Materials," U.S. Atomic Energy Commission, 1964, Chapter 1.
2. W. J. Maeck, G. L. Booman, M. C. Elliot, J. E. Rein, Anal. Chem., 1958, 30, 1902.
3. P. G. Whitkop, Anal. Chem., 1982, 54, 2475.
4. J. C. Robbins, CIM Bull., 1978, 793, 61.
5. Y. Marcus, A. S. Kertes, "Ion Exchange and Solvent Extraction of Metal Complexes," Wiley-Interscience, 1969, Chapter 8.

INSTRUMENTATION FOR REMOTE SENSING OVER FIBER OPTICS

T. Hirschfeld, G. Haugen, and F. Milanovich
Lawrence Livermore National Laboratory, CA 94550

ABSTRACT

We have developed a way to extend the sensing and analytical abilities of the laser fluorescence spectrometer beyond the physical confines of the laboratory by means of communications-grade optical fibers. These fiber probes are extremely rugged, compared with sensitive laboratory equipment, and also extremely inexpensive. They make it possible to perform sensitive chemical analyses in hostile environments without risking damage to the laser and the spectrometer.

We have produced special-purpose optrodes that are sensitive to selected chemicals. With multiplexing, we can scan a number of fibers whose terminals are at widely scattered locations, gathering information in one central instrument without the expense and delay involved in manual sample-gathering.

We have begun development of a remote analyzer for monitoring rare-earth-ion migration in a nuclear-waste repository, an environment too hostile for any previous remote sensing device. We are also developing optrodes sensitive to a wide variety of non-chemical stimuli.

INTRODUCTION

Improvements in optical fibers now make it possible to send laser pulses and receive fluorescent light signals over distances of up to a kilometer. We have been developing laser fluorimetry as a laboratory tool of great versatility and sensitivity for chemical analyses, and these improved optical fibers enable the fluorimeter to remotely sense in environments too hostile, hazardous, or inaccessible for normal instrumentation. Besides having a wide range of possible applications in industry and research, remote fiber fluorimetric analysis may also be applied in multiple-site monitoring from a central location over a switchable fiber-optic network. We are developing an instrument based on remote analysis to be used as an _in situ_ monitor for rare-earth-ion transport in a nuclear-waste repository. We are also exploring further

---

*Work performed under the auspices of the U. S. Department of Energy by the Lawrence Livermore National Laboratory under contract number W-7405-ENG-48.

applications as advances in the technique are made possible by new developments in laser-induced fluorescence and in fiber-optical technology.

Modern technology is crucially dependent on analytical monitoring and control, but there are many cases in which this analysis is best done from a distance[1,2]. Locations such as an underground nuclear-waste-disposal site, or the working zone of a coal-liquifaction reactor are much too hostile for most in situ analytical devices.

Hitherto, these situations called for the development at considerable expense and delay of specially durably made instruments or when possible, of equally rugged sampling systems that could collect a representative sample and deliver it to the instrument without alteration. Instead of these very often extensively development programs, the possibility of designing instruments to perform analyses remotely appeared as a welcome alternative.

Analytical fluorescence spectroscopy

A fluorescence spectroscope functions by exposing a chemical sample to laser light and measuring the wavelengths re-radiated. The specific wavelengths are characteristic of the emitting substance, and their intensity is proportional to its concentration in the sample. Thus, in many cases the fluorescence spectroscope can simultaneously tell us what is in the sample and how much is present[3].

Existing fluorescence spectrometers require the sample to be present in an optical cell within the instrument. To convert the instrument for remote analyses, we first needed to find a way to transmit the fluorescent light back to the spectroscope efficiently. The simultaneous development of optical fibers for the telephone and computer industries and of laser-induced fluorescence for high-sensitivity chemical analysis made it possible for us to build such a remote analysis device.

The communications industry, faced with a need to transmit signals of greater bandwidth than can be accommodated on electrical cables, developed fiber-optic strands that can transmit light over distances of up to a kilometer with remarkably little alteration[4]. They also created a whole new technology of special connectors, multiple couplers, and input-output optics that made such "light wires" no harder to use than coaxial cable. Successive technological advances have enlarged the wavelength range of such fibers, until they now cover most of the visible and a good portion of the near infrared spectrum.

We have recently improved the sensitivity of analytical fluorescence spectroscopy by using lasers as an illumination source that is extremely bright, spectrally pure, and controllable[3]. We have so far demonstrated parts-per-trillion sensitivities in the detection of many chemicals. A further virtue of the laser, its ability to be focused within an extremely small spot, has made it possible to illuminate the small entrance aperture of a long-distance optical fiber and has suggested the present work.

Remote fiber fluorimetry

We first demonstrated the feasibility of using these technological developments by coupling separate fibers to a laser source and to a photomultiplier in the spectrometer[5]. We brought the far ends of the fibers together at an angle. The sensitive volume is defined by the overlap of the fields of view of the two fibers.

A sample of water containing 10 ppm of rhodamine-6G surrounding the fiber, was illuminated with several milliwatts of green light from a laser 100 m away. The fluorescence signal at the photomultiplier, also 100 m away, was several hundred thousand counts per second. Surprisingly, sunlight made only a minor background contribution, because the fiber has a very low acceptance for any light source less intense than a laser.

This system, however, is far from optimal. Calculations show that only a small fraction of the fluorescence from the overlap volume will be collected by the fiber. To improve the collection efficiency, we added a beamsplitter at the instrument end to permit us to send the laser light and receive the fluorescence signal through the same fiber.

To be useful for chemical analyses, such a system should have a linear response function. The response to a set of rhodamine-6G solutions at different concentrations; a 100:1 beam attenuator is used to avoid saturating the detector. The calibration curve is linear from at least 0.1 to 10 ppm, and a detection limit of about 10 ppb was confirmed by actual measurement.

A mathematical analysis of the behavior of this fiber-optical sensor probe showed that its signal level is proportional to its diameter. Since we were using commercial fibers developed for other purposes, there was no simple way to increase the fiber diameter. We achieved almost the same effect, however, by adding a lens in front of the fiber.

A particularly simple version of this idea is to cement a glass or sapphire ball in the end of a glass tube. During assembly we send laser light through the fiber, observe the resulting beam, move the fiber lengthwise to find the best focus, and then hold it in that position until the cement sets. The same ball lens collects the fluorescence light and focuses it into the fiber for transmittal to the photomultiplier of the fluorescence spectrometer.

With such a standardized probe, it becomes possible to compare the response of the system to a variety of fluorescent compounds. Table 1 lists signal levels from a number of materials, each at a concentration of 10 ppm, measured at the end of a fiber 100 m long.

The technology of remote fiber fluorimetry

One of the by-products of the fiber-optic technology developed by the communications industry is the ability to connect many fibers in parallel or in sequence to a single source. This in turn allows a remote fiber fluorimeter to monitor a number of measurement points with a single instrument.

Since each of the fibers may be up to 1 km long, this scheme allows a single instrument to provide continuous or rapid sequential analysis of many points, throughout an entire waste disposal site or refinery, for example.

For routine _in situ_ sampling, it would be necessary to provide special lead-through holders for the probes. For occasional or temporary hand-held optrode (the optrode analogue of an electrode) would suffice. Either way, the probe is quite insensitive to position. All the critical alignments are inside the probe, locked in place during assembly.

We are currently developing such a multipoint sensor system in conjunction with fluorescent rare-earth tracers to monitor groundwater movement in an experimental nuclear-waste repository[6]. The glass fibers are extremely resistant to radiation damage, which gives them a unique advantage over other kinds of sensors in this application. They can also be permanently buried with the waste and monitored from the surface with no threat of being incapacitated by moisture or voltage surges.

TABLE 1

Representative fluorescent dyes and the signals they generate at a concentration of 10 ppm in the fluorescence spectrometer with a 100 m optical fiber.

	Photons/Sec
Acridine red	88 000
Rhodamine-B	380 000
Rhodamine-6G	460 000
Fluorescein	170 000
Dichlorofluorescein	90 000
Brilliant sulphoflavine	205 000
Ethidium bromide	14 000
3,3' diethyloxodicarbocyanine Iodide	50 000
Th-1-amino-4-hydroxy-anthraquinone	9 000

Because of its very high sensitivity, fluorimetry has become more and more widely used in recent years. About 90% of all chemical compounds are non-fluorescent, however, and hence not directly detectable by fluorimetry. In many cases, we have been able to extend the fluorimetric technique to include these compounds by finding reagents that render them fluorescent or whose fluorescence they affect.

We have incorporated this technology into remote fiber fluorimetry by coating the surface of the probe with an insoluble or covalently bound reagent. It is important, of course, to choose a reversible equilibrium reaction to avoid depletion of the reagent. Such a chemically specific optrode behaves much like an ion-specific electrode, detecting only its target compound.

Future developments

As advances in laser fluorimetry and fiber communications become available, we plan to incorporate them to extend the capabilities of this analytical technique. There are also new glasses, now under development and soon to be marketed, with extended spectral ranges that will allow us to make observations in the near ultraviolet and the intermediate infrared. These developments will broaden the range of fluorescent compounds that are suitable as indicators.

We are also looking into ways to make optrodes sensitive to a wide variety of non-fluorescent stimuli. Additional parameters for which we may be able to develop optrodes include temperature, pressure, and radiation.

NOTES AND REFERENCES

1. T. Hirschfeld, "The Instrumentation of Environmental Optics", Optical Engineering (in press).

2. T. Hirschfeld, "Methods the Laser Made Possible", 32nd Annual Symposium on Analytical Chemistry, Purdue University, West Lafayette, Indiana, 1979.

3. Laser fluorescence spectroscopy for chemical analyses was described in Energy and Technology Review for January 1978 (UCRL-52000-78-1), 9 pp.

4. J. E. Midwinter, Optical Fibers for Transmission, Wiler, New York, 1979.

5. T. Hirschfeld, G. R. Haugen, D. C. Johnson, and L. W. Hrubesh, "Remote Techniques Based on Long-Distance Fiber Optics", 179th National Meeting, American Chemical Society, Houston, Texas, 1980.

6. S. Klainer and T. Hirschfeld, "Trace Analysis of Ground Waters", Earth Sciences Division Annual Report, Lawrence Berkeley Laboratory, ESD-10686, 1979.

FIBER OPTICS TEMPERATURE AND PRESSURE PROBE

T. Hirschfeld, F. Wang, G. Haugen, G. Hieftje*
Lawrence Livermore National Laboratory, Livermore, CA
*Indiana University, Bloomington, IN

ABSTRACT

The half life, the line width, and the ratio of peak intensities of the R_1 and R_2 ruby lines are only a function of temperature, while the shift in line position and the absolute intensities are functions of both the pressure and temperature. The emission from an optical fiber with a ruby tip encodes both the temperature and pressure of the fiber tip. Both the temperature and the pressure can be deduced by analyzing this spectroscopic information.

INTRODUCTION

The development by the communications industry of very high transmission long range fiber optics has already had an impact on remote sensing. Here the huge information bandwidth of optical fibers, their low loss, and their immunity to electromagnetic noise have been used for data transmission between the transducer and electro-optical transmitter system at the sensing locations and the monitoring location.

These optical fibers can be used, not merely for the transmission of data but, also as a component of the sensor system itself. Such a purely optical system has most of the hardware at the monitoring location, reducing the cost of the system and making the use of multiple sensors practical as well as providing an optical method for remote sensing in hostile or in accessible environments. A remote optical sensor would have a laser located at the monitoring station coupled to an optical fiber. This fiber transmits the laser emission to an optical material at the fiber tip, which emits radiation (reflection, scatter and/or fluorescence) that is returned by the same fiber to the monitoring station. The spectroscopic properties of the tip material can be designed to be a function of the parameter to be measured, and the value of this parameter can be reduced from the analysis of the spectrum back scattered from the tip. Such a purely spectroscopic sensor is simple, low in cost, rugged, and has a potential for quite extreme rapidity of response.

---

*Work performed under the auspices of the U. S. Department of Energy by the Lawrence Livermore National Laboratory under contract number W-7405-ENG-48.

Tiny ruby splinters have been used for pressure sensors in gasketted diamond anvil cells for the last 20 years. The pressure shift of the fluorescent wavelength has been observed to be linear to 200 kbars with a resolution better than 0.2 to 0.3 kbars (Ref. 1). Since sapphire fibers are commercially available, it is possible to construct an optical remote sensor for temperature and pressure by doping the tip of the fiber with Cr_2O_3. The temperature and pressure information would be encoded onto the fluorescent signal back-scattered from the fiber tip. There are three fluorescent phenomena that could assess the temperature and pressure of the environment surrounding the fiber tip; intensity, emission line frequency shifts, and lifetime. Intensity, when measured alone, leads to the least sensitive measurement, because it varies with excitation beam intensity. The inaccuracy of this intensity measurement can be circumvented by internally calibrating the fluorescent signals against the raman scattered light generated by the fiber optic transmitting the information - bearing optical signals. Emission line frequency shifts do not depend on excitation beam intensity, but they are difficult to measure since the shifts are typically small and can be associated with considerable broadening. Fluorescent lifetime is not dependent on excitation beam intensity either and is related to quantum efficiency of the transition process occurring in the crystal which can be sensitive to environment surrounding the crystal. This article describes the preliminary study to develop a fiber optic sensor for measuring the temperature and pressure of shock waves.

RESULTS

Pressure Measurements

The spectroscopic properties of ruby are responsive to the hydrostatic stress applied to the crystal lattice by the pressure of the surrounding environment. Both emission lines are red shifted the same amount, 0.365 Å/kbar. Note that the pressure shift of the absorption spectrum of ruby (-2.96 Å/kbar) is nearly an order-of-magnitude larger than that observed for the emission spectrum (Ref. 2). This corresponds to an increase in the molar extinction coefficient, ϵ, of about 0.23 units per kbar at 5145 Å (Ref. 3). Using the familiar relationship between fluorescent intensity, I, and molar extinction coefficient, ϵ,

$$I = K I_o \phi \epsilon \ell C,$$

it is predicted that the shift in absorption peak should increase the intensity by about 1.2% per kbar. However, the quantitative observation of the variation of the intensity with pressure is not possible because of the non-reproducible transmission characteristics of the gasketted diamond anvil cell. The ratio of peak intensities is observable and it is independent of the pressure, as well as the widths of each emission line. Conditions producing non-hydrostatic stresses in the crystal lattice causes the lines to broaden and decreases their wavelength separation achieving poor correlation between spectral properties and pressure (Ref. 4). The 6943 Å, R_1, and 6929 Å, R_2, ruby lines have the same temporal behavior, i.e., the half lives are the same within the experimental measurement error (standard deviation of ± 5.5%). Measurements at twelve pressures ranging from 1 bar to 15 kbars demonstrated that within the experimental error the half life is independent of pressure. This also strongly suggests that the non-radiative processes in the crystal are independent of the pressure. Consequently, the emission intensity should increase at a rate of 1.2% per kbar because of the blue shift of the absorption peak with pressure. However, it is not possible to measure these intensity changes reliably with the bassett cell. Contrarily, the optical characteristics of the fiber probe is extremely stable with respect to collection and transmission of the fluorescence. Only possible instability in the fiber optic system is due to fluctuations in laser beam intensity internal or external to fiber. As an example, any divergence in the alignment of the laser beam and the extremely small aperture of the fiber causes fluctuations in the internal excitation intensity, but the fiber's raman back-scattering can be used to correct for these variations by normalizing the intensity of the ruby fluorescence to the intensity of the raman back-scattering.

Temperature Measurements

The emission at 23°C is a sharp line (doublet) superimposed upon a broad background emission. Increasing the temperature decreases the ruby line emission severely, while increasing and broadening the background emission. Above about 500°C the intensity of the ruby line is completely quenched, only the background emission remains. The thermal quenching of ruby line emission is non-linear. The slope is about 1.5 at low temperatures and increases continuously to a value of 26 at the high temperature end. Temperatures at which the ruby line emission is completely quenched, the black body radiation emitted from the ruby cylinder is prominent. The thermal quenching of the background emission increases as temperature is increased. Turning off the laser allows the observation of only the black body radiation. At 550°C, the background emission is 30-40% of the black body radiation, increasing the temperature to 600°C reduces this ratio to 5%. Experimentally, there is a linear relationship between the logarithm of integral of the black body emission curve and reciprocal of the temperature, which agrees with the black body radiation theory. A 65°C increase in temperature results in a 450% increase in black body emission.

The thermally induced red-shift and broadening of the ruby emission line is observed. Quantitatively, the rate of shift of the ruby line is 0.068 Å/°C (Ref. 5). The rate for thermally induced line broadening is about the same as that for the line shift (Ref. 5). This behaviour is certainly a contrast to the pressure independence of the line width.

The half-life, $t_{1/2}$ of the ruby emission is extremely sensitive to temperature. A 320°C increase in temperature causes about a 1600% decrease in the half-life. The dependence of half-life, $t_{1/2}$ on temperature, T (ok) is not represented by a simple function, but it is approximated by $t_{1/2} \simeq \frac{1}{T^6}$ the high temperature limit.

To test the feasibility of remotely measuring temperature, a Cr_2O_3 dopped sapphire fiber was affixed to the end of a standard glass optical fiber cable. The thermal and mechanical stability of the bonding agent limits the temperature range that can be studied.

The variation of the spectral intensity of the ruby tip with respect to temperature is about 1.3×10^4 counts/sec. °C. Signal levels $\sim 2 \cdot 10^6$ counts/sec are obtained at 200 m with just 12 mW of laser power in this configuration. To normalize the measurement, an internal reference was found necessary, as even with a stabilized laser the short term photometric drift approached 2%. This was reduced to 0.3%, the accuracy of the quantum counting electronics, by ratioing the measured fluorescence to the intensity of the Raman line of silica produced in the long distance fiber itself. However, the splitting of the R_1 and R_2 lines of ruby allows much simpler internal referencing by comparing their relative intensity as a function of temperature. The relative response of the peak ratio as a function of temperature is a 0.42% reduction per °C. In this range $\sim 1°C$ reproducibility is possible.

An improvement in sensitivity at long range is possible by going to longer wavelength excitation, where an He-Ne laser is used for excitation. Despite the lower power used here, and the smaller absorption cross section available, the sensitivity at this wavelength is superior for ranges > 1 Km, when using the present fiber.

A drawback of this system is its relatively long lifetime, limiting its time resolution to tens of microseconds. However, given the very favorable signal levels available, the resolution can be improved by adding a quenching ion (such as Fe^{+3}) to the crystal, giving half lifes in the tens of microsecond range and still ample signal.

REFERENCES

1 K. R. Hirsch and W. B. Holzapfel, J. Phys. E; Sci. Instr., 16 (1983), pp. 412-417.
2 D. R. Stephens and H. G. Drickamer, J. Chem. Phys., 35 (1961), 427 pp.
3 Solid State Phys., Adv. in Res. & Appl., 9, Ed. F. Seitz and D. Turnbull; D. S. McClure, "Electronic Spectra of Molecules and Ions in Crystals", 1959, 488 pp.
4 D. M. Adams, R. Appleby, and S. K. Sharma; J. of Phys. E Sci. Instr., 9 (1976), pp. 1140-1144.
5 D. E. McCumber and M. D. Sturge; J. Appl. Phys., 34 (1963), 1682 pp.

ON-LINE URANIUM DETERMINATION USING REMOTE FIBER FLUORIMETRY

R. A. Malstrom[1] and T. Hirschfeld[2]

[1] E. I. du Pont de Nemours & Co., Savannah River Laboratory, Aiken, SC 29808
[2] Lawrence Livermore National Laboratory, University of California, Livermore, CA 94550

ABSTRACT

An analytical technique has been developed for the remote, on-line measurement of uranium. The technique, Remote Fiber Fluorimetry, has four major components: a laser, an optical fiber, an optrode, and a detector. Since the laser and detector can be remote from the sampling point, this technique is well suited for analysis in hazardous or inaccessible environments.

The application of the technique to on-line uranium analysis in a nuclear fuel processing facility at the Savannah River Plant is described. Laboratory experiments have been made to determine the effects of nitric acid, temperature, and quenchers. Preliminary experiments in a mock-up of the sampling system used in the radiochemical separations buildings are discussed.

INTRODUCTION AND SUMMARY

Remote Fiber Fluorimetry (RFF) is being developed jointly at Savannah River Laboratory and Lawrence Livermore National Laboratory. This technique has four major components: a laser, an optical fiber, an optrode, and a detector. An optical fiber carries laser radiation to a remote sampling point. An optrode (the optical analog of an electrode) couples the laser light to a sample solution. Fluorescence from the sample is collected by the optrode and goes back through the same fiber to a detector.

At the Savannah River Plant, radiochemical separations are carried out in special separations buildings. Due to the hazardous nature of the solutions, direct human contact is not permitted. Analytical samples, when needed, are lifted to a sample aisle where they are collected and sent to the Laboratories Department for analysis. The sampling and analysis of these solutions are complicated by the need for shielding from radiation. Using an RFF/optrode system, sampling and analysis can be made both simpler and safer. RFF is a remote technique so no human interaction with the sample is needed. Since it is an on-line technique, analysis speed is increased and manpower requirements can be reduced when compared to off-line techniques. In addition, it could easily be incorporated into a process control system.

DISCUSSION

The initial work at Savannah River has been with uranium. Fig. 1 shows a uranyl absorption spectrum in 0.5M HNO_3. There are absorption bands at 325 nm and 416 nm. Since the optical fiber transmits best in the IR with decreasing transmission towards the blue, 416 nm laser light was used. This gives a good compromise between fiber transmission and uranyl absorption.

We have determined that in SRP process streams uranyl fluorescence depends upon uranium concentration, temperature, quenching (Fe, Al,...), and nitric acid concentration. Each of these factors will first be described individually and then as they relate to one another.

In Fig. 2 fluorescence intensity at 513 nm versus uranyl ion concentration is plotted. The concentration range is from 200 ppm to almost 10 g/L. This concentration range brackets the 2 to 7 g/L range expected in plant streams targeted for RFF analyses.

Uranyl fluorescence intensity versus temperature is shown in Fig. 3. As the temperature increases, the intensity drops off exponentially.

Fluorescence intensity at a given uranyl concentration is also a function of quenching. Column 1 in Table 1 shows the fluorescence intensity measured for nine different uranyl solutions in H_3PO_4. These measurements were taken over several days, and average 7750 cps. Phosphoric acid was used because the effects of quenching should be greater than those found with nitric acid, due to the longer fluorescence lifetime in H_3PO_4. Therefore, any quenching correction which works for phosphoric acid solutions will certainly work for nitric acid solutions.

Between 1 and 5 drops of 1 g/L $FeCl_3$ (a strong quencher) were added to each of these samples. Column 2 shows that the resulting fluorescence intensity dropped significantly. This difference is caused by quenching.

Fig. 4 shows similar data plotted as two fluorescence decay curves. In each case the fluorescence intensity was monitored at various times using a gate from a boxcar integrator. The top curve shows the single exponential decay seen from the uranyl ion in H_3PO_4. The lower curve shows the fluorescence decay from the same uranyl solution except that two drops of the $FeCl_3$ solution were also added. Quenching caused the fluorescence signal to decay more rapidly. Fluorescence lifetimes can be calculated, and for the uranyl ion without quenching the lifetime is 197 μsec. The quenched solution has a lifetime of 97 μsec.

Using a matrix lifetime technique, Hieftje and Haugen (ref. 1), it was possible to compensate for the effects of quenching. The fluorescence intensity of the solution is measured along with its fluorescence lifetime. The lifetime of an unquenched uranyl solution is constant over our concentration range. By multiplying the fluorescence intensity of the quenched solution by a lifetime normalizing factor, it is possible to calculate what the fluorescence intensity would be if no quenching were taking place. The right column in Table 1 shows the calculated fluorescence intensity using this matrix lifetime correction. Within our experimental error we were able to compensate for quenching. Using this technique, it is possible to determine what the fluorescence intensity of a solution would be if quenching were not taking place, even if we do not know what the quenchers are or what their concentrations are.

The final factor is nitric acid concentration. As the nitric acid concentration is increased from 0.01M to 1M, the fluorescence intensity increases. As before, the fluorescence maximum occurs at about 513 nm. A plot of fluorescence intensity at 513 nm versus nitric acid concentration is shown in Fig. 5.

The fluorescence relationships just described are shown in Fig. 6. As just shown, the fluorescence intensity is a function of the uranium concentration, acid concentration, quencher concentration, and temperature. The uranium concentration dependence can be described by a linear equation. The acid concentration can be fit to a quadratic function. Quenching can be fit to an exponential function. The temperature is also an exponential function. We are now doing experiments to determine an empirical function which best describes all of these relationships. This will enable us to calculate the uranium concentration directly knowing the fluorescence intensity, temperature, fluorescence lifetime, and the acid concentration. An on-line conductivity monitor is being developed by E. W. Baumann (SRL), which will be used to determine the acid concentration. Temperature can be easily measured with a thermocouple. Using gating detection, fluorescence lifetime and intensity information can also be obtained.

Part of our program is to demonstrate that this technology can be used remotely under plant conditions. We have installed an optical fiber between my laboratory and an analyzer demonstration facility (165 meters away). This facility is an exact duplication of the sampling system used in the canyon facilities I described earlier. In this facility, we will be able to determine if there are any engineering problems to be solved in moving up from a laboratory system to a plant system.

ACKNOWLEDGEMENT

The information contained in this article was developed during the course of work under Contract No. DE-AC09-76SR00001 with the U.S. Department of Energy.

REFERENCE

[1] G. M. Hieftje and G. R. Haugen, Analytica Chemica Acta, 123 (1981) 255-261.

TABLE 1

Quenching Effect and Correction

10^{-3} M UO_2^{++} (cps)	+ 1-5 DROPS $FeCl_3$	QUENCHING CORRECTED
7900	3800	7720
7730	3900	7370
7680	2020	7380
8000	2750	7180
7500	3000	7470
7730	2700	7050
7800	3400	7450
7700	4300	7360
7700	2700	7790
AVG = 7750 ±140		AVG = 7420 ±230

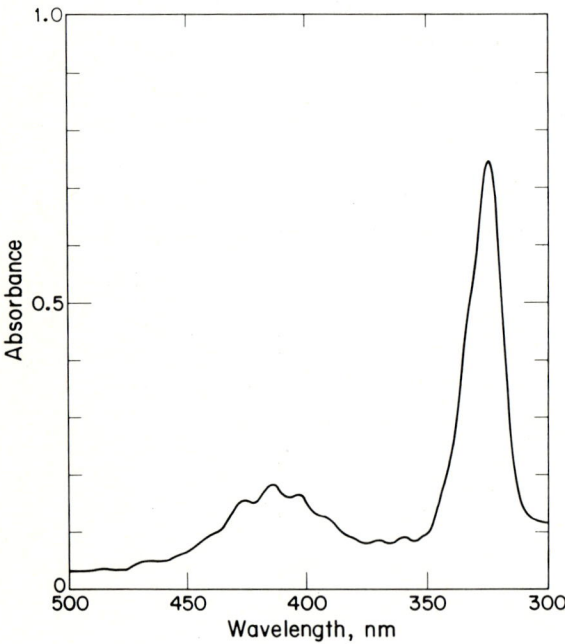

Fig. 1. Uranyl Absorption Spectrum

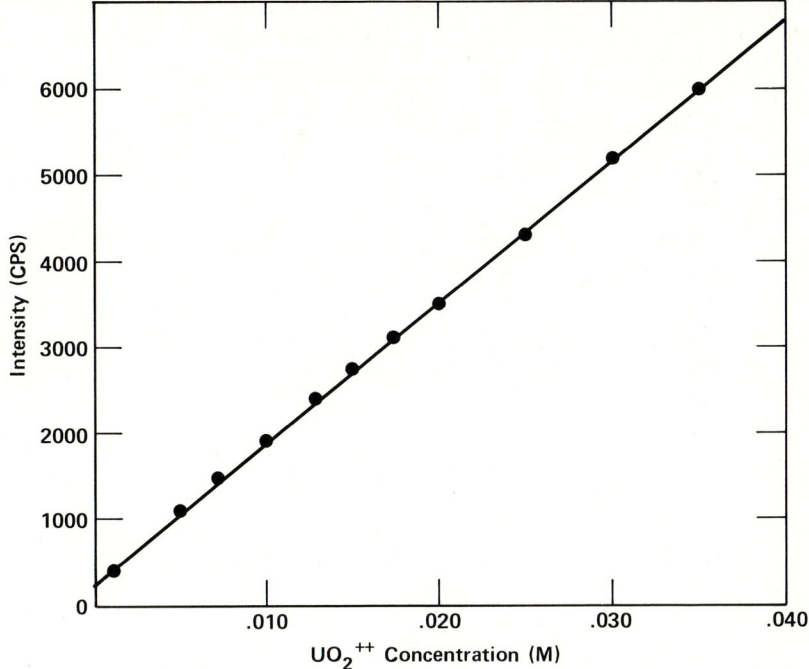

Fig. 2. Uranyl Calibration Curve

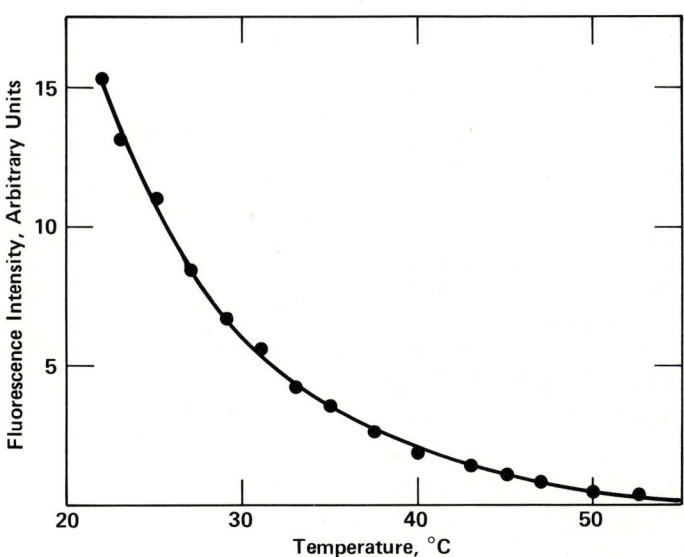

Fig. 3. Effect of Temperature on Uranyl Fluorescence Intensity

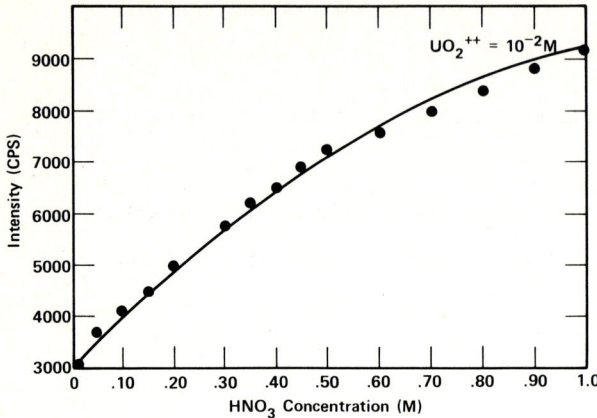

Fig. 4. Fluorescence Decay Curves

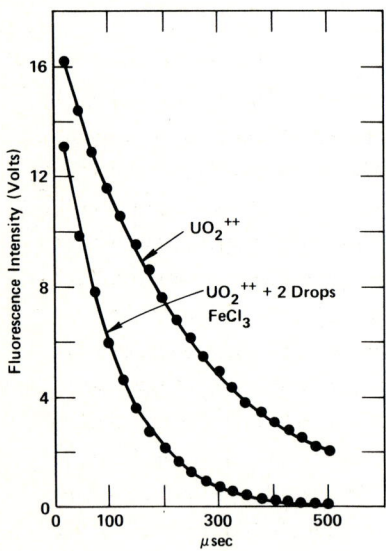

Fig. 5. Effect of Nitric Acid on Uranyl Fluorescence Intensity

- $I = F(U, A, Q, T)$
  - URANIUM CONCENTRATION $\quad I(U) = A_0 + A_1 U$
  - ACID CONCENTRATION $\quad I(A) = B_0 + B_1 A + B_2 A^2$
  - QUENCHER CONCENTRATION $\quad I(Q) = C_0 E^{-C_1 Q} + \ldots$
  - TEMPERATURE $\quad I(T) = D_0 E^{-D_1 T}$

- $U = F'[I(A, Q, T)]$

Fig. 6. Fluorescence Relationships

VERSATILITY OF LASER INDUCED FLUORESCENCE LINE NARROWING IN CHEMICAL ANALYSIS

M. J. MCGLADE[1], J. M. HAYES[1], G. J. SMALL[1], V. HEISIG[2] and A. M. JEFFREY[2]
[1]Ames Laboratory-USDOE and Department of Chemistry, Iowa State University, Ames, IA 50011, U.S.A.
[2]Institute Cancer Research, Cancer Center, Columbia University, New York, NY 10032

ABSTRACT

Laser excited fluorescence line narrowing spectroscopy (FLNS) in organic glasses at low temperatures has previously been shown to be a high sensitivity and resolution technique for the analysis of polycyclic aromatic hydrocarbons (PAH) in real samples. In this paper the versatility of FLNS is underscored by application to amino-PAH, PAH metabolites and aromatic carcinogen-DNA adducts.

INTRODUCTION

During the past several years three <u>high</u> resolution fluorescence based techniques have been developed which have been used for the <u>direct</u> analysis of PAH in real samples such as solvent refined coal. They are fluorescence line narrowing spectroscopy (FLNS) in organic glasses (1,2); laser excited Shpol'skii spectroscopy or LESS (3) and; rotationally cooled-laser induced fluorescence/GC (4). The latter (RC-LIF/GC) is a gas phase approach which employs a supersonic expansion while the other two are low T solid state techniques. Matrix isolation laser excited fluorescence spectrometry (MIFS) is also a promising approach and has been used, with minimal preseparation, for the analysis of compounds in very complex samples (5). Each method has its advantages and disadvantages. For example, the selectivity of RC-LIF/GC is significantly higher than that of FLNS, LESS and MIFS or capillary column GC/MS and, for nonfluorescent species, the supersonic jet can be combined with multi-photon ionization. On the other hand, FLNS and LESS can be applied to thermally labile compounds or compounds with very low vapor pressures. A great deal of additional work is required in order to establish the attributes and limitations of each of these methodologies.

The underlying physics of FLN has recently been discussed in detail (6) but for completeness the following brief discussion is presented. A molecule imbedded in an amorphous solid such as glass or polymer typically exhibits at 4.2 K vibronic absorption bandwidths of about 300 cm^{-1}. This is referred to as site inhomogeneous broadening and is a direct result of the host disorder. Thus, the molecule adopts essentially an infinite number of energetically inequivalent sites resulting in a broad distribution of excitation energies for each and every vibronic transition. Classical broad band excitation results in excitation of all sites and a fluorescence spectrum as broad as in absorption. However, if a narrow band laser (in our experiments about 1 cm^{-1}) is used for excitation, only a narrow subset (isochromat) of sites is excited. At low temperatures and sufficiently low concentrations, to avoid intermolecular electronic energy transfer, only the isochromat excited emits. Dramatic narrowing of the fluorescence spectrum results and this phenomenon is commonly referred to as FLN. At temperatures near 4.2 K the thermal broadening contribution to the linewidth is small, <1 cm^{-1}.

It should be emphasized that FLN is a general phenomenon and, therefore, operative with crystalline hosts such as the Shpol'skii matrix in LESS where inhomogeneous line broadening can be as large as ~10-20 cm^{-1}. Thus when narrow band laser excitation is utilized, the maximum selectivity of any low temperature solid state fluorescence technique is obtained only when the appropriate (1,6) FLN conditions are employed.

Organic glasses were chosen over other matrices for our studies for a number of reasons (1). Perhaps the most important is that "universal" glass solvents which accept a wide variety of compounds exist. We have usually employed glycerol:water based glasses because they form readily without cracking and with high optical quality for minimization of laser light scatter. In this paper we establish that FLNS in water containing glasses can be applied to amino-PAH, PAH metabolites and aromatic carcinogen-DNA adducts. The other solid state techniques cannot be applied to the first and latter compounds.

EXPERIMENTAL

The experimental apparatus utilized in these experiments is identical to that described earlier (1,7) except that the home built gated detection electronics (7) was replaced with a Quanta-Ray DGA-1 dual gated amplifier with a fixed gate width of 100 ns. For excitation wavelengths in the range 300-340 nm the frequency doubled output of a Quanta-Ray PDL-1 dye laser pumped by the 530 nm second harmonic of a DCR-1 Nd:YAG laser was used. For excitation wavelengths greater than 360 nm a Molectron TDL-200 dye laser pumped by a Molectron UV-14 nitrogen laser was used. The sample temperature for all FLN spectra was 4.2 K. Fluorescence linewidths are monochromator limited at ~6 cm^{-1}.

RESULTS AND DISCUSSION

<u>Amino Derivatives of Polycyclic Aromatic Hydrocarbons</u>

Based on earlier work on 1-aminopyrene, 1- and 2-aminoanthracene it was concluded that the S_1 fluorescent state of amino-PAHs possesses significant charge-transfer (CT) character (8). As a result the assertion was made that low T high resolution fluorescence techniques will not generally be applicable to amino-PAHs (8). The reason is that the electron-phonon interaction associated with a CT state is very strong. Thus, even under FLN conditions, only very broad (structureless) CT fluorescence is observed. However, it was found that this problem could be circumvented with FLNS for the above molecules by acidification of the glycerol:H_2O glass. Acidification leads to protonation of the $-NH_2$ group and, thus, elimination of the CT-type state; the fluorescent state of the protonated species is essentially the S_1 state of the parent PAH. FLN was observed for all three protonated species and the selectivity of the technqiue restored.

In more recent studies (9) we have examined the following molecules for FLN: naphthalene, 1-methylnaphthalene, 2-methylnaphthalene, 1-aminonaphthalene, 2-aminonaphthalene, 2,7-diaminonaphthalene, 1,5-diaminonaphthalene, 1,8-diaminonaphthalene, 2,3-diaminonaphthalene, fluorene, 2-aminofluorene and 2,7-diaminofluorene. We note that the first three exhibit ideal FLN spectra and that a simple mixture of them was readily resolvable by selective excitation and FLN (9). Apparently, the Shpol'skii effect is not operative for the naphthalenes. In accordance with earlier work (8) no FLN was observed for any of the amino-PAH (only broad CT-like fluorescence) when aprotic glasses are used. However, when the glycerol/HCl (2N) glass was employed FLN was observed for all the amino-PAHs except 2,3-diaminonaphthalene and 1,8-diaminonaphthalene. A possible reason for the absence of sharp FLN emission for the latter two is that the two amino groups are sufficiently close to each other so that the basicity of the second amino group is markedly reduced by protonation of the first amino group. Again, the absorption and FLN spectra of the protonated amino-PAH bear a close resemblance to that of the parent. Details are given elsewhere (9).

Typical FLN spectra are shown in Figs. 1 and 2 for 1-aminonaphthalene and 2,7-diaminofluorene. The sharp features in Fig. 2 beginning at ~3100 Å are reproducible and are zero-phonon lines whose energies relative to the L,(0,0) feature yield ground state vibrational frequencies in close agreement with those of fluorene (9). We note that the phonon side bands which build on the low energy side of the zero-phonon lines are quite intense for the fluorenes (in contrast to the naphthalenes).

Based on a careful comparison of the FLN spectra and excitation wavelengths for PAH, their alkylated and amino-derivatives we conclude that FLNS cannot generally be used to <u>directly</u> distinguish between all three types of compounds in real complex samples. Relatively simple preseparation schemes are available which yield neutral, basic and acidic fractions. With improvements in our FLNS apparatus we are hopeful that the characterization and determination of closely related species (isomers) in the individual fractions will be possible.

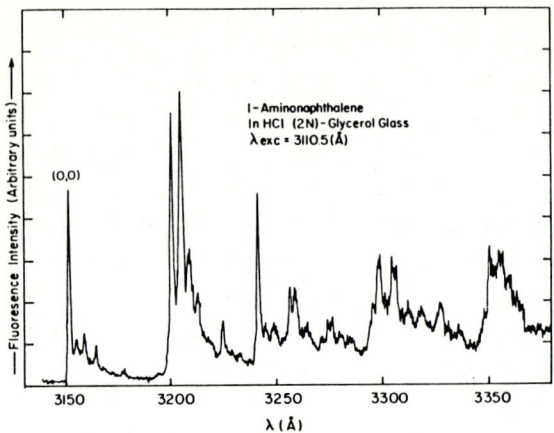

Figure 1. FLN spectrum of 1-aminonaphthalene imbedded in a HCl (2N)-glycerol glass at 4.2 K. Excitation is at 3110.5 Å.

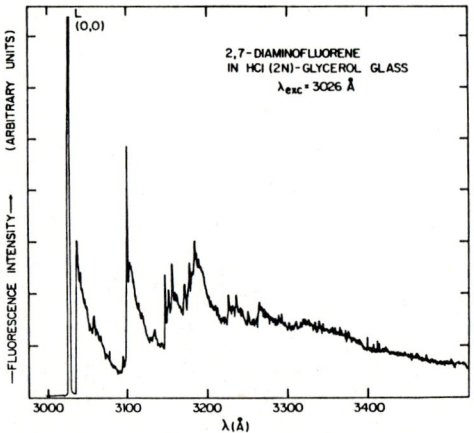

Figure 2. FLN spectrum of 2,7-diaminofluorene imbedded in a HCl (2N)-glycerol glass at 4.2 K. Excitation is at 3026 Å.

PAH Metabolites and Aromatic Carcinogen-DNA Adducts

Despite the development of several high resolution spectroscopic techniques for the analysis of carcinogens in real samples such as solvent refined coal and shale oil, no comparable technique exists for the analysis of DNA from cells exposed to carcinogens, toxins, etc. The development of such would be significant in many respects including that high selectivity chemical dosage measurements would become feasible. The analysis of DNA for aromatic carcinogen-DNA adducts by any jet spectroscopic technique, MIFS or LESS is not possible either because of the vanishingly small vapor pressure of the adduct or lack of solubility. These are not difficulties for FLNS in water containing glasses.

The first question to be addressed is whether FLN is generally operative for PAH metabolites and macromolecules as large as carcinogen-DNA adducts. A concern is that the electron-phonon interaction may be so strong that broad phonon side bands dominate the spectrum. In addition to DNA modified (.4%) with (±)-r-7,t-8-dihydroxy-t-9,10-epoxy-7,8,9,10-tetrahydrobenzo[a]pyrene (BPDE) [the "ultimate" carcinogenic metabolite of benzo[a]pyrene (BP)], we have studied the following compounds: 7,8,9,10-tetrahydrobenzo[a]pyrene (THBP), 7,8,9,10-tetrahydroxy THBP, 7-hydroxy THBP, 10-hydroxy THBP, deoxyuanosine-BPDE adduct and 9-methoxy-BP 4,5 dihydrodiol. Importantly, ideal FLN spectra are observed for all compounds. As expected the spectra of BPDE-DNA and the deoxyuanonsine-BPDE are very similar, Figs. 3 and 4. For all compounds except the 4,5-dihdrodiol the fluorescent chromophore is pyrene. It is chrysene for the 4,5-dihydrodiol.

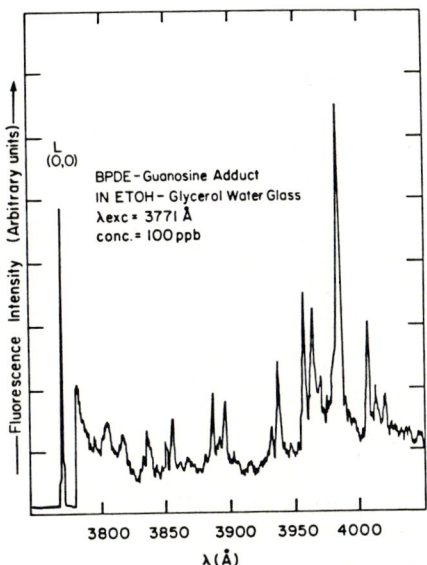

Figure 3. FLN spectrum of the BPDE-guanosine adduct imbedded in an ethanol-glycerol-water glass at 4.2 K.

Figure 4. FLN spectrum of BPDE-DNA adduct imbedded in an ethanol-glycerol-water glass at 4.2 K. Excitation is at 3790 Å.

In addition to BPDE, 9-hydroxy-BP 4,5-oxide has been reported as being a metabolite responsible for the binding of BP to DNA heptocytes. As a first test of the selectivity of FLNS, we have shown that (following isolation of nucleic acids from cells exposed to BP) the BPDE and 9-hydroxy-BP 4,5 oxide adducts can be distinguished (10). This is a considerable simplification over the best current method for this characterization which involves extensive enzyme digestion of the DNA, derivatization and HPLC analysis.

In conclusion, FLN in polar glasses is a promising approach for high selectivity analysis of DNA obtained following in vivo exposure to PAH mixtures. Such analyses will require improvements in sensitivity and selectivity which is possible at this time and is discussed elsewhere (9).

ACKNOWLEDGEMENT

This research was supported by the Office of Health and Environmental Research, Office of Energy Research of the U.S. Department of Energy under contract No. W-7405-Eng-82 and by DHS, NCI Grant CA 021111.

REFERENCES

1. J. C. Brown, J. A. Duncanson, Jr. and G. J. Small, Anal. Chem. 52, 1711 (1980).
2. J. M. Hayes, I. Chiang, M. J. McGlade, J. A. Warren and G. J. Small In "Laser Spectroscopy for Sensitive Detection"; J. A. Gelbwachs, Ed.; SPIE: Bellingham, WA, 286, 117 (1981).
3. Y. Yang, A. P. D'Silva and V. A. Fassel, Anal. Chem. 53, 894 (1981).
4. J. M. Hayes and G. J. Small, Anal. Chem. 55, 565A (1983).
5. E. L. Wehry, Trends Anal. Chem. 2, 143 (1983).
6. R. I. Personov In "Spectroscopy and Excitation of Condensed Molecular Systems"; V. M. Agranovich and R. M. Hochstrasser, Eds.; North Holland, New York, NY 1983, Chapter 10.
7. J. C. Brown, Ph.D. Dissertation, Iowa State University, 1982.
8. I. Chiang, J. M. Hayes and G. J. Small, Anal. Chem. 54, 315 (1982).
9. M. J. McGlade, M.S. Thesis, Iowa State University, 1983.
10. V. Heisig, A. M. Jeffrey, M. J. McGlade and G. J. Small, submitted for publication.

MATRIX-ISOLATION PHOTOACOUSTIC SPECTROSCOPY*

R. W. SHAW,[1] H. E. HOWELL,[2] G. MAMANTOV,[2] and E. L. WEHRY[2]
[1]Analytical Chemistry Division, Oak Ridge National Laboratory, Oak Ridge, TN 37831
[2]Department of Chemistry, University of Tennessee, Knoxville, TN 37996

ABSTRACT

We have recorded the first photoacoustic (PA) spectra of polycyclic aromatic hydrocarbons and n̄-heterocycles isolated in rare gas matrices at 5-10K. Matrix-isolation sampling was utilized in the hope of improving PA spectral resolution so as to provide greater analytical selectivity for the analysis of mixtures. The most sensitive transducer found was an undamped PZT disc, yielding detection limits ranging from 50 to 500 ng, depending on the molar absorptivity of the individual compound measured. Matrix isolation spectra of n̄-heterocycles in argon were two- to three-fold sharper than their room temperature solution counterparts. Internal and external heavy atom effects were observed for naphthalene.

INTRODUCTION

Spectroscopy of polycyclic aromatic compounds (PACs) at cryogenic temperatures has been of great interest recently (ref.1). Improved spectral resolution under these conditions (e.g., Shpol'skii frozen solutions or rare gas matrices) creates the possibility of identifying and quantifying individual compounds in the complex mixtures found in real world samples. One method, fluorescence spectroscopy, has been particularly successful in this regard. We have developed techniques for photoacoustic (PA) spectroscopy of PACs in rare gas matrices (ref.2). Matrix-isolation photoacoustic (MI-PA) spectroscopy could conceivably serve as a companion technique for fluorescence spectroscopy, finding particular utility for compounds that are only weakly fluorescent or are nonfluorescent. In order to combine MI sampling with PA spectroscopy, we have extended the use of piezoelectric PA transducers (ref.3) to temperatures as low as 1.5K (ref.4).

METHODS

Our technique for MI-PAS will be described in detail elsewhere (ref.2). Briefly, the compound of interest was co-deposited with a rare gas on a transparent window attached to the cold finger of a liquid helium cryostat (operating

---

*Research sponsored by the Office of Energy Research, U.S. Department of Energy, under Contract W-7405-eng-26 with the Union Carbide Corporation and by the National Science Foundation (Grant CHE-8025282).

at 10-20K). The matrix isolation ratio (rare gas:analyte) was approximately 1000:1. Because the compounds we examined are relatively volatile at room temperature, a sidearm sample delivery tube (ref.5) was used (Fig. 1).

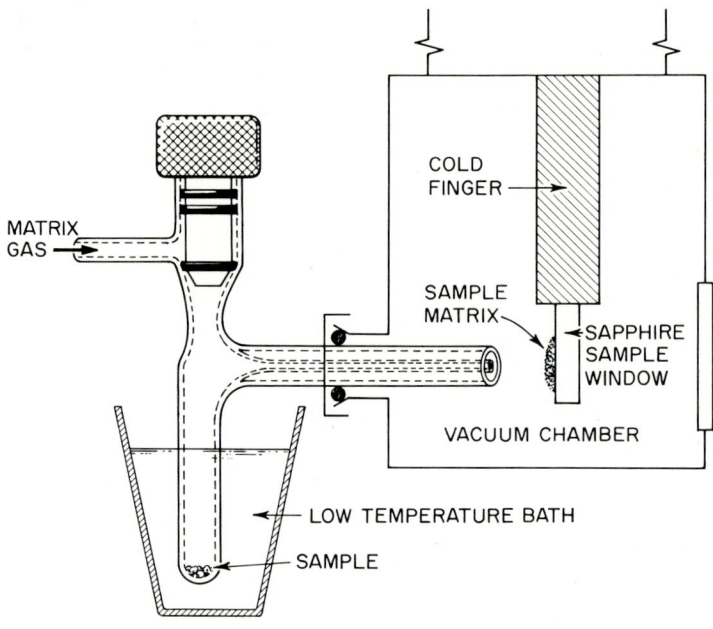

Fig. 1. Apparatus for preparing matrix-isolation samples for volatile compounds.

With this apparatus, the analyte (usually 1-10 μg) in a suitable solvent was added to the sidearm tube, the majority of the solvent evaporated, and the last traces of solvent removed at reduced pressure while the sample tube was maintained at a point below the analyte sublimation temperature. The cryostat cold finger temperature was then reduced from ambient to 10K; finally, the low temperature bath surrounding the sample tube was removed to allow sublimation of the analyte with the rare gas.

A lead zirconate titanate (PZT) ceramic disc transducer was bonded to the cold finger window to detect the PA signal. Alternately a slab of X-cut quartz served as both the sample substrate and piezoelectric transducer. The former approach was found to be approximately ten-fold more sensitive, largely due to cable loading of the X-cut quartz device (very low capacitance). Ringing photoacoustic response was generated upon absorption of 1 μs pulses from a frequency-doubled, flashlamp-pumped dye laser (0.02-0.2 mJ/pulse). The PA signal was amplified, filtered, processed with a boxcar averager, and finally normalized relative to the laser pulse energy using a photometer and ratiometer.

RESULTS AND DISCUSSION

An example of a MI-PA spectrum--5,6-benzoquinoline (5 μg) in xenon at 18 K--is shown in Figure 2A. For reference, the corresponding room temperature solution (10^{-4} $\underline{M}$, hexane) spectrum of 5,6-benzoquinoline is presented in Fig. 2B.

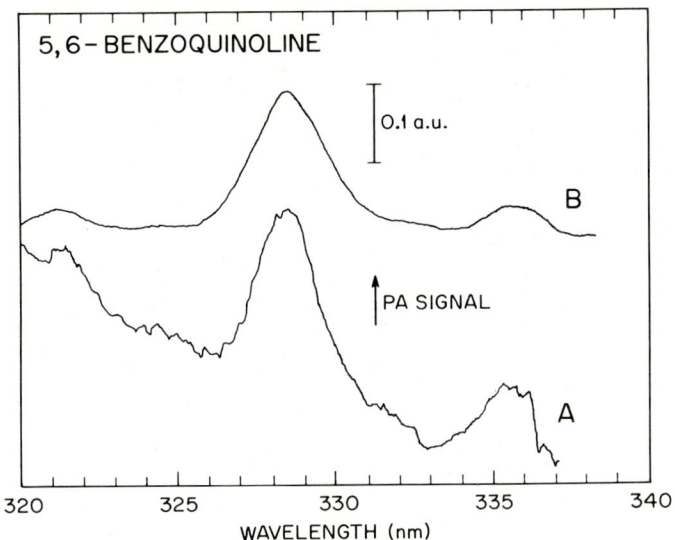

Fig. 2. 5,6-Benzoquinoline spectra. A. MI-PA spectrum, 5 μg in Xe at 18K; B. Room temperature solution absorption spectrum, 10^{-4} $\underline{M}$ in hexane.

Fig. 2 is only a portion of the complete 5,6-benzoquinoline spectrum, that part that lies within the scan range of rhodamine 640 laser dye. One additional absorption maximum (344 nm) is present at longer wavelength for both the PA and absorption spectra. For this particular compound the spectral resolution enhancement is only 1.4 for the 328.5 nm band: 240 cm^{-1} full width at half fraction absorbed for the MI-PA spectrum and 330 cm^{-1} for its room temperature solution counterpart. No spectral shift is apparent for the xenon matrix data relative to the solution spectrum. Both of these features are changed substantially for spectra of PACs isolated in argon matrices. For the latter, spectral sharpening is typically two- to three-fold (120-250 cm^{-1} full width at half fraction absorbed for the MI compounds) and a blue shift of approximately 100 to 400 cm^{-1} is observed (both relative to room temperature solution spectra). These factors point to a smaller analyte-matrix interaction for argon. Even when resolution enhancements of two-to-three-fold can be attained, the spectral linewidths of matrix-isolated PACs typically are too large to eliminate overlap of the absorption bands of isomers. We have examined several mixtures of quinolines (methyl- and dimethyl-isomers) and of benzoquinolines and find residual overlap at low temperature for all cases. By way of example, spectra of a 1:1 mixture of quinoline and isoquinoline are presented in Fig. 3. Trace 3A is the room temperature solution spectrum (10^{-4} M each in heptane) and spectrum 3B is a MI-absorption spectrum (3 µg each in nitrogen at 15K). The spectral bands are labeled as Q (quinoline) and I (isoquinoline), as identified in spectra of the individual compounds. Even though a substantial resolution improvement is observed for the (0,0) bands of these isomers, the (0,1) band of isoquinoline underlies the (0,0) band of quinoline (311 nm) and would result in an error of approximately 15% for the determination of quinoline in this 1:1 mixture. This error could of course be minimized by spectral subtraction if accurate spectra of the individual compounds are available, but that procedure becomes increasingly difficult for higher ratios of isoquinoline-to-quinoline. Site selection techniques (ref.1) provide a means to extricate MI-fluorescence spectroscopy from this poor spectral resolution dilemma. That option is of course not open to MI-PA spectroscopy because a continuum of host sites would each be probed in turn as the laser excitation wavelength is scanned to record spectra.

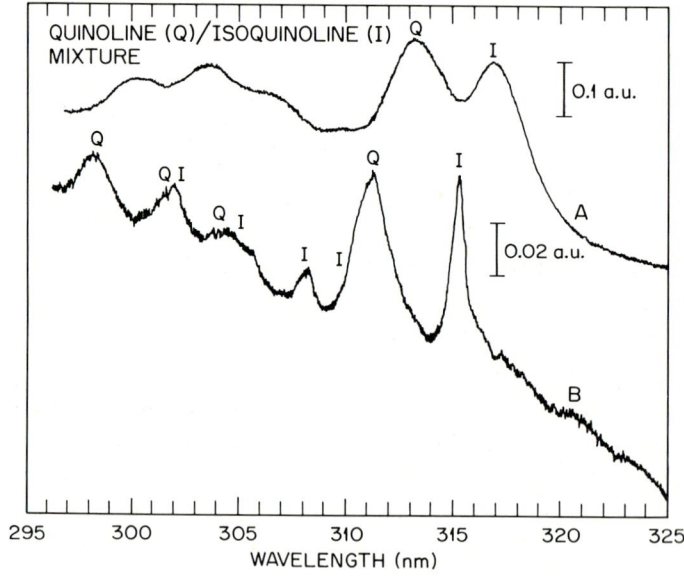

Fig. 3. Spectra of 1:1 quinoline:isoquinoline mixtures. A. Room temperature solution absorption spectrum, 10^{-4} M each in heptane; B. MI-absorption spectrum, 3 µg each in N_2 at 15K.

One solution to this problem of limited spectral resolution for matrix-isolated PACs would be to employ Shpol'skii frozen solutions, wherein much narrower linewidths (≤ 15 cm^{-1}) are observed (ref.1). We have attempted PA spectra in such opaque hosts and find that the PA background signal due to scattered light overwhelms any true analyte signal. Lai and coworkers (ref.6) have recorded a PA spectrum of 9,10-dimethylanthracene in an n-hexane Shpol'skii solution at 77K, but their PA signal is only barely above the noise level.

Our detection limit (S/N ~ 2) for detection of matrix-isolated PACs ranges from 50 ng (pyrene) to 500 ng (quinoline), depending on the molar absorptivity of the particular transition studied. In absorbance terms, those detection limits correspond to 10 mAU. It is conceivable that these limits could be reduced if greater UV laser pulse energies were available. Table 1 is a summary of the compounds for which we have recorded MI-PA spectra, including the various hosts employed.

TABLE 1
PACs and matrices for which laser MI-PA spectra have been recorded.

Compound	Matrices	Compound	Matrices
Quinoline	N$_2$, Ar, Xe	3-Methylisoquinoline	Ar
3-Bromoquinoline	Ar	Pyrene	Ar
2,6-Dimethylquinoline	Xe	Naphthalene	Ar, Xe
3,4-Benzoquinoline	Ar	1-Bromonaphthalene	Ar, Xe
5,6-Benzoquinoline	Ar, Xe	Dibenzothiophene	Ar
7,8-Benzoquinoline	Ar	Dibenzofuran	Ar

For quantities of naphthalene (N) and 1-bromonaphthalene (BrN) isolated in argon that yield equal absorbance, BrN yields an enhanced PA signal: BrN (314 nm band)/N (310 nm band) = 2.3. This observation is most likely due to an internal heavy atom effect (ref.7) in BrN that promotes a higher yield of intersystem crossing for BrN than for N. For PACs isolated in rare gas hosts, the nonradiative energy release that creates the PA effect probably stems from intersystem crossing; singlet nonradiative decay is probably hindered due to poor coupling to the low energy lattice phonons, and triplet nonradiative decay is too slow ($\tau > 1$ ms) to contribute to the PA signal. In xenon (an external heavy atom effect host) the BrN/N PA enhancement is even greater. These heavy atom effect arguments must be viewed as tentative because the quantum yields for fluorescence, intersystem crossing, and phosphorescence are unavailable for these PAC/matrix systems.

CONCLUSIONS

We have recorded PA spectra for several matrix-isolated PACs in solid nitrogen, argon, and xenon at 10-20K. Modest sharpening of spectral features was observed relative to room temperature solution spectra. Some overlap of spectral bands of isomeric compounds remains for these matrix-isolated PACs. Analytical detection limits ranged from 50 to 500 ng. Our detection limits are comparable with those of matrix-isolation/Fourier transform infrared spectroscopy, a technique exhibiting somewhat better spectral resolution.

REFERENCES

1 E.L. Wehry and G. Mamantov, in E.L. Wehry (ed.), Modern Fluorescence Spectroscopy, V. 4, Plenum Press, NY, 1981, Chapter 6.
2 H.E. Howell, R.W. Shaw, G. Mamantov and E.L. Wehry; submitted for publication in Anal. Chem.
3 C.K.N. Patel and A.C. Tam, Rev. Mod. Phys. 53 (1981), 517.
4 R.W. Shaw and H.E. Howell, Appl. Optics, 21 (1982), 100.
5 D.M. Hembree, E.R. Hinton, Jr., R.R. Kemmerer, G. Mamantov, and E.L. Wehry, Appl. Spectrosc., 33 (1979), 477.
6 E.P. Lai, A. Jurgensen, E. Voightman and J.D. Winefordner, Appl. Spectrosc., 36 (1982), 589.
7 S.P. McGlynn, T. Azumi and M. Kinoshita; Molecular Spectroscopy of the Triplet State, Prentice-Hall, Englewood Cliffs, NJ, 1969.

SELECTIVE FLUOROMETRIC DETERMINATIONS BY LASER-INDUCED MATRIX ISOLATION FLUORESCENCE SPECTROMETRY

E.L. WEHRY

Department of Chemistry, University of Tennessee, Knoxville, Tennessee 37996 (USA)

ABSTRACT

The preparation of samples by matrix isolation for low-temperature spectrometric examination is described, and the specific characteristics of matrix isolation molecular luminescence spectra are considered. The advantages of laser excitation of matrix isolation fluorescence are outlined, and applications of the technique to characterization of complex samples are surveyed.

MATRIX ISOLATION SPECTROSCOPY

Matrix isolation (hereafter abbreviated "MI") is a cryogenic sample preparation technique initially developed in the 1950's for obtaining spectra of transient species (ref.1,2). Although MI was not originally devised as a technique for the spectroscopy of stable molecules, it has proven very useful for that purpose (ref.3). In an MI experiment, the compound(s) of interest are vaporized and then mixed with a substantial excess of a "matrix gas". Any substance having an appreciable vapor pressure at room temperature can be used as a matrix gas. Thus, although "classical" MI experiments normally employ rare gases (Ne, Ar, Xe) as matrices, MI of organic molecules may be most advantageously achieved if organic compounds (such as n-alkanes or fluorocarbons) are used as matrix materials. The gaseous mixture so produced is then collected as a solid on a cryogenic surface, usually at temperatures of 15 K or less, and the spectroscopic behavior of the solid is examined.

The purpose of MI is to form a cryogenic solid in which solute molecules are isolated from each other and have only matrix atoms or molecules as near neighbors. Except for molecules which are strongly aggregated in the gas phase, this objective is ensured by "dissolving" the sample in the "solvent" (matrix) in the gas phase under conditions wherein the matrix is present in large excess.

The principal reasons for using MI as a sample preparation technique in analytical fluorescence spectroscopy are:

(a) To achieve sharp, structured, highly characteristic fluorescence and excitation spectra, rather than the broad, featureless spectra characteristic of the fluorescence and absorption spectra of most organic molecules in liquid solution. This is the principal objective of any type of cryogenic luminescence spectroscopy (ref.4), and is of importance for both qualitative and quantitative analytical applications of fluorescence spectrometry. It is desirable to obtain luminescence spectra which contain sufficient information to be employed for identification of specific sample constituents (i.e., "molecular fingerprinting") (ref.5). It is particularly desirable to have the capability of exciting selectively the fluorescence of individual sample constituents, preferably without fractionation or other prior treatment of the sample (ref.5,6).

(b) To eliminate errors in quantitative analyses caused by quenching or intermolecular electronic energy transfer (either short-range or long-range). In a properly prepared deposit, the average distance between analyte molecules exceeds the "Förster radius" for long-range energy transfer.

(c) To obviate the need to dissolve the analytes in a liquid solvent. In MI, the analytes are vacuum sublimed or distilled from the sample. This is particularly advantageous if the original sample is an intractable solid (such as coal or solvent-refined coals). Moreover, the "best" spectroscopic solvent for a particular analyte often is a poor "chemical" solvent for that analyte. For example, the highest spectral resolution for polar aromatic compounds often is obtained in fluorocarbon matrices, yet most of these compounds are exceedingly insoluble in fluorocarbons (ref.7).

(d) To facilitate coupling of the luminescence measurement with chemical separation techniques for those samples for which even the very high selectivity of low-temperature fluorescence is insufficient. Inasmuch as preparation of a matrix-isolated sample requires vaporization of the analyte, coupling of MI spectrometry directly to gas chromatographic separations is an obvious idea, and coupling of MI infrared (ref.8) and fluorescence (ref.9) spectrometry to GC separations has been reported. Interfacing of MI (or any other fluorometric measurement technique requiring vaporization of analytes) to HPLC is a great deal more problematical and has yet to be reported.

Experimental techniques in analytical MI spectrometry have been described elsewhere (ref.10,11,12). The required cryogenic apparatus is commercially available and can be used readily by workers lacking prior experience in cryogenics.

LASER-INDUCED FLUORESCENCE

Although lamp-excited fluorescence in vapor-deposited cryogenic matrices can produce much useful analytical information (ref.10,13), the full capability of MI spectrometry is most readily exploited by use of a tunable dye laser as the excitation source. Although "laserphobia" remains a potent factor limiting the application of laser spectroscopy to real analytical problems, it is now possible to purchase dye laser systems that (while expensive) are reliable and capable of being used by persons having no previous experience with lasers. One obvious advantage of a laser is the spatial coherence of the output, which permits samples of very small area or volume to be examined. In conjunction with "microsampling" techniques for preparation of matrix-isolated samples (ref.12), this characteristic of laser sources permits very small quantities (subpicogram) of fluorescent analytes to be detected by MI fluorometry.

A more important advantage of lasers is the high spectral purity of the radiation which they produce. The objective of cryogenic fluorometric techniques is to produce sharp excitation spectra. If that goal is achieved, it becomes possible to excite fluorescence from individual fluorescent constituents of very complex samples without prior separation (ref.5,6,7,14); perhaps the most spectacular example thus far reported is the individual selective excitation of several polycyclic aromatic hydrocarbons and their deuterated analogs in a shale oil sample by frozen-solution laser-induced fluorometry (ref.15).

Proper choice of matrix may be crucial in achieving selectivity of excitation. Different molecules of a particular solute in a solid may be trapped in slightly different microenvironments. This situation is manifested spectroscopically by broadening of both absorption and emission spectra, termed "inhomogeneous broadening" (ref.4,16). To achieve the maximal selectivity in MI fluorometry, the effects of inhomogeneous broadening must be minimized. One way to achieve this is to incorporate the solute species in a matrix in which solutes can be incorporated substitutionally into the matrix by displacement of a small number of matrix molecules, such that one (or at most several) types of "matrix sites" are occupied by the solute. The classical example of this approach is the use of n-alkanes as matrices for polycyclic aromatic hydrocarbons (the "Shpol'skii effect"), wherein exceedingly sharp, characteristic fluorescence excitation and emission spectra can be obtained via both frozen-solution (ref. 14,15) and matrix isolation (ref.5,6,17) sample-formation techniques. The use of MI fluorometry in vapor-deposited Shpol'skii matrices to achieve identification and quantification of polycyclic aromatic hydrocarbons in an untreated intractable solid solvent-refined coal (ref.5) represents a good example of an analytical problem for which laser-induced matrix isolation fluorometry is especially well suited.

Unfortunately, many fluorescent compounds do not exhibit the Shpol'skii effect. In many cases, however, the extent of inhomogeneous broadening for such compounds can be decreased dramatically by selectively exciting a subset of molecules of a particular solute which exhibit nearly identical electronic transition energies (ref.18). This procedure, referred to by various workers as "fluorescence line narrowing", "site selection", or "transition energy selection", has been discussed elsewhere in this volume and in numerous research publications (ref.19,20) by Small, Hayes, and co-workers, who have employed glass-forming organic solvents and frozen-solution sample-preparation techniques. The technique also has been used to achieve selective excitation of aromatic hydroxyl compounds and nitrogen hetero aromatics in vapor-deposited rare gas and fluorocarbon matrices (ref.7). It is virtually impossible to achieve fluorescence spectral narrowing by this technique under realistic analytical conditions unless a laser is used for excitation (ref.4,16).

An additional advantage of laser excitation in MI fluorometry is the possibility of employing time-resolved fluorescence measurement to distinguish spectrally similar compounds (ref.21) and to reduce spectral background (ref.5).

FUTURE TRENDS

Continued development of selective laser-induced MI fluorometric determination of fluorescent compounds (especially using "site-selective" excitation) is expected. Extensions to determination of nonfluorescent compounds also are anticipated. The classical procedure for using fluorescence spectrometry to determine nonfluorescent analytes is to convert them to fluorescent species by chemical derivatization; such procedures are especially widely used in HPLC detection. We are currently examining the use of electron impact (ref.21) and UV laser photolysis as techniques for fragmenting nonfluorescent molecules in the gas phase to form fluorescent fragments which can then be detected by high-resolution matrix isolation fluorometry. Many small molecular fragments (e.g., CH, CN, NH, CO^+, Cl, H, etc.) fluoresce with reasonable quantum efficiency in low-temperature matrices. The imposition of a "fragmentation" step between vaporization (required in any MI experiment) and measurement of laser-induced fluorescence promises to expand greatly the number of important chemical compounds than can be identified and quantified by matrix-isolation fluorescence spectrometry. Perhaps paradoxically, this most recent approach to "analytical" MI fluorometry represents a return to the classical reason for performance of matrix isolation experiments--to stabilize highly reactive species for spectroscopic examination at leisure under controlled conditions.

ACKNOWLEDGMENT

This work has been supported by the U.S. National Science Foundation (Grant CHE-8025282).

REFERENCES

1 H.E. Hallam, Vibrational Spectroscopy of Trapped Species, Wiley, London, 1973.
2 S. Cradock and A.J. Hinchcliffe, Matrix Isolation, Cambridge University Press, New York, 1975.
3 B. Meyer, Low Temperature Spectroscopy, American Elsevier, New York, 1971.

4 E.L. Wehry and G. Mamantov, in E.L. Wehry (Ed.), Modern Fluorescence Spectroscopy, Plenum, New York, 1981, Vol.4, p.193.
5 M.B. Perry, E.L. Wehry, and G. Mamantov, Anal. Chem. (in press).
6 V.B. Conrad and E.L. Wehry, Appl. Spectrosc., 37 (1983) 46.
7 J.R. Maple and E.L. Wehry, Anal. Chem., 53 (1981) 266.
8 D.M. Hembree, A.A. Garrison, R.A. Crocombe, R.A. Yokley, E.L. Wehry, and G. Mamantov, Anal. Chem. 53 (1981) 1783.
9 V.B. Conrad, W.J. Carter, E.L. Wehry, and G. Mamantov, Anal. Chem., 55 (1983) 1340.
10 R.C. Stroupe, P. Tokousbalides, R.B. Dickinson, Jr., E.L. Wehry, and G. Mamantov, Anal. Chem., 49 (1977) 701.
11 E.L. Wehry and G. Mamantov, Anal. Chem., 51 (1979) 643A.
12 D.M. Hembree, E.R. Hinton, Jr., R.R. Kemmerer, G. Mamantov, and E.L. Wehry, Appl. Spectrosc., 33 (1979) 477.
13 P. Tokousbalides, E.R. Hinton, Jr., R.B. Dickinson, Jr., P.V. Bilotta, E.L. Wehry, and G. Mamantov, Anal. Chem., 50 (1978) 1189.
14 Y. Yang, A.P. D'Silva, and V.A. Fassel, Anal. Chem., 53 (1981) 894.
15 Y. Yang, A.P. D'Silva, and V.A. Fassel, Anal. Chem., 53 (1981) 2107.
16 B.E. Kohler, in C.B. Moore (Ed.), Chemical and Biochemical Applications of Lasers, Academic Press, New York, 1979, Vol.4, p.31.
17 J.R. Maple, E.L. Wehry, and G. Mamantov, Anal. Chem., 52 (1980) 920.
18 F.A. Burkhalter and U.P. Wild, Chem. Phys. Lett., 66 (1982) 327.
19 J.C. Brown, M.C. Edelson, and G.J. Small, Anal. Chem., 50 (1978) 1394.
20 J.C. Brown, J.A. Duncanson, Jr., and G.J. Small, Anal. Chem., 52 (1980) 1711.
21 R.B. Dickinson, Jr. and E.L. Wehry, Anal. Chem., 51 (1979) 778.
22 M.L. Robin, G.K. Schweitzer, and E.L. Wehry, Appl. Spectrosc. Rev., 17 (1981) 165.

METAL SURFACE ENHANCED FLUORESCENCE

T. D. Harris, A. M. Glass and D. H. Olson
Bell Laboratories, Murray Hill, New Jersey 07974

ABSTRACT

The report in 1977 by Jeanmarie and Van Dyne of greatly enhanced Raman scattering from silver electrodes triggered a worldwide flurry of activity in surface perturbations of spectroscopy. This activity has centered on the nature and application of the phenomenon for Raman scattering. However, it has become increasingly clear that the origin of the high intensity is a simple electrodynamic effect due to the peculiar spectroscopy of small metal particles. It has also become clear that other forms of spectroscopy (absorption, emission, photochemistry, etc.) will be similarly affected. Several recent reports have demonstrated the large perturbation of the apparent quantum yield and the relaxation rate for fluorophore near both smooth metal surfaces and arrays of small metal particles.

The proven and potential applications of these dramatic changes in fluorescence parameters are discussed. The preparation of appropriate substrates is described. The generality of this phenomenon and the consequences of surface chemistry is outlined. Finally the application of these substrates for reduction of quenching interference in fluorometric analysis is demonstrated.

INTRODUCTION

The report in 1974 of anomalously high Raman scattering is from silver electrodes[1] and its subsequent characterization by Van Duyne[2] has led to many groups investigating and exploiting the phenomenon. While the controversy over the origin of the mechanism has subsided there remains many observational details to be explained. It is now widely agreed that a major part of the enhancement in Raman intensity obtains from simple electromagnetic effects from the response of the metal to incident radiation. The result of this interaction is to create very high electric fields near the surface of the metal particle. For the purposes of the model, the rough silver electrodes are treated as particles on or near a smooth metal surface.

The anomalous Raman scattering obtains from large electric fields and many other field sensitive spectroscopic processes also exhibit similar perturbations. These processes include, absorption,[3] fluorescence,[4] second harmonic generation,[5] photochemistry,[6] two photon spectroscopy,[7] and four wave mixing.[8] The processes that include a molecular excited state, fluorescence and photochemistry must also include the non-radiative parameters of the particle while the prompt interactions, such as Raman scattering and four wave mixing do not.

To exploit the properties of these systems, the nature of their resonance must be understood. The frequency and magnitude of the resonance is primarily a property of the material. The frequency can be "tuned" to some extent by controlling the size and shape of the particles. This tuning is done at a sacrifice in the magnitude of the effect. A useful rule of thumb is that increased size or ellipticity moves the resonance to lower frequencies while increasing the refractive index of the surrounding medium moves the resonance to higher frequencies. The calculated half-width of the resonance for a perfect spherical silver particle is about 20 nm. A typical experimentally accessible distribution of size, shape and spacing exhibits a resonance more than 100 nm wide.

The nature of the interaction between particles and molecular excited states is complex but can be summarized by two observations. The radiative relaxation rate is increased and energy transfer from the molecule to non-radiative modes in the particle is possible. The fluorescence process is then perturbed in three ways; enhanced absorption, increased rate of emission, and increased rate of non-radiative relaxation. For experiments reported, the net effect of these three interactions has been a modest increase or decrease in fluorescence intensity and a dramatically shortened excited state lifetime.

This combination of properties leads to an important potential application, reported here for the first time. Since the excited state lifetime is shortened, the effect of competing non-radiative channels for de-excitation is diminished. One very troublesome non-radiative pathway in fluorescence analysis is quenching from species in solution with the fluorophore. These quenching effects become important for a quencher concentration at which the rate constant for quenching approaches the rate constant for fluorescence. The only reported lifetimes measurement show an increase of 10^3 for the fluorescence rate when the fluorophore is located near silver particles of 75 Å diameter. This implies that for molecules near the particles quenching effects would become important at quencher concentrations 1000 times higher than for that same fluorophore in the absence of the particle. For many sample preparation schemes this change effectively eliminates quenching as a significant impediment to the analysis.

EXPERIMENT

The metal particle arrays used in this work were identical to those used in several previously reported studies.[4] They are known as island films and are shown schematically in Figure 1. Silver particles with dimensions of 50-75 Å will result if a film of that same nominal thickness is evaporated onto a glass substrate. To ensure that the optimum size particles are available, a thickness wedge was used. In this way particles sizes from 10 Å up to approximately 100 Å can be studied on a single substrate. The maximum particle size is determined by the ability of silver to wet the glass. A thickness of greater than 100 Å yields a continuous film rather than islands.

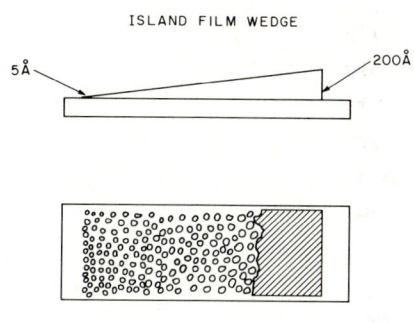

FIGURE 1

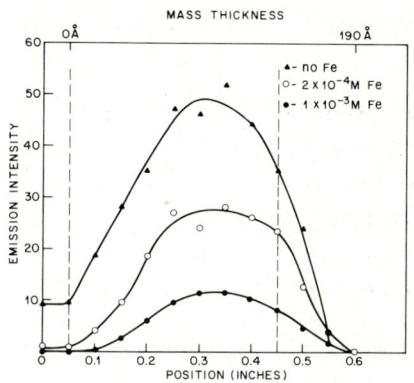

FIGURE 2

The films were coated with dye-methanol solutions by placing the substrate on a spin coating apparatus. Several drops of solution were placed on the silver film and the substrate was spun dry. Solutions of Rhodamine B in methanol were used. Some solutions were quenched by the addition of varying amounts of ferrous sulfate. Fluorescence spectra were recorded by illuminating the coated substrates with a chopped, unfocussed beam of 514.5 nm radiation from an argon ion laser. A slit mask was used to illuminate only one film thickness area at a time. The emitted radiation was collected with an f2 lens and focussed onto the slit of a double monochromater. Detection was with a photomultiplier-lock-in amplifier arrangement.

The purpose of this initial study was to demonstrate the reduction of quenching as outlined in the introduction. To this end fluorescence spectra were recorded as a function of silver film thickness and quencher concentration. If the hypothesis were correct the ratio of fluorescence intensity from the silver particle array to that from the bare glass slide should increase significantly with increasing quencher concentration. Data are shown in Figure 2 and Table I. It is clear that the effect of the quencher has been reduced. The extent to which any change in the intensity of emission from silver islands is predicted is a complex and speculative problem. Many factors including dye coverage, dye molecule distribution, silver particle shape, coverage and dye quencher-interactions must be considered. Control to these parameters must be much finer for even a general interpretation. Work to this end is in progress. These outstanding questions should not detract from the results reported here.

TABLE I

Quencher Conc.	Intensity Ratio[*]
5×10^{-3} M	>250:1
1×10^{-3} M	230:1
2×10^{-4}	25:1
0	5:1

[*] For 3×10^{-5} M Rhodamine B with $FeSO_4$ quencher. Ratios are relative intensities of bare glass slide to maximum emission from silver particles.

CONCLUSION

It has been demonstrated that the effects of quenchers on fluorescence intensity can be greatly diminished with the use of silver particle arrays. The potential to simplify and generalize many fluorometric determinations is considerable. The effect is general to all Raman enhancing substrates and quenchers. An appropriate match between analytic emission and metal optical properties is required. Reductions of quenching effects from 10^3 to 10^4 can be expected.

REFERENCES

[1] M. Fleischman, P. J. Hendra, and A. J. McQuillan, *Chem. Phys. Lett.*, **26**, (1974) 163.

[2] R. P. Van Duyne, in C. B. Moore (Ed), *Chemical and Biochemical Applications of Lasers,* Vol. 4, Academic, New York, (1978)

[3] H. G. Craighead and A. M. Glass, *Opt. Lett.*, **6**, (1981) 248-250.

[4] A. M. Glass, P. F. Liao, J. G. Bergman, and D. H. Olson, *Opt. Lett.*, **5**, (1981) 368-370.

[5] C. K. Chen, A. R. B. de Castro, and Y. R. Chen, *Phys. Rev. Lett.*, **46**, (1981) 145-148.

[6] A. Nitzan and L. E. Brus, *J. Chem. Phys.*, **75**, (1981) 2205-2214.

[7] A. M. Glass, A. Wokaun, J. P. Heritage, J. G. Bergman, P. F. Liao, and D. H. Olson, *Phys. Rev. B,* **24**, (1981) 4906-4909.

[8] D. S. Chemla, J. P. Heritage, P. F. Liao, and E. D. Isaacs, *Phys. Rev. B,* **27**, (1983) 45-4558.

[9] D. A. Weitz, S. Garoff, C. D. Hanson, T. J. Gramila, and J. I. Gersten, *Opt. Lett.,* **7**, (1982) 89-91.

PROSPECTS FOR SINGLE-MOLECULE DETECTION IN LIQUIDS BY LASER-INDUCED FLUORESCENCE

M. TRKULA,[1] N. J. DOVICHI,[2] J. C. MARTIN,[3] J. H. JETT,[3] AND R. A. KELLER[1]

[1] Chemistry Division, Los Alamos National Laboratory, Los Alamos, NM 87545
[2] Chemistry Department, University of Wyoming, Laramie, WY 82071
[3] Life Sciences Division, Los Alamos National Laboratory, Los Alamos, NM 87545

ABSTRACT

A laser-induced fluoresence determination of aqueous solutions of rhodamine 6G resulted in a detection limit of 18 attograms, or 22,000 molecules, of rhodamine 6G. These results allow the projection to single-molecule detection with reasonable improvements in the experimental apparatus.

INTRODUCTION

In recent years single-atom detection for several atoms has been reported (ref.1,2). To date there are no analytical techniques for molecular species which even approach this limit of sensitivity. Analytical techniques capable of detecting single-molecules would be quite useful in applications where sample size is severely limited; hence, there are inherently few signal "events" produced by the procedure. We report here a flow cytometry based technique which has improved the laser-induced fluorescence detection limit of molecular species by two orders of magnitude. Projections based upon these experiments indicate that single-molecule detection should be possible with reasonable improvements in the experimental apparatus.

EXPERIMENTAL

The technique of flow cytometry was utilized in this analysis (ref.3). Briefly, a sample fluid and a sheath fluid are introduced into the sample cell under laminar flow conditions. In the tapered region of the cell hydrodynamic focusing takes place and a sample stream diameter on the order of micrometers is produced. A tightly focused laser beam intersects the sample stream with its waist centered in the sample stream. In this manner, a probe volume of a few picoliters results. There are two important advantages of the sheath protected flow system. The first is that there is an insignificant index of refraction change at the sample/sheath interface, therefore background scatter generated in the immediate area of the probe volume is reduced. Secondly, the sheath prevents sample adsorption onto the cell walls, an important consideration at low concentrations.

In this work (ref.4,5) the sample stream radius was 15 μm and the laser beam diameter was 11 μm, resulting in an 11 pℓ probe volume. The excitation source was an argon-ion laser operated at 514.5 nm in the light-stabilized mode. The output of the argon-ion laser was 100% amplitude modulated at 10 kHz by an electro-optic modulator with a 50% duty cycle. Peak power out of the modulator was 500 mW. Approximately 6% of the fluorescent and scattered light from the interaction region was collected with a 32X, 0.6 NA microscope objective. A 0.75-mm pinhole was placed at the image plane of the microscope objective such that only light from the probe volume was passed to the detector. After the spatial filter spectral filtering consisting of a 550-nm, 40-nm FWHM bandpass interference filter, a 530-nm longwave passing interference filter and a 530-nm longwave passing colored glass filter, occurred. An S-20 photomultiplier with an estimated quantum efficiency of 12% at 550 nm was used as a detector. The output of the phototube was preamplified and sent to a lock-in amplifier. The signal and noise were determined by voltage to frequency converting and counting the lock-in amplifier output. Typically five to ten one-second readings were recorded and the average and standard deviation were computed.

RESULTS AND DISCUSSION

A calibration curve for aqueous solutions of rhodamine 6G was found to be linear over a range of five orders of magnitude. The lower bound is determined by the sensitivity of the technique and the upper bound by saturation of the detection electronics. The two-sigma concentration detection limit, determined from the measured noise in the blank and the signal from a 3.5×10^{-11} M rhodamine 6G solution, is 8.9×10^{-14} M. The detection limit corresponds to 18 attograms, or 22,000 molecules, of rhodamine 6G flowing through the probe volume during the one-second integration period of the electronics. This represents a factor-of-100 improvement in terms of number of molecules and a factor-of-30 improvement in terms of mass for molecular laser-induced fluorescence determination. The 11-pℓ probe volume and the 0.42 μℓ sampled volume are the smallest ever reported for molecular fluorescence detection of flowing samples.

At the detection limit there is a probability of 0.6 that a molecule is present in the probe volume at any instant in time. However, due to the one-second integration period of the detection electronics, this does not represent true single-molecule detection. In order to determine if true single-molecule detection is possible a complete characterization of the apparatus and the experimental parameters was performed.

With our present experimental arrangement we measured 2.9×10^5 photoelectron/s from background scatter, therefore in the 37 μs required for a molecule to traverse the laser beam an average of 10.7 photoelectrons would be counted in a photon-counting experiment. At our present detection limit, this corresponds to an average of 0.07 photoelectrons per molecule being counted. A set of reasonable improvements in the experimental apparatus, such as improved photon collection efficiency, reduced probe volume, slower linear flow velocity, and improved spectral and spatial filtering, indicates that we can increase the photoelectrons counted from fluorescence by a factor of 747 while increasing the background counts by only a factor of 6.5. Thus we predict that an improved version of the experiment would result in an average of 70 photoelectrons from background scatter and 52 photoelectrons from a single molecule's fluorescence being counted in a molecular transit time through the laser beam. The noise in the background is given by the square root of the number of background counts, hence a single molecule should be detectable with a signal-to-noise of about 6. Consideration of the Poisson distributions associated with the background scatter and the background scatter plus fluorescence in our projected apparatus shows that a discriminator level of 105 counts gives a 0.95 probability of detecting a single-molecule as it passes through the laser beam with only a 6×10^{-5} probability of mistaking the background for a molecule (ref.4). Thus it should be possible to detect single molecules with few errors. An apparatus which incorporates our improvements is now under construction and will soon be ready for testing.

Applications of this analytical technique which we envision are: (1) a sensitive and selective detector for high-performance liquid chromatography, (2) tracer systems where fluorescent micro-beads are used to demonstrate the connection or independence of two water systems, and (3) a fluorescent analog for radioimmunoassay.

REFERENCES

1 C. L. Pan, W. M. Prodan, W. M. Fairbank, Jr. and C. Y. She, Opt. Lett. 5 (1980) 459-461.
2 G. W. Greenless, D. L. Clark, S. L. Kaufman, D. A. Lewis, J. F. Tonn and J. H. Brodhurst, Opt. Commun. 23 (1977) 236-239.
3 M. R. Melamed, P. F. Mullaney and M. L. Mendelson, Flow-Cytometry and Sorting, 1979, Wiley, New York.
4 N. J. Dovichi, J. C. Martin, J. H. Jett, M. Trkula and R. A. Keller, Analytical Chem. (in press).
5 N. J. Dovichi, J. C. Martin, J. H. Jett and R. A. Keller, Science 219 (1983) 845-847.

KINETIC ANALYSIS OF LASER INDUCED PHOSPHORESCENCE IN URANYL PHOSPHATE FOR IMPROVED ANALYTICAL MEASUREMENTS*

B. A. BUSHAW

Chemical Methods and Kinetics Section, Pacific Northwest Laboratory, P.O.Box 999, Richland, WA 99352

ABSTRACT

Pulsed dye-laser excitation with multichannel scaler photon counting is used to obtain time resolved emission spectra of uranyl ions in aqueous solution. Kinetic analysis of this data corrects for matrix quenching and temperature effects which reduce the quantum yield of the uranyl ion luminescence. The method gives accurate measurements without separative prechemistry or the use of internal standards. Detection limits of one part-per-trillion (pptr) have been demonstrated, and in samples with concentrations greater than 100 pptr, relative standard deviations of less than 3% are achieved routinely.

INTRODUCTION

The phosphorescent emission of the uranyl ion has long been recognized as a sensitive means of analyzing for trace quantities of uranium (ref.1). Microgram quantities are easily determined with standard fused-salt and ultraviolet lamp excitation methods (ref.2). Recent interest in lower detection limits for environmental and industrial monitoring, geological survey, and bioassay applications has prompted a number of improvements in these methods. These have included using pulsed lamp (ref.3) or laser (ref.4) light sources with time-gated detection, taking advantage of the long emission lifetime of the uranyl ion to discriminate against scattered light and fluorescence from short-lived species. Addition of phosphates (ref.5) or fluorides have allowed analyses in aqueous solution. These "enhancers" complex the uranyl ion, preventing quenching by water (ref.6). Detection limits of a few picograms have been demonstrated (ref.7) using these improvements.

Phosphorescence quantum yield is very sensitive to quenching by a variety of species, particularly in solution (ref.8,9,10). The quantum yield is also sensitive to temperature (ref.7,10), ionic strength, and pH (ref.11). In the analysis of real samples these effects have generally limited the precision and accuracy of measurements. To control these effects, fused-salt methods have relied upon separative prechemistry to remove the quenchers, while the gated detection methods, with their greatly improved detection limits, have used

---

*This work was performed for the U.S. Department of Energy under contract DE-AC06-76RLO 1830.

sample dilution to reduce quenching. In practice, accurate measurements with either method requires the use of internal standards to compensate for any residual quenching. Also, care must be taken to prevent the internal standards themselves from significantly altering the quantum yield.

This paper reports a new technique using pulsed, dye-laser excitation with time resolved photon-counting detection to measure uranium concentrations in a variety of samples. The technique is unique in that multichannel scaler (MCS) photon counting provides discrimination against short-lived emitting species and scattered light while simultaneously correcting quenching effects. This correction, based upon kinetic analysis of the uranyl phosphorescence, has been found accurate for up to 80% quenching, even at sub-part-per-billion levels.

EXPERIMENTAL

Instrumentation

The instrumentation used for the measurements reported here is shown schematically in Fig. 1. A nitrogen laser pumped dye laser provides 0.3 millijoule pulses of 5 nanosecond duration at a 60 Hz repetition rate. The dye laser operates broadband using stilbene 420 dye, matching the excitation spectrum profile of the uranyl phosphate complex (see Fig.2). The sample cell is a standard ultraviolet grade fused silica fluorimetry curvette with a 1 cm path length. Emitted photons are collected at right angles to the laser path with a f1.0 lens. The collected light passes through a 10 nm bandpass interference filter, centered at 520 nm, to isolate a single vibronic band in the emission

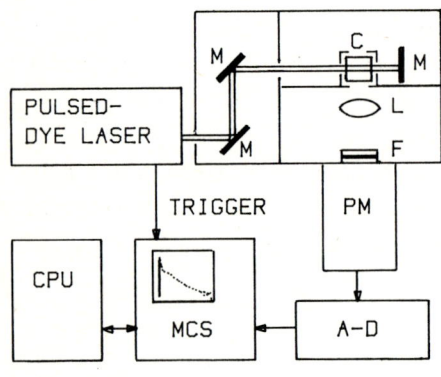
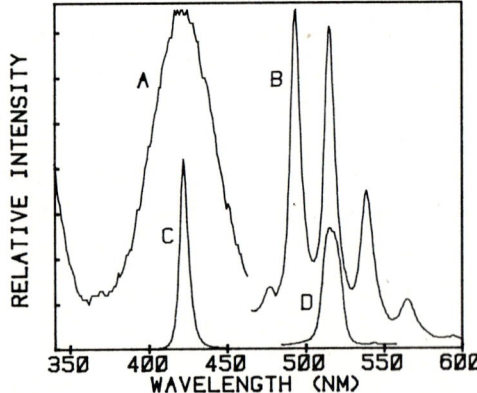

Figure 1. Instrumentation; see text for description. Abreviations: M, mirrors; C, sample cell and holder; L, lens; F, interference filter; PM, photomultiplier; A-D, amplifier - discriminator; CPU, computer.

Figure 2. Excitation (A) and emission (B) spectra of 20 ppm uranyl nitrate in 1 M phosphoric acid recorded with a conventional spectrofluorimeter. Dye laser spectral output (C) and bandpass of the interference filter (D) are also shown

spectrum (see Fig.2). A cooled photomultiplier detects the filtered emissions and resultant current pulses are processed by a fast amplifier and discriminator. A multichannel analyzer is operated in the MCS mode with a dwell of 10 microseconds per channel to provide time resolved counting of the discriminator output. Each laser pulse triggers an MCS sweep of ten millisecond duration. Typically, a measurement consisted of summing 6000 sweeps, corresponding to a 100 second aquisition period. Data is transfered to a minicomputer at the end of the aquisition period for background subtraction and kinetic analysis.

Sample preparation

Doubly deionized water (18 megohm/cm) was used in the preparation of reagents and where water is called for in the procedures. Perchloric and nitric acids used for wet ashing samples were J. T. Baker ULTREX grade. 1 M phosphoric acid was prepared by slowly adding 25 g. singly sublimed phosphorous pentoxide (J. T. Baker, ULTREX) to 100 ml water. After cooling, the solution was brought up to a total volume of 350 ml. A standard uranium stock solution was prepared by dissolving U_3O_8 (New Brunswick Laboratory standard reference #114) in concentrated nitric acid. Low level standards were prepared by dilution of the stock solution. All glassware was preleached by soaking in 6M nitric acid at 60°C for several days, then rinsed thoroughly with water.

Preparation of samples depended upon their origin. Water samples required no pretreatment other than adding a drop of nitric acid to assure that the uranium was in the +6 oxidation state, and to prevent adsorption onto surfaces. Samples of biological origin were wet ashed as follows. 2 ml sample, 0.5 ml nitric acid and 0.5 ml perchloric acid were placed in a 10 ml pyrex beaker and taken to dryness on a hot plate. The solid residue was dissolved in 2 ml water with a drop of 1 M nitric acid, quantitiatively transferred to a 5 ml volumetric flask, and brought to volume with water. A 1 ml aliquot of this solution was mixed with 1.5 ml of 1 M phosphoric acid in the fluorimetry cell. Brine water samples, which contain up to 25% salts, required second and third wet ashing with 1 ml perchloric acid to complete the removal of halide ions. Blanks used for background subtraction were prepared from 1 ml water and 1.5 ml 1M phosphoric acid.

RESULTS AND DISCUSSION

Kinetic analysis

The basis for kinetic analysis as a correction for quenching is as follows. If a population of uranyl ions is excited to the emitting state, decay of that population may be described by the first order differential equation:

$$d\,U^*{}_t/dt = -(k_p + k_q)[U^*]_t \tag{1}$$

where $[U^*]_t$ is the excited state population at time t, k_p is the intrinsic rate constant for phosphorescent decay, while k_q represents a sumation of the rate constants for all modes of radiationless deactivation. This includes intermolecular relaxations and external quenching effects discussed in the introduction. The validity of assuming first order kinetics may be seen in Fig. 3, showing the time resolved emission spectrum from a 7 ppb standard. The single component exponential decay can be observed clearly for more than seven lifetimes. The integrated form of (1) is a linear equation:

$$\ln([U^*]_t) = \ln([U^*]_0) - (k_p + k_q)t \qquad (2)$$

Since the number of detected photons is proportional to $[U^*]_t$, least squares fitting of a logarithmic plot of count rate versus time yields an intercept proportional to $\ln([U^*]_0)$ independent of any quenching effects. The effectiveness of this correction for quenchers may be seen in Fig. 4 which shows decay curves for a 1 ppb uranium standard containing varying amounts of chloride ion to alter the quenching rate. Note that the integrated emission intensity for the most highly quenched sample is less than 19% of the unquenched case, yet there is no apparent loss in the accuracy of the measurement - only a loss of statistical precision commensurate with the reduction in total counts recorded. Similar results have been obtained for cationic quenchers and for real samples where the source of quenching is unknown.

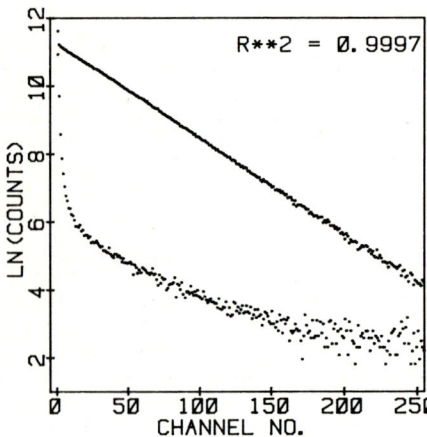
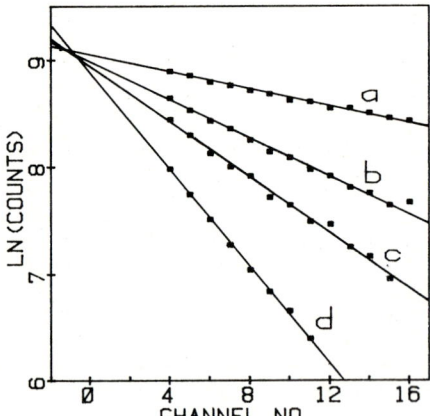

Fig. 3. Time resolved emission of a 7 ppb uranium standard. Background, which has been subtracted, is also shown.

Fig. 4. Time resolved emission from a 1 ppb uranium standard to which chloride has been added as a quencher. Millimolar chloride concentrations and observed lifetime in microseconds are: a) 0, 267; b) 0.17, 120; c) 0.33, 81; d) 0.66, 50.

Detection limits

Background due to luminescence from the cell, equivalent to approximately 0.1 ppb uranium, limits lower levels of detection. This background is relatively constant and detection limits of about 5% of its value, 5 pptr, have been routinely achieved. Detection limits of 1 pptr can be reached using a longer signal integration period, matched cells, and by paying particular attention to cell cleaning proceedures. The upper limits of detection, governed by count rate saturation in the detection electronics, are about 10 ppb. Higher level samples can be handled instrumentally a number of ways, however, in practice the simplest approach is to dilute the sample.

Sample analyses

Kinetic phosphorimetry has been successfully applied to the analysis of a wide variety of samples including urine, tissue digestions, ore dissolutions, brines, and ground and surface waters. Due to space limitations only those examples which have been verified by other means will be discussed here.

Table 1. shows the results for two deep groundwater samples and a series of brine waters derived from salt dome structures. The uranium concentration in these samples were also measured by U-233 tracer, filament source mass spectrometry (ref.12). Analyses were carried out in triplicate for both methods. The agreement is excellent: within the combined standard deviations in all cases.

TABLE 1

Comparison of kinetic phosphorimetry and U-233 tracer mass spectrometric analyses for low level uranium in brines and ground water.

Sample	Kinetic phosphorimetry		Mass spectrometry	
	vol. (ml)	U (ppb) (a)	vol. (ml)	U (ppb) (a)
ground water #1	1	0.010 ± .003	8000	0.007 ± .001
	25(b)	0.0078 ± .001		
	100(b)	0.0085 ± .001		
ground water #2	50(b)	0.025 ± .002	4000	0.025 ± .001
zone 1 brine	1	0.14 ± .01	500	0.22 ± .16(c)
zone 2 brine	1	0.23 ± .02	500	0.16 ± .03
zone 4 brine	1	0.36 ± .02	500	0.34 ± .05
zone 5 brine	1	0.43 ± .05	500	0.34 ± .04

(a) Error values are the standard deviation for three replicate samples.
(b) Samples were preconcentrated by acidifying, boiling and filtering off any silicate precipitates.
(c) The triplicate analyses yielded values of 0.12, 0.14 and 0.40. The latter sample probably suffered contamination during preparation.

Table 2. shows the results for repetitive analyses of a blind urine standard provided by an interlaboratory comparison program (Ref.13). Simple integration of counts over channels 5 to 45 provides results equivalent to that obtained by time-gated methods. These are included for comparison to the kinetic analysis measurements. The 46% average recovery for gated detection shows the need for internal standards with this approach. The direct accuracy and improved precision obtained with kinetic analysis is apparent.

TABLE 2
Comparison of time-gated detection and kinetic analysis of a blind, simulated urine standard. Uranium concentrations are reported in ug/Kg. The given value for the standard was 7.024 ug/Kg. Errors for individual kinetic analyses are the standard deviation derived from the intercept calculation.

Aliquot #	U time-gated	U kinetic
1	2.81	7.15 ± .03
2	3.24	7.11 ± .03
3	3.27	7.25 ± .02
4	3.94	7.34 ± .02
5	2.92	7.01 ± .03
mean	3.24 + .44	7.17 + .13
RSD	14%	1.8%
% of given	46	102

Conclusions
A new approach to phosphorescent analysis for uranium has been developed. The method is characterized by part-per-trillion detection limits and a dynamic range of over three orders of magnitude. Kinetic analysis of time resolved emission provides unprecedented precision and accuracy, automatically correcting for quenching effects. Minimal sample preparation is required and internal standards are not needed. Simplified chemistry and small sample volumes make the low cost analysis of large numbers of samples practical.

REFERENCES
1 E.L. Nichols and K.M. Slattery, J. Opt. Soc. Amer., 12 (1926) 449-66.
2 F.S. Grimaldi, I. May and M.H. Fletcher, U.S. Geological Survey Circular No. 199 (1952).
3 R. Kaminski, F.J. Purcelli and E. Russavage, Anal. Chem., 53 (1981) 1093-96.
4 J.C. Robbins, CIM bulletin, 793 (1978) 61-67.
5 C.W. Sill and H.E. Peterson, Anal. Chem., 19 (1947) 646-51.
6 H.D. Burrows and T.J. Kemp, Chem. Soc. Rev., 3 (1974) 139-65.
7 A.C. Zook, L.H. Collins and C.E. Pietri, Mikrochim. Acta, 2 (1981) 457-68.
8 Y. Yokoyama, M. Moriyasu and S. Ikeda, J. Inorg. Nucl. Chem., 38 (1976) 1329-33.
9 ibid., 39 (1977) 2205-09.
10 ibid., 39 (1977) 2211-14.
11 M.D. Marcantonatos, Inorg. Chim. Acta., 25 (1977) L101.
12 J.C. Laul, et. al., in preparation for submission to Geochim. Acta.
13 Technical evaulation of Draft ANSI Standard N13.30: "Performance Criteria for Radiobioassay".

PERFORMANCE APPRAISAL STUDIES OF LASER-ENHANCED IONIZATION IN FLAMES --
THE DETERMINATION OF NICKEL IN PETROLEUM PRODUCTS

G.C. Turk[1], G.J. Havrilla[1,3], J.D. Webb[2], and A.R. Forster[2]

[1]Center for Analytical Chemistry, National Bureau of Standards, Washington, D.C. 20234

[2]Shell Development Company, Westhollow Research Center, Houston, TX 77001

[3]Present Address: The Standard Oil Company (Ohio), Research Center, Cleveland, OH 44128

ABSTRACT

Laser-enhanced ionization (LEI) in flames is an ultrasensitive atomic flame spectrometric technique based on the efficient thermal ionization of atomic species which have been selectively excited by tunable laser radiation. The performance of LEI for real sample analysis is presently being evaluated. A successful determination of trace Ni concentrations in heavy oil flash distillate and Standard Reference Material Fuel Oil has been performed. One gram samples were diluted into 100-700 mL volumes of a xylene/n-butanol solvent mixture and aspirated directly into an air-acetylene flame. Stepwise laser excitation of Ni was performed using a Nd:YAG pumped dual-dye laser system. Accurate and reproducible results were obtained.

INTRODUCTION

Laser-enhanced ionization (LEI) in flames is an application of optogalvanic spectroscopy which is being developed as a method of trace metal analysis (ref.1-10). The method uses tunable laser radiation to specifically produce excited analyte atoms from a sample which has been aspirated into a flame and atomized. The rate of collisional ionization of the excited analyte atoms is enhanced as a consequence of their reduced energy barrier to ionization. This can be detected with appropriate electrodes as an increase in an electrical current conducted through the flame. Limits of detection have been measured for over 30 elements, and generally fall in the 0.1 ng/mL range (ref.5,6). Over the past few years, most of the underlying physical mechanisms of LEI have been the subject of several theoretical and experimental investigations (ref.11-15). The present emphasis in the development of LEI is on maintaining high levels of accuracy and sensitivity over the widest possible variety of sample matrices. Toward that end the performance of LEI for real sample analysis is being evaluated at the National Bureau of Standards. This paper will describe LEI performance when applied to the determination of Ni in heavy oil flash distillate (HOFD) and Standard Reference Material Fuel Oil (SRM

1634a).

The determination of Ni in petroleum process streams is important since Ni can act as a poison when deposited upon the surface of a catalyst. The heavy oil flash distillate sample used in this study is typical of petroleum process streams that might be used in hydrotreating reactors. The determination of Ni before and after hydrotreating is a useful monitor of catalytic activity.

METHOD

The LEI instrumentation is described in ref. 7. Briefly, the laser system is composed of an oscillator-only frequency doubled Nd:YAG laser which is beam split to pump two transverse-pumped oscillator-amplifier tunable dye lasers. The wavelength drives of the dye lasers are under computer control. A servo-tracking frequency doubling crystal is used to generate tunable ultraviolet radiation from one of the dye lasers. The flame is a premixed air-acetylene flame supported on a 5-cm single-slot burner. A 0.635 cm diameter stainless steel water cooled cathode is immersed in the flame with a 12 mm separation between the bottom of the cathode and the burner head. A -1500 V potential is applied to the water cooled cathode. The burner head is utilized as the anode. Direct current caused by natural flame ionization passes from the anode to ground through a 10 kΩ resistor. A 0.03 µF capacitor connects the anode to a 10^6 V/A pre-amp (ref. 16) for amplification of the pulses of LEI current. Further amplification and filtering is performed by a 1 kHz - 1 MHz filter-amplifier. A gated integrator, with a 1 µs aperture duration, is used for signal averaging of the filtered and amplified pulses. A laboratory micro-computer allows further data processing and storage.

For the best sensitivity and selectivity for Ni, stepwise laser excitation was performed. The two laser beams were aligned 1-2 mm below the cathode. The first step laser wavelength at 300.249 nm excites the Ni atom from the low-lying energy level at 205 cm^{-1} to 33501 cm^{-1}, and the second step wavelength at 561.479 nm further excites the Ni atoms from 33501 cm^{-1} to 51306 cm^{-1}, 10273 cm^{-1} below the ionization potential of Ni. Laser energies were approximately 0.1 mJ for the 1st excitation step, and 2 mJ for the second. These energies are about half of what the system is capable of delivering. The unfocussed laser beams were approximately 2 mm in diameter. Laser bandwidth was approximately 0.01 nm.

The samples were diluted with a 50/50 (v/v) xylene/n-butanol solvent mixture. Aspiration of neat xylene into the flame produced sooting, with a non-specific laser ionization background signal, and for this reason was not used. No such problems were encountered with the xylene/butanol mixture. Acetylene flow was reduced to keep the flame burning in a fuel-lean condition. The sample uptake rate was 1.9 mL/min.

Standard Reference Materials 1084 and 1085 (Wear Metals in Lubricating Oil), with certified Ni concentrations of 101±4 µg/g and 303±7 µg/g respectively, were used to prepare calibration standards from 50 to 300 ng/mL in the xylene/butanol solvent. The HOFD samples were diluted to approximately 1g/100 mL, and the SRM 1634a Fuel Oil samples were diluted to approximately 1 g/666 mL.

Wavelength scan data were collected with the gated integrator set for a 0.1 s (1 shot) real-time constant (i.e. 1 µs time constant with 1 µs aperture duration at a laser pulse rate of 10 Hz) and a signal value measured and stored at 0.001 nm intervals for uv scans, and at 0.002 nm intervals for visible scans. A 4 time constant pause was taken at each wavelength point before data collection. A Gaussian data smoothing routine was performed on each stored spectrum to enhance the signal to noise ratio. For the uv scans, a 0.007 nm halfwidth gaussian function was used; a 0.014 nm halfwidth gaussian was used for the visible scans.

Data collection for quantitative analysis was performed as follows. The gated integrator was set for a 1 s (10 shot) real-time constant. Five measurements were recorded at 5 s intervals with both wavelengths fixed on the analysis line centers. Background correction was then performed by tuning the second step laser off-resonance to 561.1 nm and again 5 replicate measurements were taken. The wavelength was then returned to line center, where 5 more measurements were recorded. The average of the total of 10 measurements taken on resonance, less the average of the background reading, was used as the analytical measure.

A bracketing run order of standard-sample-standard-sample ... was used to compensate for laser power intensity drifts which occured during the course of data collection. The results for each sample were calculated using the average calibration value from the preceeding and following standard.

RESULTS

Wavelength scans near the Ni analysis line were performed to measure spectral background and check for interfering lines. Figure 1 shows a series of three spectra recorded while aspirating the HOFD solution, and also the blank xylene/butanol spectra. Fig 1(a) was taken using single photon LEI near the Ni 300.2 nm line. Stepwise excitation LEI is shown in figure 1 (b) and (c). The first step wavelength was scanned in 1(b) with the second step wavelength fixed on the Ni second step resonance. The second step laser was scanned in 1(c), with the first step laser fixed on the Ni first step line.

Five lines are observed in the single photon scan shown in fig. 1(a). They are two Ni lines at 300.249 and 300.363 nm, two Fe lines at 300.095 and 300.303 nm, and an unidentified line at approximately 300.28 nm, which partially overlaps the Ni analysis line. The Ni signal intensity appears to be adequate, but line overlap interference is observed.

A better situation is seen in figs. 1(b) and 1(c), where stepwise excitation has selectively enhanced the Ni analytical signal by a factor of 35. Besides the improved sensitivity, the line overlap is no longer a problem. Single photon signals from the other nearby lines are still observed in fig. 1(b), but at a lower intensity relative to the analytical line. No nearby lines are observed in the fig. 1(c) second step scan. A small broadband background interference was observed, as can be seen in fig. 1(b) from the offset between the sample and blank baseline levels. The offset in fig. 1(c) includes this broadband background signal, together with the single photon Ni LEI signal from the first step laser.

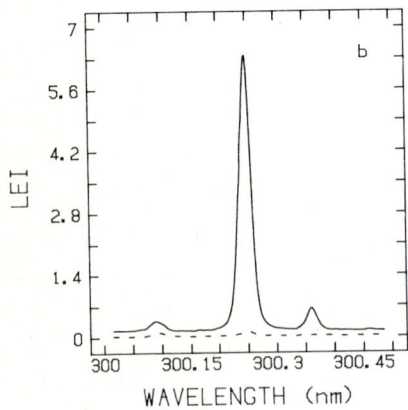

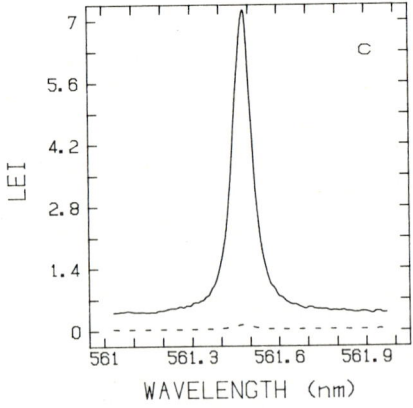

Fig. 1. LEI spectra for the determination of Ni in heavy oil flash distillate. Solid lines are sample signals while dashed lines represent the blank signals. (a) Single photon excitation near Ni analysis line at 300.249 nm. (b) Stepwise excitation with 2nd step wavelength fixed at 561.48 nm to enhance Ni signal. (c) Stepwise excitation with 2nd step wavelength scanned and 1st step wavelength fixed at 300.25 nm. LEI signal units are arbitrary, but consistent between spectra.

Background correction was performed by detuning the second step laser to 561.1 nm. A small (1/35th) loss in Ni signal is incurred by this method of background correction, since the Ni single photon signal is subtracted from the total signal together with the interfering background signal. No error occurs if background correction is performed in the same manner for samples and standards. Background correction can also be performed by detuning the first step laser to some point between the nearby atomic lines. In this case no analytical signal would be lost.

Table I presents results for the determination of Ni in the samples studied. Accuracy of the results is demonstrated by the agreement between the LEI result and the certified value for SRM 1634a. The largest component of the total uncertainties reported in Table I for the LEI results is the uncertainty in the certified values of Ni in the SRMs which were used to prepare the calibration standards. The total measurement reproducibility (excluding uncertainties in standards) for the analysis was approximately 1.5% at the 95% confidence limit.

TABLE I
Determination of Nickel in Petroleum Products

Sample	LEI Result[*] (µg/g)	Certified Value (µg/g)
SRM 1634a (Fuel Oil)	28.2 ± 1.2	29 ± 1
Heavy Oil Flash Distillate	12.6 ± .35	------

[*]Uncertainty at 95 % Confidence Limit

CONCLUSION

LEI spectrometry has been demonstrated to be an accurate and sensitive method for the determination of Ni in petroleum products. Although the Ni concentrations in the samples studied were high enough to be determined using conventional analytical methods (ref. 17), the very high signal to noise ratios observed indicate that LEI would perform well at much lower levels. The ratio of the net Ni signal to the standard deviation of the off-resonance background signal was approximately 700 for the HOFD sample and 150 for SRM 1634a. In addition, the very high sensitivity of LEI allows the use of much higher levels of dilution than is typical for conventional flame analysis. This minimizes any possibility of interferences resulting from viscosity differences between samples and standards.

REFERENCES

1 R.B. Green, R.A. Keller, P..K. Schenck, G.G. Luther, and J.C. Travis, J. Am. Chem. Soc., 98 (1976) 1517-1518.
2 G.C. Turk, J.C. Travis, J.R. DeVoe, and T.C. O'Haver, Anal. Chem. 50 (1978) 817-821.
3 G.C. Turk, J.C. Travis, J.R. DeVoe, and T.C. O'Haver, Anal. Chem. 51 (1979) 1890-1896.
4 G.C. Turk, Anal. Chem., 53 (1981) 1187-1190.
5 G.C. Turk, J.R. DeVoe, J.C. Travis, Anal. Chem., 54 (1982) 643-645.
6 J.C. Travis, G.C. Turk, and R.B. Green, Anal. Chem., 54 (1982) 1006A-1018A.
7 G.C. Turk, J.C. Travis, and J.R. DeVoe, in Proc. of Colloque International C.N.R.S. No. 352 on Optogalvanic Spectroscopy and its Applications, Aussois, France, June 20-25, 1983, J. Phys. Colloq., in press.
8 G..J. Havrilla, S.J. Weeks, and J.C. Travis, Anal. Chem. 54 (1982) 2566-2570.
9 A.S. Gonchacov, N.B. Zorov, Y.Y. Kuzyakov, O.I. Matveev, Anal. Lett. 12 (1979) 1037-1041.
10 C.A. Van Dijk, F.M. Curran, K.C. Lin, and S.R. Crouch, Anal. Chem. 53 (1981) 1275-1279.
11 J.C. Travis, P.K. Schenck, G.C. Turk, and W.G. Mallard, Anal. Chem. 51 (1979) 1516-1520.
12 K.C. Smyth, P.K. Schenck, and W.G. Mallard, in D.R. Crosley (ed.), Laser Probes of Combustion Chemistry, ACS Symposium Series 134, American Chemical Society, Washington, D.C., 1980, Ch. 12.
13 C.A. Van Dijk, C.Th.J. Alkemade, Comb. and Flame 38 (1980) 37-49.
14 T.Berthoud, J. Lipinsky, P. Camus, J.L. Stehle, Anal. Chem. 55 (1983) 959-962.
15 P.K. Schenck, J.C. Travis, G.C. Turk, and T.C. O'Haver, J. Phys. Chem. 85 (1981) 2547-2557.
16 G.J. Havrilla, and R.B. Green, Chem. Biomed. Environ. Instrum. 11 (1981) 273-280.
17 R.J. Brown, Spectrochim. Acta, 38B (1983) 283-259.

LASER SPECTROSCOPIC STUDIES OF SOLVATION PHENOMENA

M. J. Wirth

Department of Chemistry, University of Wisconsin, Madison, Wisconsin 53706 USA

ABSTRACT

Two-photon spectroscopic measurements have been shown to provide solvation structural information (ref. 1). The principles of this method are outlined and applications under current study are given. Future spectroscopic measurements on the picosecond time scale promise to provide independent information about the dynamics of solvation interactions.

INTRODUCTION

The measurement of solvation structures in liquid solution is of pervasive chemical interest. The disordered and dynamic nature of liquids has been a considerable challenge to both theoretical and experimental progress in the fundamentals of solvation. Improvements in computer technology have allowed great advances in the theory of liquids, showing, for example, that short range structure is important. On the other hand, experimental methods to prove solvation structures have been lacking entirely. This work describes the original development at Wisconsin of laser spectroscopic techniques to generate solvation structural information.

TWO-PHOTON SYMMETRY PERTURBATIONS AS SOLVATION STRUCTURAL PROBES

Background

In order to discuss solvation structural probing by two-photon spectroscopy, it is useful to outline the principles of the spectroscopic process. What two-photon spectroscopy uniquely offers is the capability of assigning excited state symmetry through the use of polarization measurements (ref. 2). Consider an oriented naphthalene molecule with the long axis aligned with the laboratory x direction, and the molecular plane as the laboratory xy plane, shown in Fig. 1. Absorption of one photon polarized along the x-axis gives an excited state polarized along the x-axis, when the wavelength of the photon is at a resonance. One can think of this as conservation of symmetry. Analogous behavior occurs for y axis excitation; however, for z axis excitation there is no absorption because the planar aromatic molecules have no states polarized in the z direction. For an oriented single crystal, one-photon polarization experiments can thus be used to assign the symmetries of the excited states.

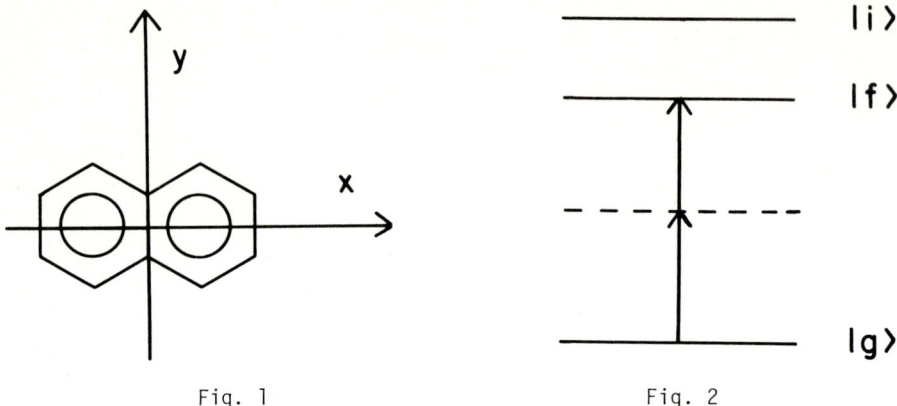

Fig. 1   Fig. 2

Absorption of two photons simultaneously proceeds according to the energy level diagram shown in Fig. 2. If the two photons are polarized along the x-axis, they excite to a totally symmetric state. In the conservation of symmetry argument, the polarizations of the two photons cancel, resulting in an excited state having the same symmetry as the ground state, which is totally symmetric. The response of an oriented molecule to two x-polarized photons is designated S_{xx}. Likewise, two y-polarized photons give rise to a totally symmetric excited state, but by a different amount, denoted S_{yy}. $S_{xx} \neq S_{yy}$ because the intermediate states are different for the two processes. It is also possible to make the two photons perpendicularly polarized, where the x-polarized photon connects the ground state to the intermediate state, and the y-polarized photon connects the intermediate state to the final state. The corresponding response is denoted S_{xy}, and interaction in the reverse order gives S_{yx}.

It is convenient to write the responses in the form of 2x2 matrices called tensor patterns. McClain (ref. 3) has shown that form of the tensor pattern is characteristic of the excited state symmetry, given the molecular point group. For the D_{2h} point group, of which naphthalene is a member, the tensor patterns for the A_g and B_{1g} states, which are the only two excited state symmetries allowed in a two-photon process, are as shown in Fig. 3. The diagonal elements, S_{xx} and S_{yy}, are nonzero for a totally symmetric excited state and zero for an antisymmetric state. The off-diagonal elements are zero and nonzero for totally symmetric and antisymmetric excited states, respectively. Experimental determination of the tensor pattern thus reveals the excited state symmetry. Monson and McClain have shown that the tensor patterns can be determined even for materials that are randomly oriented, such as liquids and gases (ref. 4). For excitation by a single laser, where both photons have the same wavelength, the tensor element information is obtained from measuring the two-photon excited

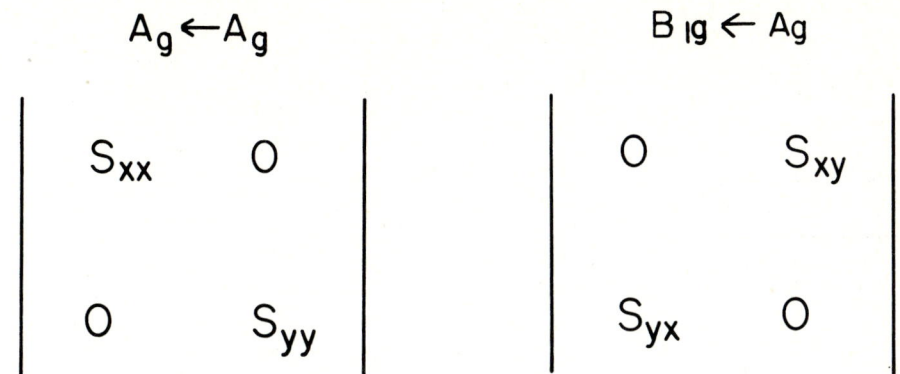

Fig. 3. Tensor patterns for unperturbed planar molecules of the D_{2h} point group.

fluorescence from linearly polarized excitation $F_{//}$, and that from circularly polarized excitation, F_{cc}. These are related to the tensor elements as follows.

$$\delta_f = [3F_{//} - 2F_{cc}]/k = [S_{xx} + S_{yy}]^2 \qquad (1)$$

$$\delta_g = [1\ F_{//} + 1\ F_{cc}]/k = S_{xx}^2 + S_{yy}^2 + S_{xy}^2 + S_{yx}^2 \qquad (2)$$

Thus, a determination of δ_f alone can distinguish between the A_g and B_{1g} excited states of naphthalene since a nonzero value arises only from an A_g state and a zero value from a B_{1g} state. Determination of δ_g provides a useful normalization factor to scale δ_f for oscillator strength and laser characteristics. Koskelo (ref. 5) has shown that the utility of δ_f in the single laser experiment is diminished by interference effects among wavefunctions, resulting in occasional observation of inordinately small δ_f values for totally symmetric excited states. This problem can be circumvented by the use of two laser experiments to determine the tensor patterns.

Solvation Symmetry Perturbations

The interaction of a molecule such as naphthalene with a nearby molecule slightly changes the distribution of electron density in the naphthalene rings. The spatial properties of the electron density distribution are related to the potential energy surface for the interaction between naphthalene and its neighbor. If the spatial arrangement is of lower symmetry than the molecular point group, then the electron density distribution of the ground state is no longer totally symmetric in the molecular point group. The effect on the quantum states of naphthalene is that new states of lower symmetry are mixed into the ground state, where these new states are other states of isolated naphthalene. New states are also mixed into the excited states. First order perturbation theory can be used to describe the contributions of new states $|n>$ upon perturbation of a state $|k^o>$, by the solvation Hamiltonian H^1.

$$|k> = |k^o> + \sum_j \frac{<n_j|H^1|k^o>}{E_{n_j} - E_{k^o}} |n_j> \qquad (3)$$

For the integral $<n|H|k^o>$ to be nonzero, the product of the symmetry representations of $|n>$, H^1 and $|k^o>$ must be totally symmetric. For example, if the state $|k^o>$ is totally symmetric, and if H^1 has B_{1g} symmetry, then only states $|n>$ having B_{1g} symmetry can be contributed to $|k>$. The two-photon experiment then determines the symmetry of H^1 by sensing the symmetry of $|n>$.

The new states $|n>$ contribute to the magnitudes of the tensor elements according to the symmetries of the $|n>$ states. In the previous example, the B_{1g} symmetry of $|n>$ would contribute to the off-diagonal tensor elements, which are nominally zero for the A_g excited state. Denoting the contributions to the tensor elements from the solvent perturbation as S^1_{xy} and S^1_{yx}, the tensor pattern for an A_g state perturbed by a B_{1g} solvation environment is shown in Fig. 4. The magnitude of either of the off-diagonal elements in the $A_g \leftarrow A_g$ case can serve as a measure of the magnitude of B_{1g} solvation perturbation. Such a unique relationship connecting $|k^o>$, H^1 and $|n>$ exists for all states, thus the contribution of any symmetry representation in this solute point group can be determined by selecting the appropriate tensor element for measurement. The fact that these relationships are unique is crucial to the success of the technique because it allows for only one possible symmetry component of H_1 for a detected symmetry of $|n>$.

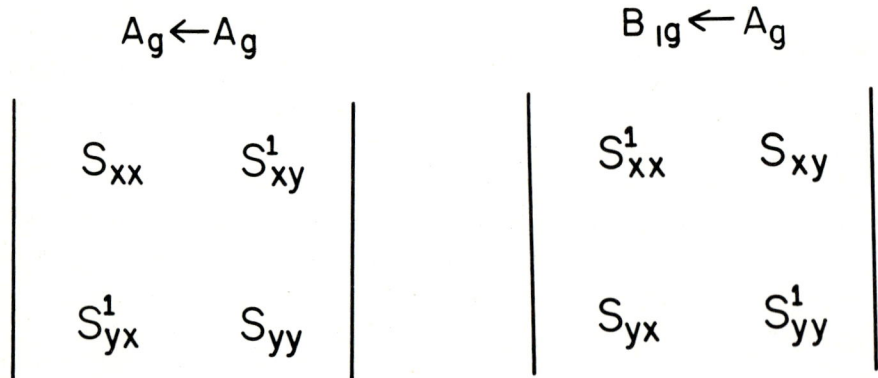

Fig. 4. Example tensor patterns of perturbed, planar molecules of the D_{2h} point group, where the solvation symmetry is described by only one representation.

In liquid solutions, there is sufficient disorder that one would expect H^1 to have components of all possible representations of the point group symmetry. It is thus desired to determine the amount that each point group representation contributes to the representation of H^1, which can be expressed for the D_{2h} point group as follows.

$$\Gamma_{H^1} = a\,\Gamma_{Ag} + b\,\Gamma_{B_{1g}} + c\,\Gamma_{B_{2u}} + d\,\Gamma_{B_{3u}} \tag{4}$$

The solvation structural information obtainable from two-photon spectroscopic measurements is in the magnitudes of a, b, c and d. A solvation structure that is highly disordered with respect to the molecular point group would have approximately equal values of a, b, c and d. A significant amount of order is indicated by a dominance of one or several of the coefficients. As with other complex media such as surfaces or polymer solutions, the solvation measurements are most insightful when interpreted in conjunction with theoretical studies.

In principle, measurement of the diagonal and off-diagonal tensor elements both for a transition to a gerade state and a transition to an ungerade state, which entails four measurements in all, provides the values of a, b, c and d. Since the point group representations are orthogonal, these four values provide four degrees of freedom regarding the solvation structure. For planar solutes of lower point group symmetry, such as C_{2v}, there are two degrees of freedom. Two degrees of freedom may be sufficient to distinguish among possibilities in some cases. For nonplanar solutes, additional information is obtainable from the representations containing components in the out-of-plane direction. More sophisticated experiments are needed to gather this information because the number of independent polarization experiments must be as large as the number of degrees of freedom.

The single laser measurement of δ_f and δ_g does not allow direct quantitative determination of the symmetry components of H^1 because the experiment does not isolate individual tensor elements. Less direct information is nonetheless available because relative amounts of S^1_{aa} and S^1_{ab} can be estimated from δ_f/δ_g. Such information has been valuable in developing and demonstrating the concept of two-photon measurements of solvation structures. More quantitative experiments will involve: 1) two color laser excitation to isolate tensor elements, 2) single crystal studies to quantify the relation between the magnitudes of the tensor elements and the coefficients a, b, c, and d, and 3) theoretical studies to predict solvation structures which can be tested by the two-photon experiment.

APPLICATIONS

Even in this early stage of development, two-photon studies of solvation have already provided information regarding the degree of structural order in solvation environments. For the aromatic solutes naphthalene, phenanthrene, chrysene and fluorene, it has been shown that the solvents cyclohexane, benzene and benzyl alcohol give disordered solvation environments, while significant order exists for solvation by methanol (ref. 1). The latter is currently under detailed investigation involving other aromatic solutes and several transitions for each molecule. The nature of the ordering in methanol is important in understanding the origin of the relatively large solubilities of aromatics in this solvent, and in explaining the chromatographic retention behavior of aromatics with methanol as the mobile phase.

DYNAMICAL METHODS

Two two-photon method detects the symmetry of the environment on the sub-picosecond time scale, avoiding complications due to solvation dynamics. Independent information about the structure can be obtained from direct studies of the dynamics because, as the solute randomly moves, it samples the structure of its environment. A rotational diffusion study has shown that the solvation structure can dramatically affect the directions of the motions of the solute molecule (ref. 6). On the picosecond time scale, spectroscopic bandwidths have been shown to reveal the packing interactions controlling the chromatographic retention mechanism in reverse phase chromatography (ref. 7). For the latter, dephasing measurements, i.e., lineshape studies in the time domain, can potentially reveal exchange rates of solvent molecules in the solvation environment. Rapid progress on the fundamental understanding of a phenomenon as complex as solvation can be made through the use of such a variety of experimental tools to examine the constituent processes.

REFERENCES

1. M.J. Wirth, A.C. Koskelo and C.E. Mohler, J. Phys. Chem., in press.
2. W.M. McClain and R.A. Harris, in Excited States V. 3, E.L. Lim (Ed.), Academic Press, New York, 1977, pp. 1-56.
3. W.M. McClain, J. Chem. Phys. 55 (1971) 2789.
4. P.R. Monson and W.M. McClain, J. Chem. Phys. 53 (1970) 29.
5. A.C. Koskelo and M.J. Wirth, J. Chem. Phys., submitted for publication.
6. M.J. Sanders and M.J. Wirth, Chem. Phys. Lett., in press.
7. M.J. Wirth, D.A. Hahn and R.A. Holland, Anal. Chem. 55 (1983) 787.

LASER SPECTROSCOPIC STUDIES OF VIBRATIONAL OVERTONES. INFLUENCE OF MOLECULAR CONFORMATION ON OVERTONE ENERGIES.

HOWARD L. FANG and ROBERT L. SWOFFORD
The Standard Oil Co., Corporate Research Center, Warrensville Heights, OH 44128

ABSTRACT

Intracavity cw dye laser photoacoustic spectroscopy is an extremely sensitive way of studying vibrational overtones in gas-phase samples. New details of the influence of conformation on bond strength are revealed.

INTRODUCTION

We have been using photoacoustic spectroscopy (PAS) to study the near-ir and visible absorption of molecules. Here can be seen the high vibrational overtones ($\Delta v = 3,4,5,6,7...$) of X-H stretching (X = C,N,O,etc.) The spectra are not readily explainable in terms of normal modes built on the coupled motions of individual atomic vibrations in the molecule. On the other hand, the local mode (LM) model can explain the spectra rather simply (ref. 1). The LM model treats a molecule as a set of loosely-coupled anharmonic oscillators localized on the individual X-H bonds. Thus the simple anharmonic progressions are described as the overtones of a single X-H stretching oscillator belonging to one of the types of X-H bonds in the molecule.

The overtone spectra reveal the presence of nonequivalent X-H oscillator sites in the molecule. Primary, secondary, and tertiary alkyl C-H and aromatic C-H oscillators are easily distinguished from one another. More subtle differences between the local environments among otherwise "equivalent" C-H can be seen, e.g., in cyclohexane, where distinct bands are assigned as overtones of the equatorial and axial C-H oscillators (ref. 2). The interconversion rate of the "chair" cyclohexane is not sufficiently rapid to motionally average the C-H sites. More recently, it has been shown that the conformationally nonequivalent C-H bonds in a methyl group can be resolved (ref. 3). For the present report we show several examples to demonstrate the information available from an overtone spectral study.

EXPERIMENT

The overtone absorption spectra of the gas phase, room-temperature samples were measured by intracavity cw dye laser PAS (ref. 4). The spectral range was covered by a number of laser dyes pumped by a krypton-ion or argon-ion laser. Samples were transferred into all-glass PA cells at a pressure of approximately 20 torr. The PA signal was corrected for variation in intracavity dye laser power.

RESULTS AND DISCUSSION

Figure 1 shows the spectrum of gas-phase ethanol. The relative sensitivity is indicated for the individual segments. The most prominent features are the O-H stretching overtones, $\Delta v(OH) = 3,4,5$ and 6. The weaker, broader features are the C-H stretching overtones of primary and secondary C-H oscillators, $\Delta v(CH) = 4,5,6$ and 7. Our earlier studies of alkanes showed that secondary C-H oscillators have the lower overtone transition energies (ref. 2). The remaining weak bands are associated with combinations involving an X-H stretch overtone and some lower frequency motion in the molecule. The present discussion will concentrate only on the prominent O-H overtones.

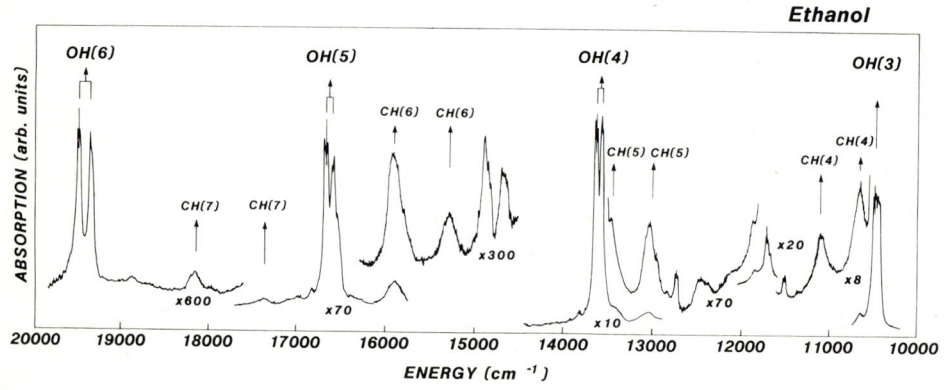

Fig. 1. Photoacoustic spectrum of gas-phase ethanol (T=296K, P=20 torr).

Figure 2 shows an expanded plot of the O-H overtones. Two bands are observed at each overtone and are attributed to rotational conformers. The torsional potential of the O-H group about the C-O axis gives rise to three equilibrium positions: one <u>trans</u> conformer (absolute energy minimum) and two equivalent <u>gauche</u> conformers. The O-H overtones show rotational structure which can be partially resolved with the present cw dye laser (nominal linewidth 0.5 cm^{-1}). The characteristic appearance of the rotational contour within each band provides firm evidence for the conformational assignments. The well-defined P,Q,R structure of the higher-energy band at each overtone is evidence of a nearly "parallel-type" transition while the low-energy band is

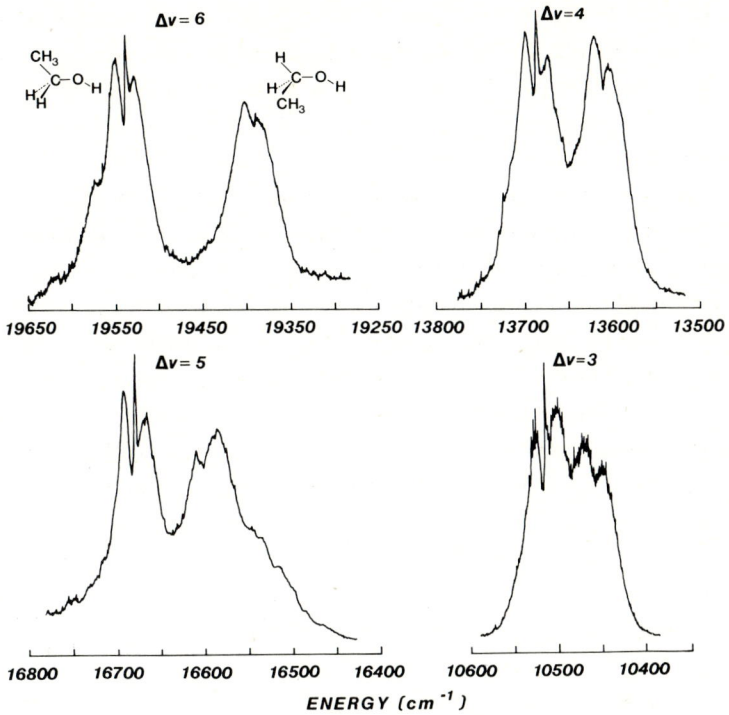

Fig. 2. Expanded plot of ethanol O-H overtones from Fig. 1.

attributed to a "perpendicular-type" transition. In the former, the O-H transition moment is essentially parallel to the figure axis, while in the latter, the transition is essentially perpendicular to the figure axis. The energy separation of the two bands increases with quantum number, but the bandwidths are essentially unchanged. The asymmetry on the low-energy side of the perpendicular band at $\Delta v(OH)=5$ is not understood. Similar behavior has been seen for methanol (ref 3). It is possible that a combination built on the next lower O-H overtone is responsible for the appearance of the band.

We note that in the _gauche_ conformer of ethanol, the O-H is _trans_ to hydrogen. The influence which a _trans_ hydrogen has on the strength of an X-H bond has been reported by many workers, but a thorough explanation of the effect has not been proposed. We have seen consistently in our overtone studies that the transition frequencies of X-H _trans_ to hydrogen are significantly lower.

It is also interesting to note that the lower-energy transition at each overtone does not correspond to the lower-energy rotational conformer. In work which we are reporting elsewhere, we have measured the temperature dependence of the relative overtone intensities of the two conformers in order to determine the enthalpy difference (ΔH). Our results indicate ΔH = 0.7 kcal/mole, in good agreement with previous measurements.

Recently we reported the overtone spectroscopy of nonequivalent methyl C-H oscillators and the influence of conformation on the overtone energies (ref. 3). There it was seen how the strength of a suitably oriented methyl C-H bond was influenced by the presence of an adjacent anisotropic moiety such as lone-pair electrons. An adjacent C=C bond also leads to nonequivalent methyl C-H oscillators, as was seen in an earlier study of propene (ref. 4). The Δv(CH)=5 overtone in propene is shown in Fig. 3. In previous work we used deuterium substitution to assign the two highest energy peaks as overtone transitions of the olefinic terminal C-H, the middle peak as olefinic nonterminal C-H, and the two lowest energy peaks as methyl C-H. At that time we were unable to assign the olefinic terminal C-H as <u>cis</u> or <u>trans</u> to methyl. Recently we recorded the Δv(CH)=5 overtone of propene-cis-1,2-d2, which is also shown in Fig. 3. We can thus assign the highest energy band in propene-cis-1,2-d2 as the olefinic terminal C-H <u>cis</u> to methyl. This band corresponds to the second-highest energy band in propene. The assignment is consistent with our previous experience that C-H bonds <u>trans</u> to C-H have lower overtone energy than C-H bonds <u>trans</u> to C-C.

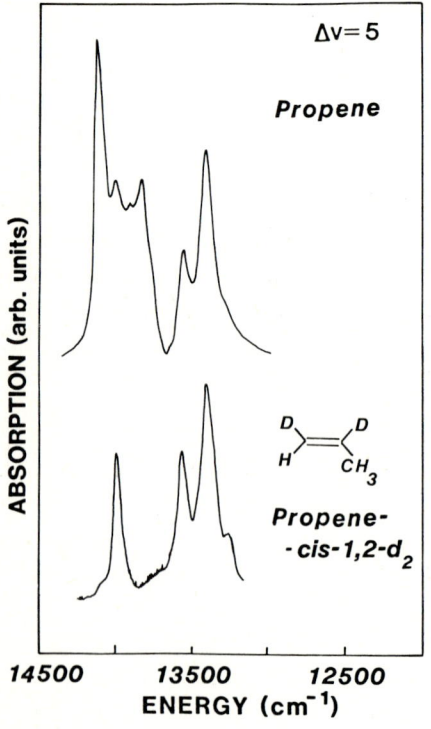

Fig. 3. Comparison of the photoacoustic spectra of propene and propene-cis-1,2-d2 at Δv(CH)=5. The use of selective deuteration allows the assignment of the highest-energy peak in propene to the olefinic terminal C-H <u>trans</u> to methyl.

We are extending our alkene studies to more complicated molecules and will report on these studies in detail elsewhere. As an example of that work, we show in Fig. 4 the overtone absorption spectrum of 1,3-butadiene. As can be seen, the spectrum is quite similar to that of propene except for the absence of methyl C-H. In Fig. 5 we show the spectrum of 2,3-dimethyl-1,3-butadiene, which is quite similar to that of propene except for the absence of the vinyl nonterminal C-H. However, detailed studies have shown that in the latter case the energy order of the vinyl terminal C-H is reversed, with the C-H <u>trans</u> to the backbone C-C being lower than the C-H <u>cis</u> to C-C. The two lowest-energy peaks at each overtone are the methyl C-H, the lower energy band being the two out-of-plane C-H. There is an intriguing increase in the bandwidths of the methyl C-H peaks at $\Delta v(CH)=6$. This may be due to the influence of methyl group torsion in the molecule. Other evidence indicates that these bandwidths are sensitive indicators of steric crowding of the methyl groups.

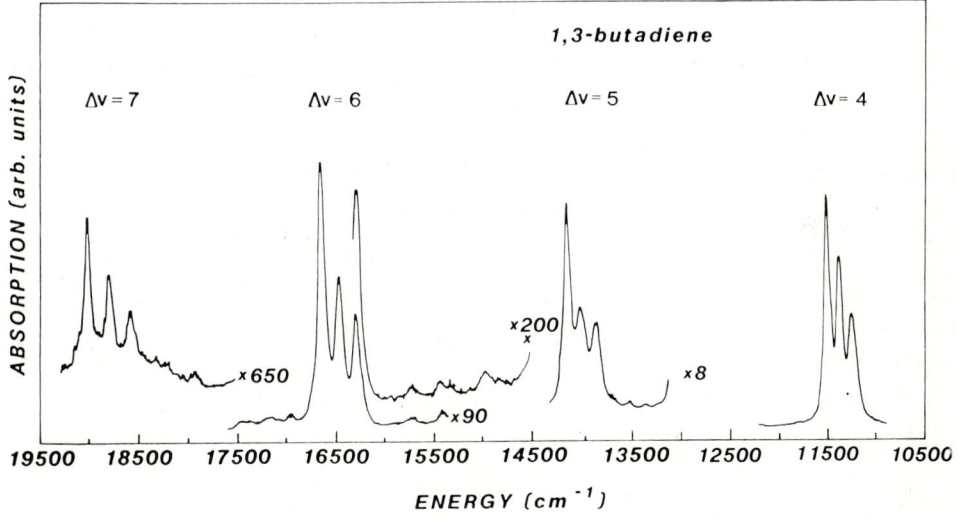

Fig. 4. Photoacoustic spectrum of gas-phase trans-1,3-butadiene (T=296K, P=20 torr).

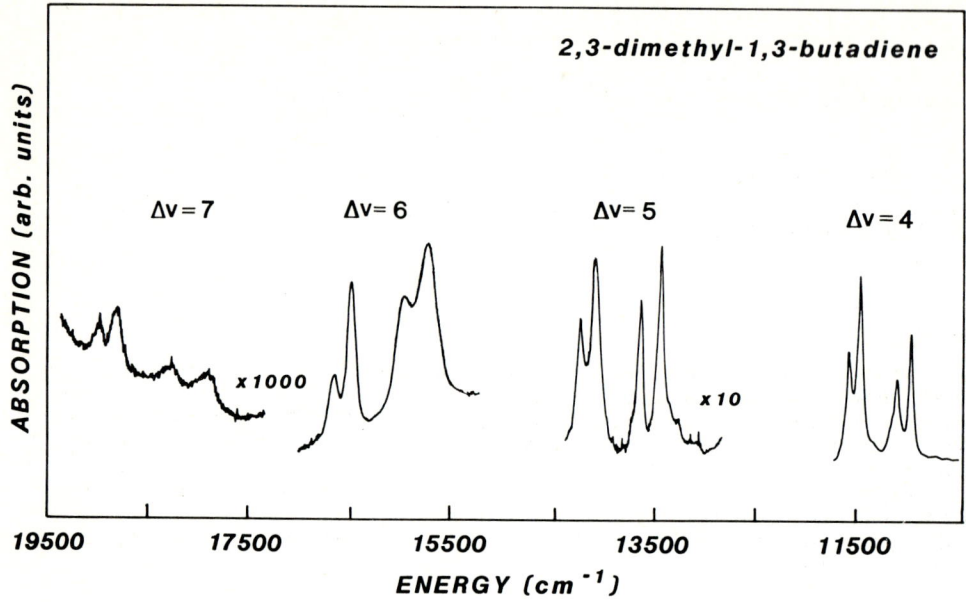

Fig. 5. Photoacoustic spectrum of gas-phase 2,3-dimethyl-1,3-butadiene (T=296K, P=20 torr).

CONCLUSION

We have attempted to demonstrate the potential of this new approach to vibrational spectroscopy. We expect that these studies will have an important impact on our understanding of the influence of molecular structure on the strengths of individual chemical bonds. The prospect of high-resolution vibrational spectroscopy on polyatomics isolated by supersonic expansion techniques is truly exciting, and we are now in the process of setting up these experiments.

REFERENCES

1. B.R. Henry, Acc. Chem. Res. 10, 207 (1977);
   R.G. Bray and M.J. Berry, J. Chem. Phys. 71, 4909 (1979);
   J.S. Wong and C.B. Moore, J. Chem. Phys. 77, 603 (1982).
2. H.L. Fang and R.L. Swofford, J. Chem. Phys. 73, 2607 (1980).
3. H.L. Fang, D.M. Meister and R.L. Swofford, J. Phys. Chem., in press.
4. H.L. Fang and R.L. Swofford, Appl. Opt. 21, 55 (1982).

W.S. Lyon (Editor), *Analytical Spectroscopy*
Elsevier Science Publishers B.V., Amsterdam — Printed in The Netherlands

INFRARED LASER SPECTROSCOPY USING A LONG PATHLENGTH ABSORPTION CELL*

K. C. Kim and R. A. Briesmeister
University of California, Los Alamos National Laboratory, P. O. Box 1663,
Los Alamos, NM 87545 (U.S.A.)

ABSTRACT

   We developed two very long pathlength absorption cells to be used in conjunction with diode lasers. They were designed to operate at controlled temperatures with the optical pathlength variable up to ~1.5 Km. Not only very low sample pressures are used for studies with such cells but also the spectroscopic sensitivity is enhanced over conventional methods by a factor of 10^3 to 10^4. In this paper we present some analytical aspects of the diode laser spectroscopy using the long pathlength absorption cells in the areas of absorption line widths, pressure broadening coefficients, isotope composition measurements and trace impurity analysis.

INTRODUCTION

   The development of tunable ir laser sources in recent years has played a major role in laser spectroscopy. [1] The high resolution capability greatly adhanced the study of rovibrational transitions in molecules. Diode lasers with the tunability and the narrow linewidth have been responsible for striking advances in infrared spectroscopy. [2]

   Several years ago, we recognized the need for an absorption cell which utilizes fully the high resolution capability of diode lasers. The absorption measurements in an ordinary experimental cell may require typically a few torr pressure of sample gas. At these pressures the absorption lines are usually pressured-broadened and, therefore, closely spaced transitions are poorly resolved even at diode laser resolution. This situation is greatly improved if spectroscopic measurements can be made on molecules before they have time to collide at extremely low sample pressures.

   We developed two very long pathlength absorption cells to be used in conjunction with diode lasers. The first cell had a physical length of 10.5 meters [3] and the second of 4.5 meters. [4] They were designed to operate at controlled temperatures in the range from ~100K to 300K with the optical pathlength variable from 16 m up to about 1.5 Km. Obviously, very low sample pressures are used for studies with such cells. The spectroscopic sensitivity is also enchanced over conventional methods by a factor of 10^3 to 10^4, improving

---

*Work performed under the auspices of the U.S. Department of Energy.

the analytical capability of measuring particle densities to the order of 1×10^{1} molecules/cm^3.

Over the past few years, we used these two spectroscopic facilities for a variety of spectroscopic studies of fundamental molecular parameters [5], vibrational intensity measurements [6] and laser photochemistry. [7] In this paper, we present some other analytical aspects of the diode laser measurements using the long pathlength absorption cells. They are measurements of absorption line widths, pressure broadening coefficients, isotopic composition, and trace impurity analysis.

DIODE LASER SPECTROMETER WITH A LONG PATHLENGTH CELL

Figure 1 shows a schematic diagram of the long pathlength absorption cell. The cell employs conventional white cell optics. We found that white cell optics are most suitable for long pathlength optical transmission in terms of operational ease and are relatively immune to vibrational interferences. The separation of the reflecting mirror surfaces is the same as the radius of the curvature. The laser beam enters the optical cavity as a point source and exits from the same side with very little distortion of the image. Thus it is relatively simple to sort out beams that have made different number of passes between the mirrors.

Figure 2 shows an experimental set-up utilizing a diode laser spectrometer. The monochromator shown is used to select a longitudinal mode of the laser output and also serves as an approximate wavelength reference. The wavenumber calibration is done using a germanium etalon. For example, for a 1 in. germanium etalon, the free spectral range at 16 μm is 0.048268 cm^{-1}. Accurate wavenumber markers are established by recording absorption lines of well-known

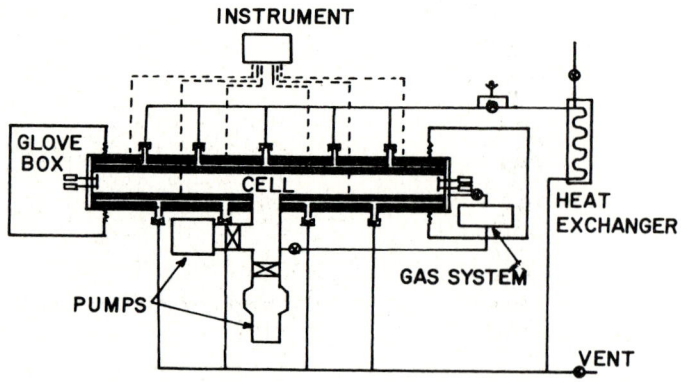

Fig. 1. A schematic diagram of the long pathlength absorption cell.

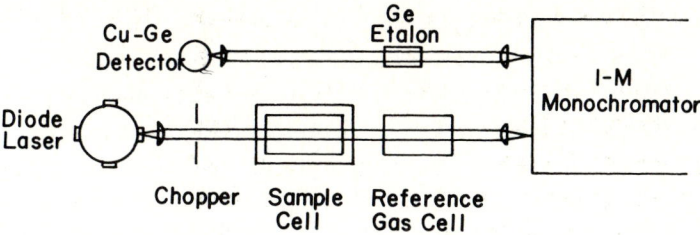

Fig. 2. An experimental set-up utilizing a diode laser spectrometer.

frequency standards (such as CO_2 and NO_2, etc). [8] The diode lasers are current- and temperature-tuned to cover approximately 0.5 cm^{-1} range. The details of diode laser operation were thoroughly reported previously [1] and will not be discussed here. Figure 3 shows a typical diode laser output tuned over a narrow spectral range, a few absorption lines and an etalon trace.

LINEWIDTH MEASUREMENTS

The amount of sample gas is typically a few mtorr, and at these low pressures it is possible to make Doppler-limited spectroscopic measurements. An example of pressure-broadened line shapes of N_2O in the 16 μm has been shown in Fig. 5 of Ref. 3. Peak absorption of a pressure-broadened absorption line does not increase with increasing sample size. Frequently there are other adjacent

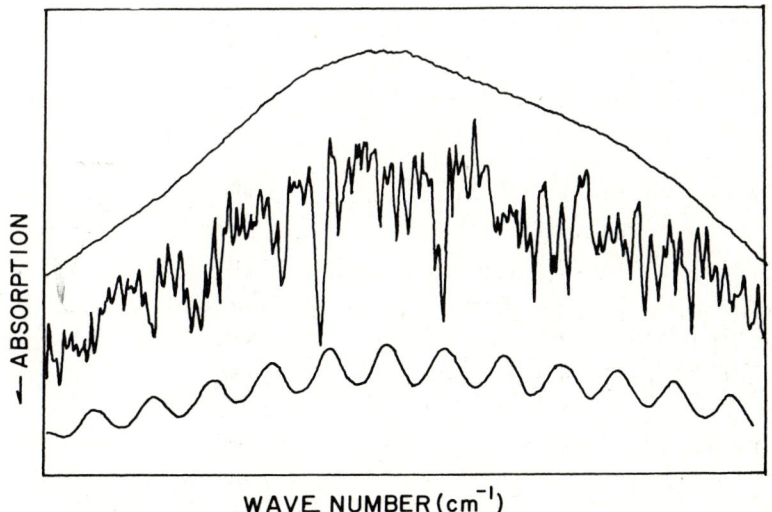

Fig. 3. A typical diode laser output tuned over a narrow spectral range, absorption lines of a heavy hexafluoride molecule and an etalon trace in the 16 μm region.

interfering absorption lines which obscure the spectroscopic identification. Spectroscopic measurements of low pressure sample gases therefore enhance the analytical sensitivity and makes maximum use of the diode laser resolution.

Although the spectral range that can be covered is small and limited within a few wavenumbers because of the very nature of the diode laser, the Doppler widths of absorption lines provide information on the molecular weight of the sample gas. In Table I, we summarize experimentally measured linewidths of several gases compared to the room temperature Doppler widths evaluated at the particular transition wavenumbers [$\Delta v_D = 2v_o (2 kT \ln 2/Mc^2)^{1/2}$]. Only those absorption lines that appear free of fine structures were used for the line width measurements. The measurement of Doppler widths alone seldom identifies a particular gas sample but approximate molecular weights combined with accurate wavenumbers narrow down the list of possible compounds. Obviously, an FTIR spectrometer which can cover a wider spectral range with an intermediate resolution will be an ideal match with the long pathlength absorption cell for trace impurity analysis. For most polyatomic molecules, however, the Doppler widths are narrower than the resolution of commercially available FTIR spectrometers.

OTHER APPLICATIONS

Another important application of the diode laser spectrometer with the long pathlength absorption cell is measurements of lineshape and pressure-broadening coefficients by foreign gases. This has been demonstrated for UF_6 and CO_2 and the pressure-broadening coefficients by various gases have been obtained. Although somewhat insensitive to temperature variations, lifetimes of the excited states can be measured as a function of temperature.

Isotope shifts of some molecular species are rather small. High resolution measurements allow isotopic composition analysis. An example is shown in Fig. 4.

TABLE 1: Summary of Linewidth Measurements

Compound	Molecular weight	Room Temp Linewidth ($cm^{-1} \times 10^3$)	Transition Wavenumber (cm^{-1})	Doppler Widths[a] ($cm^{-1} \times 10^3$)
NH_3	17	3.70	628.444	1.89
CO_2	44	1.50	617.2916	1.15
N_2O	44	1.30	585.2747	1.09
		1.35	612.3516	1.14
CF_4	88	1.18	617.2648	0.82
ClF_3	92	1.25	627.701	0.81
UF_6	352	0.57	627.5988	0.41

[a] Doppler widths evaluated at the particular transition wavenumbers [$\Delta v_D = 2v_o (2 kT \ln 2/Mc^2)^{1/2}$].

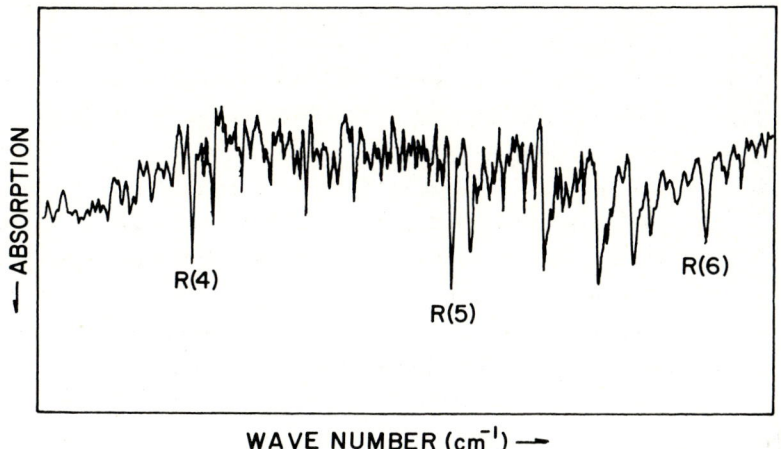

Fig. 4. High resolution absorption lines of a mixture of 5% $^{235}UF_6$ in 95% $^{238}UF_6$.

In this figure, we analyze a mixture of ~5% $^{235}UF_6$ and 95% $^{238}UF_6$. Figure 5 shows absorption lines of a mixture of CO_2 isotopic species in the 16 µm region. [9] Each species is identified by their isotopic composition. (The notation ijk at each peak corresponds to the $^iC^jO^kO$.) This method has the advantage of measuring isotopic composition in situ where there is a potential problem of isotopic scrambling.

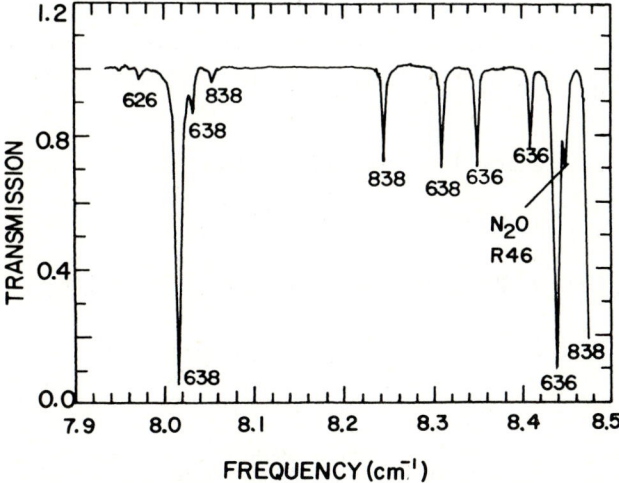

Fig. 5. CO_2 isotopic species in the 16 µm region. The notation ijk at each peak corresponds to the $^iC^jO^kO$. The abscissa has the units of +620 cm^{-1}.

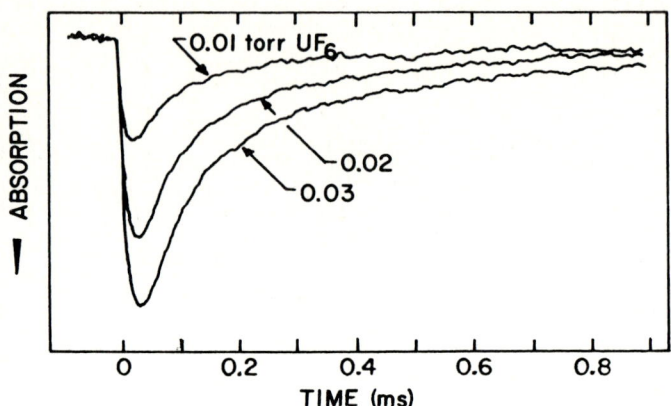

Fig. 6. Transient absorption at 16 μm by UF_5 generated from laser photolysis of UF_6.

Finally, an important area of the laser spectroscopy in conjunction with the long pathlength absorption is the study of transient absorption measurements. Using powerful lasers one can generate significant concentrations of transient species. Because of the low pressures, these reactive species have relatively long lifetimes and they can be detected with the enhanced sensitivity of the long pathlength absorption technique. Figure 6 shows an example of the absorption by UF_5 generated from laser photolysis of UF_6.

REFERENCES

1. K. C. Kim, J. P. Aldridge, H. Filip, H. Flicker, R. F. Holland, N. G. Nereson, and W. B. Person, Proceedings of The Electro-Optics International Laser 1975 Conference, Anaheim, California, November, 1975, p. 191.

2. R. S. McDowell, "Vibrational Spectroscopy using Tunable Lasers," Vibrational Spectra and Structure, Vol. 10 (J. R. Durig, ed.), Elsevier, Amsterdam, 1980.

3. K. C. Kim, E. Griggs, and W. B. Person, Appl. Opt., 17 (1978) 2511.

4. R. A. Briesmeister, G. W. Read, K. C. Kim, and J. R. FitzPatrick, J. Appl. Spectrosc. (in press) 1983.

5. K. C. Kim, W. B. Person, D. Seitz, and B. J. Krohn, J. Mol. Spectrosc., 76 (1979) 322.

6. K. C. Kim and W. B. Person, J. Chem. Phys., 74 (1981) 171.

7. K. C. Kim, M. Reisfeld, and D. Seitz, J. Chem. Phys., 73 (1980) 5605.

8. R. A. McClatchey, W. S. Benedict, S. A. Clough, D. E. Burch, R. F. Calfee, K. Fox, L. S. Rothman, and J. S. Garing, "AFCRL Atmospheric Absorption Line Parameter Compilation," Air Force Cambridge Research Laboratories Report AFCRL TR-73-9906 (1973).

9. M. J. Reisfeld, private communication.

MASS SPECTROMETRY

SOME ADVANCES IN INORGANIC ANALYTICAL MASS SPECTROMETRY

H. J. SVEC
Ames Laboratory-USDOE and Dept. of Chemistry, Iowa State University, Ames, IA 50011.

ABSTRACT

Three recent developments in mass spectrometry are applicable to analytical problems. High powered laser excitation of metals, semi-conductors, and insulators combined with mass analysis provides an absolute analytical probe. Combining an inductively coupled argon plasma with mass analysis provides facile and sensitive analysis of aqueous solutions. A cryogenic hollow cathode containing ice is a proven ion source for 20 to 50 µL of aqueous samples with prior treatment.

INTRODUCTION

Today when you consider analytical applications of mass spectroscopy you are apt to be overwhelmed with the various kinds of important organic analyses that are being reported. This is natural because of the great stress placed today on problems wherein organic compounds are involved. There are however important developments in inorganic analyses by mass spectroscopy but the problems are not of such broad general interest and the workers in the field are fewer. Indeed, during the past decade or so the most prevalent work is abroad, with Russian workers reporting the bulk of it. In the United States only a handful of laboratories are involved and then so at a limited level. However, the results of this research are exciting and we have been in the midst of the activity. At the risk of being parochial I intend to emphasize three developments, in which we've played an important role.

The first of these will involve the marriage of a laser of high radiant power with a mass spectrometer. In this work a highly focussed laser is employed along with a double focussing mass spectrometer. The apparatus can be considered to be an analytical probe, usefully applied to solids which may be metals, semi-conductors and insulators. It is also capable of bulk analysis with high accuracy and precision. Its crowning attribute is its capability of doing analytical work on solids without extensive preparation or treatment and without the use of calibration standards. The technique has been applied to a number of real analytical problems, the results of which I'll illustrate with a few examples.

The second development concerns the mass spectrometric analysis of aqueous solutions, ever a difficult application in the past. In this case an inductively coupled argon plasma, operating at atmospheric pressure, atomizes and ionizes the constituents of a solution at high temperature and the ions produced are mass analyzed using a quadrupole mass filter. The development of an ICP/Mass Spectrometer, which was initiated in the Ames Laboratory, has been successful to the point that two commercial companies are now producing them. There is a concerted effort going on at present to extend the applicability of this technology to lower and lower analytical sensitivities for inorganic solutes and also polar organic solutes.

A third development which will be described briefly concerns another approach to the mass spectrometric analysis of aqueous solutions which is applicable to small (20-50 µL) volumes. In this case the sample is introduced into the cavity of a hollow cathode discharge apparatus without any preparation whatsoever, frozen at near liquid-N_2 temperature, and then submitted to a low voltage dc discharge. To date, the feasibility of the approach has been demonstrated and the outstanding problems assessed. The promise of the technique calls for futher research and development albeit, it will have to take place in the hands of a researcher other than its originators.

Laser/Mass Spectrometry

The analysis of solids, without prior chemical treatment and with Periodic Table coverage has a long history. It began with the development of optical spectroscopy by Bunsen and Kirchhoff in Germany, continued with that of spark source mass spectrography based on the early work of Dempster in this country (Manhattan Project Days) and has now progressed to the use of laser excitation followed by mass analysis of the ions produced. While the first two techniques require special treatment or adaptations of the samples to make them into electrical conductors, LMS does not. None of the problems attending analysis of the pristine samples are present except the necessity to clean off the surface in some cases in order to avoid analytical artifacts.

The published work in the field prior to 1982 has been reviewed by Conzemius in 1980 and 1983. In 1978, Bykovskii et al. in Moscow outlined the advantages of using a laser of high radiant power in the analysis of solids. In summary these are: analysis of any solid without standards; sterile sampling; coverage of the entire periodic table; 100% to 3×10^{-7}% concentration range; 10^{-10} to 10^{-20} g absolute detection limit; accuracy at 10^{-6}% level within ±30%; accuracy at 1-100% level ~1% relative; depth resolution as low as 0.1 µm with a detection limit of 10^{-6}%; detection surface impurities at 10^{10} atoms/cm^2; lateral resolution as low as 10 µm; milligram sample consumption at lowest detection limits; and automatic sampling and recording possible.

It turns out that we in the Ames Laboratory were the first to take up these predictions seriously in this country and apply them to the analysis of solids for inorganic constituents. The approach was to utilize a Q-switched Nd doped Yttrium Aluminum Garnet laser (λ = 1.06 µm) focussed to a spot ~10 µm in diameter. The laser (Holobeam Corp.) has triocular optics, two as a binocular microscope used to choose a spot on a sample to analyze, and the third for monitoring the process via a closed television setup. The power of the laser beam is 1-6 gigawatts/cm^2 which is sufficient for example to vaporize thorium from a crater 12-15 µm in diameter and 2-3 µm deep. At this level virtually all of the vaporized species are ions, thus the relative sensitivities of all the elements are expected to be the same. Figure 1 shows the relationship between radiant power and ion fraction. Although at low radiant power there is a spread in the ion fractions of various elements due to different volatilities and ionization energies, it is patently clear that the spread appears to become unity at an extrapolated radiance of 5-10 gigawatts/cm^2. From this standpoint, if the collection efficiency for ions in a mass spectrometer is the same for all ions of different masses, a method of analysis based on this fact would be free of matrix effects and requires no calibration standards. Results to date in our laboratory indicates that this is true. I will illustrate this subsequently with examples.

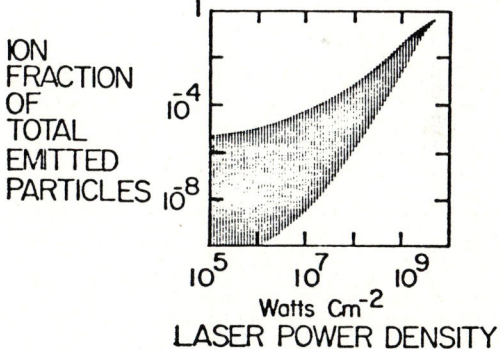

Fig. 1. Ion fraction vs. laser radiant power.

The pulsing rate of our laser is variable from 1 to 20000 Hz and the pulse length is ~100 ns. So far, the energy spread of the largely singly charged ions produced requires a double focussing ion optic system in order to get good mass resolution. With shorter pulses, TOF measurements utilizing ion reflectors would probably eliminate this need.

The focussed beam of the laser can be rastered using galvo-scanning mirrors over an area of 1 x 1 mm^2. The mode of rastering is completely optional, namely, selected laser shots, lines of laser shots, or overlapping laser shots. With a 10-12 µm focal spot, the convention of laying down laser shots 25 µm (0.001 inch) apart was adopted. No apparent cross contamination is observed.

Figure 2 is a diagrammatic representation of the apparatus. Note that the laser is outside the vacuum chamber. Because of the wavelength of the YAG laser, a glass window and prism can be used.

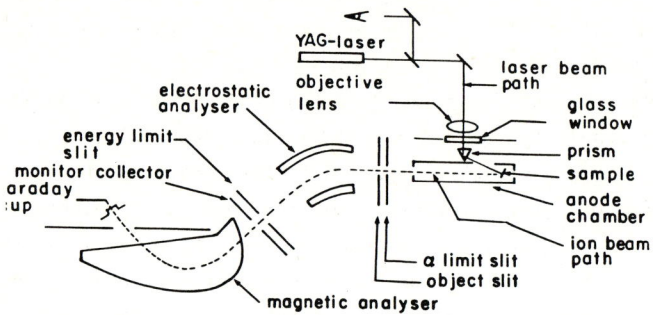

Fig. 2. Diagrammatic arrangement of the laser/mass spectrometer.

We tested the lateral resolution of the apparatus with a diffusion couple made up of 2.5 mm diameter tantalum and thorium rods welded together. Figure 3 illustrates the results. The points were obtained as Ta^+ and Th^+ signals produced by a short line of laser shots 48 µm apart. The figure shows graphically that Ta diffuses into Th and persists as a Ta^+ signal at a 10^{-3} level for ~200 µm into the Th. There is no evidence of Th diffusing into Ta.

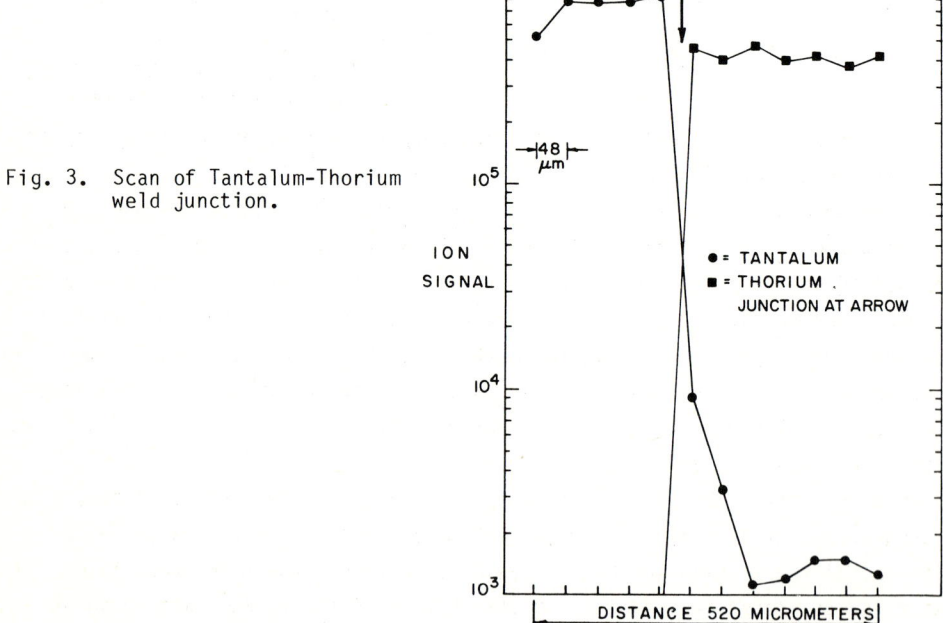

Fig. 3. Scan of Tantalum-Thorium weld junction.

We have long held that proposed analytical methods should work on real samples. Those that don't, show up on the scene and then fade away contributing mainly to the glut in the literature. Accordingly, the utility of the LMS method is best demonstrated by tests with real samples, a few examples of which will be illustrated here.

Figure 4 shows the results of analyzing four samples of thorium doped with Mo at levels from ~115 ppmw to ~25 ppmw. The samples were analyzed spectrophotometrically on the basis of make up and calibration data and then by LMS. The laser results were obtained by counting 10^6 ions due to the matrix, observing the $^{98}Mo^+$ ions, making allowances for isotopic abundance, and converting the ppma data obtained to ppmw data. No relative sensitivity coefficients were used. The correspondence between the two independent analytical methods is excellent. Similar results were obtained with Th samples doped with W, Zr, and Re.

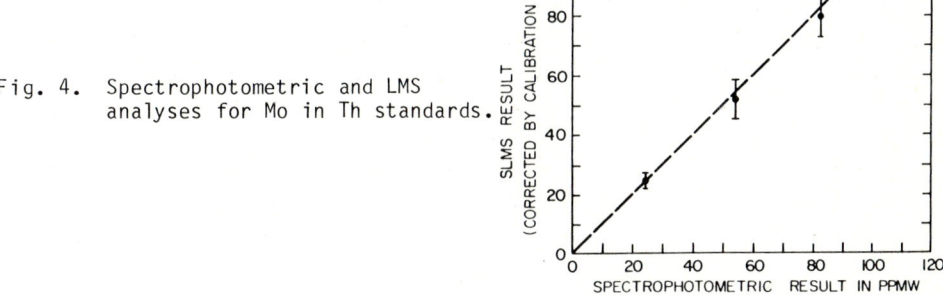

Fig. 4. Spectrophotometric and LMS analyses for Mo in Th standards.

These results were applied to electrotransport samples of Th containing 100 ppmw of these dopants. The metal is formed in 2.5 mm diameter bars 16 cm long. They are held at 1500°C for 30 or more hours with a 2.6 dc voltage applied. Figure 5 shows the analytical results for Mo which migrated toward the cathode. The points were obtained by rastering 500 laser shots perpendicular to the bar at 110 sampling sites. Note the highly purified Th at the anode end (left <1 ppmw Mo) and the relatively impure Th at the cathode end (right >500 ppmw). Because the bar was too long for the ion source chamber it was cut into lengths A, B, and C. The apparatus is now under computer control and can generate such analytical data automatically.

Another test of the LMS method concerns analysis of metals for carbon, nitrogen, and oxygen. These involve not only sampling procedures but also clean up of the surface so that the results represent the bulk composition. Rastering the laser beam over a sufficient area of the cleaned sample then insures a legitimate sample. The method was proven by determining these inter-

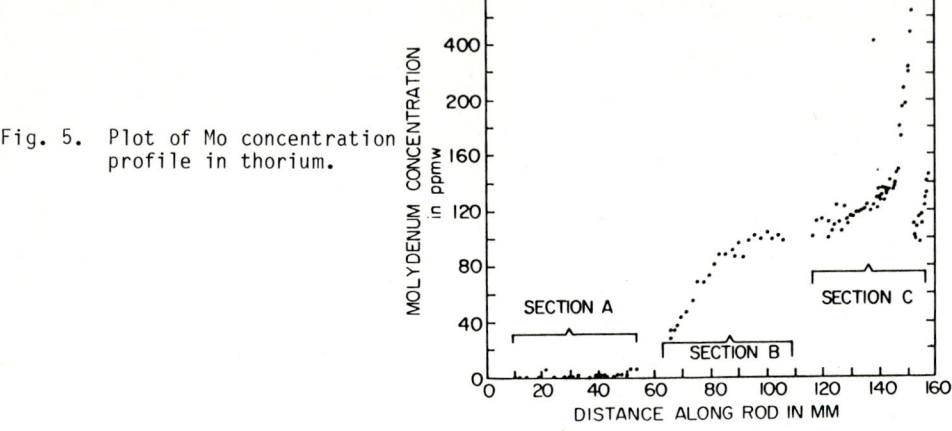

Fig. 5. Plot of Mo concentration profile in thorium.

stitial elements in NBS SRM 1040 irons. Table 1 shows the results in tabular form for eight of these materials. Two thousand laser shots were used for each of the measurements at a radiant power of 3.2 gigawatts/cm^2. Prior to the measurements the area of each specimen to be sampled was cleaned under high vacuum by laser irradiation. The results are based on ion currents due to C^+, N^+ and O^+ relative to 10 Fe^+ ions, and these ppma results were converted to ppmw using the formula given in footnote c. No independent calibration data were used! As is evident from the table, the correspondence between the NBS and experimental values is excellent, almost embarrassingly so. Table 2 shows the results of five separate measurements on NBS SRM 1094, a maraging steel. Again the correspondence between the results and the NBS values are excellent.

Table 1. Analysis of C, N, and O in NBS standard 1040 series.

Sample	C[a]		N		O	
	NBS[b] Data (wt%)	Exp.[c] (wt%)	Cert.[b] (wt%)	Exp. (wt%)	Cert.[b] (wt%)	Exp. (wt%)
1040	0.03	0.028	0.003	0.0031	0.018	0.014
1041	0.4	0.38	0.004	0.0043	0.017	0.013
1042	0.1	0.090	0.014	0.0142	0.017	0.025
1043	0.2	0.21	0.005	0.0065	0.002	0.003
1044	0.2	0.23	0.004	0.0053	0.009	0.0075
1045	0.4	0.42	0.004	0.0042	0.007	0.0074
1046	0.02	0.010	0.005	0.0056	0.106	0.112
1047	0.2	0.28	0.004	0.0060	0.017	0.015

[a] Carbon concentration is not certified.
[b] No uncertainty values given by NBS.
[c] Exp.(wt%) = Exp.(ppma) x (At. wt.$_i$) x 10^{-4}/(At. Wt.$_{Fe}$)

Table 2. Analysis of O in maraging iron, NBS 1094.

Element	Certified[a] Value	Exp.[b] (ppma)	Mean	Std. Dev.	Rel. Std. Dev.
O	4.5 ppm by wt. (16 ppma)	16.5 17.8 17.8 18.9 17.2	17.6	0.89	5.0%

[a]NBS reports a range in the determinations from 2.5 to 7.5 ppmw.
[*]5 separate determinations on same loading.

A test of the probe capabilities of the method was made by attempting to analyze niobium dendrites in a copper matrix for carbon. The composite material is part of an Ames Laboratory development aimed at a better way of making superconducting wire containing Nb-Sn alloy. The size of the dendrites varies from 5 to 20 μm in diameter. A cross section of the material was lightly etched with dilute HNO_3 and under 475X magnification positions on the dendrites and matrix were located for single laser shots. The analytical results based on ions from single craters indicated ppm amounts of carbon in both matrix and dendrites. The amount in the latter was variable and some so high as to interfere with the next step in the next step in the process of making super conducting wire.

Other tests of this analytical method have been applied to a variety of diffusion couples and electrotransport materials, to thin films on metals and insulators, to insulators like quartz and sapphire, and to minerals in geological specimens. The analytical data has been obtained by electrometric and photograph methods and covers the entire periodic table. Extensive studies have been made of the operating parameters which affect the analytical results. Most recently the entire instrumental method has been placed under computer control.

This is not a static area of research. Much is going on in Europe, particularly in Russia and we expect similar work to be initiated in the People's Republic of China. This is an exciting development.

ICP/MS

The second topic of concern is the fast developing area involving an inductively coupled argon plasma as an ion source in conjunction with a quadrupole mass filter for the analysis of aqueous solutions. The pioneering work was begun in the Ames Laboratory with financial help from the E.P.A. about 7 or 8 years ago and continues under the direction of Dr. R. S. Houk. Much of the development has concerned being able to extract directly and effectively ions from the 6000-7000K plasma. This is done now through a 0.5 mm hole in a metal cone (Cu, Ni) as shown in Fig. 6. Ions enter a differentially pumped vacuum system, are skimmed, energy filtered, focussed, mass analyzed and counted. Photons from the plasma which produce a high background are effectively handled now with a stop in the center of the ion stream.

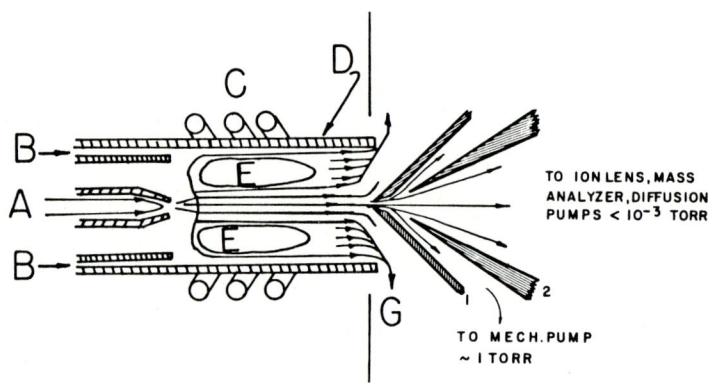

Fig. 6. Sampling arrangement for ICP/MS spectrometer.

The aqueous samples are introduced as dilute HNO_3 solutions into the plasma in the form of a nebulized stream. Prominent background ion currents due to O^+, OH^+, H_2O^+, Ar^+, and ArH^+ are observed along with other less prominent ionic species as shown in Fig. 7. When a 0.8 ppmw solution of silver is the sample material, the spectra shown in Fig. 8 is obtained. Note the intense peaks due to $^{107}Ag^+$ and $^{109}Ag^+$ ions ranging high above the background signal of 300 counts/s in the upper traces which was recorded at a sensitivity 128x that of the lower trace. Peaks at m/z 80 (Ar_2^+), 56 (ArO^+) and 52 are in the blank. An analytical sensitivity for silver in the low ppbw range without a photon stop is indicated by these data. In more recent work using a photon stop a background count approaching ~10 c/s has been achieved. Thus the expected analytical sensitivity is well below a part-per-billion by weight.

When an equimolar solution of Al, V, Nb (~1 ppmw), Cs, La, and Tm are introduced into the ICP and the background spectra are subtracted the spectrum shown on Fig. 9 results. Some of the observed variation in the peak heights is due to the way the solution was made up and also the fact that when these data were obtained nothing was done to compensate for mass discrimination in the energy analyzer and mass filter. Note the small peak due to La^{+2} at 69 1/2. This is observable because of the low second ionization energy of lanthanum which makes ionization of La^+ in the plasma possible. No peaks due to LaO^+ ions are discernible in this spectrum.

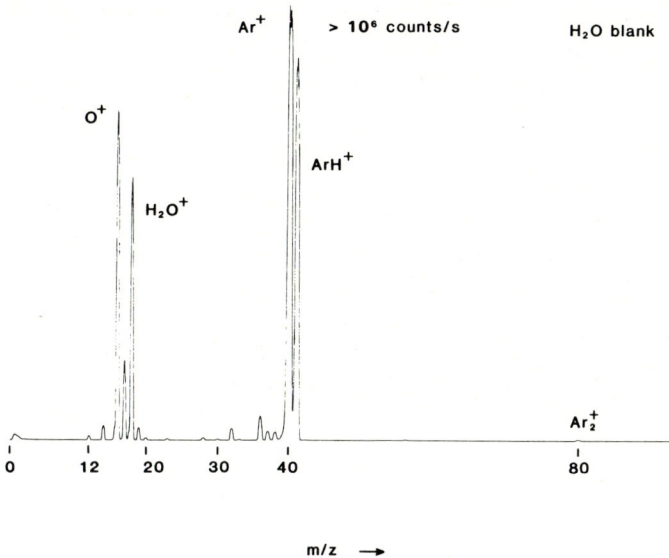

Fig. 7. Background spectra for ICP/MS.

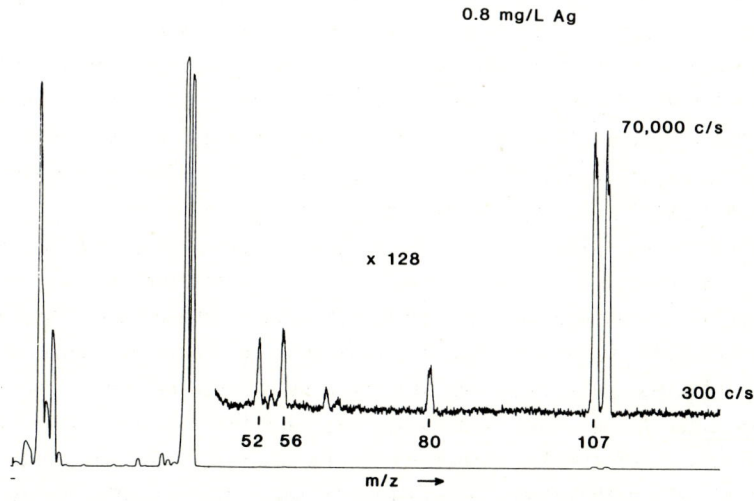

Fig. 8. ICP/MS spectra of a 0.8 ppm solution of Ag.

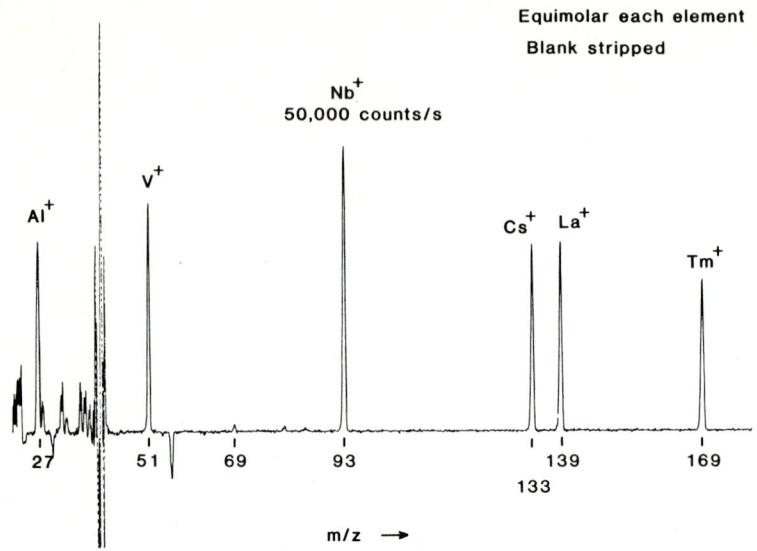

Fig. 9. ICP/MS analysis of mixture of Al, V, Nb, Cs, La, Tm solution.

As indicated earlier, a proposed analytical method that doesn't work for real samples is not fully legitimate. Accordingly, a sample of Ames tap water was analyzed. Figure 10 is the mass spectrum observed in the low m/z range. The data shown are the result of 128 scans using a signal averager. The lower spectrum is what results when distilled water is introduced into the plasma. Note the small 23^+ due to sodium, a ubiquitous impurity, the 28^+ and 32^+ due to N_2 and O_2 in the argon, the 30^+ due to NO formed in the plasma and a small 29^+ of unknown origin. After the lower spectrum is subtracted from the composite tap water spectrum, the upper spectrum representing the solute results. The much more abundant Na shows up with the Mg and Si isotopes clearly indicated. Of considerable interest are the 32, 33, and 34 peaks which come from the sulfate that is known to be in the water. Thus under the thermal conditions in in the plasma, SO_4^{-2} is atomized and the ground state S^+ ions formed are clearly discernible. A similar argument can be made for the origin of Si^+ ions due to soluble silicate in the water. These latter elements do not show up in emission spectra. An important result illustrated here is that in addition to elemental analyses one obtains relatively sensitive isotopic abundance data for elements that occur at trace levels.

A graphic example of this is shown in Fig. 11. On the left are the zinc isotopes at m/z 66, 67, 68 which were observed in a solution of normal zinc nitrate. The peak hights correspond to the normal isotopic abundances. On the right are the zinc isotopes from the sera of a rat that had zinc enriched in ^{67}Zn added to its diet. The sensitivity of the ICP/MS method now makes this a

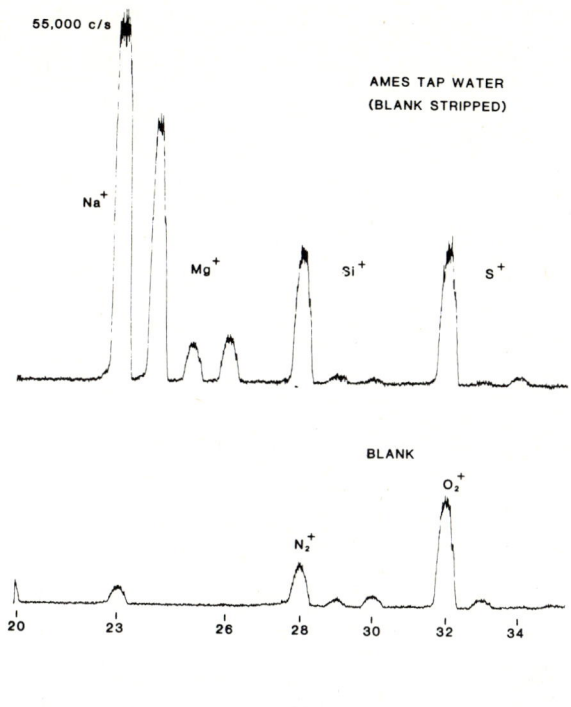

Fig. 10. ICP/MS analysis of Ames tap water.

convenient way to follow trace levels of elements in biological specimens.

Cryogenic Hollow Cathode Ion Source

The third topic to be covered concerns a feasibility study in which a hollow cathode discharge is used to generate ions from a small (20-50 μL) aqueous sample. The raw sample with no prior preparation is frozen in the cathode at liquid nitrogen temperature, placed in the ion source chamber of a mass spectrograph and then submitted to a dc discharge at ~250 V. The discharge is started under low pressure oxygen and then is maintained under the ambient vapor pressure of water. Ion currents due to water molecule clusters and impurities in the aqueous solution are observed as the ice is ablated from the cathode. No solute segregation is discernible. The ion source shown in Fig. 12 was mounted in a special vacuum housing attached to the ion optic system of a spark source mass spectrograph. Tests of the system were made with nine solutions containing an aggragate of 71 elements at various concentration levels. Simultaneous recording of ion currents was done photographically. Seven elements, F, P, S, Sc, Mn, Ni, and Ta (material for hollow cathode) were not analyzable due to interferences. The remainder of the elements could be determined at sensitivity levels ranging from sub parts-per-billion to parts-

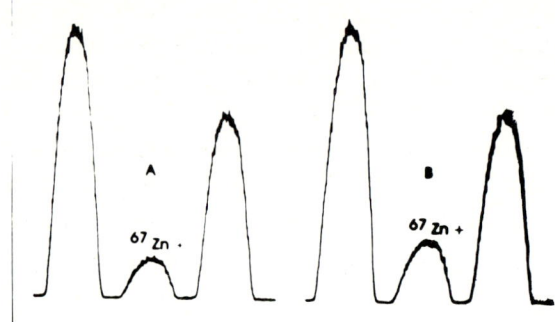

Fig. 11. A - Normal distribution of Zn isotopes from solution of $Zn(NO_3)_2$; B - Zn isotopes from sera of a rat fed zinc enriched in ^{67}Zn.

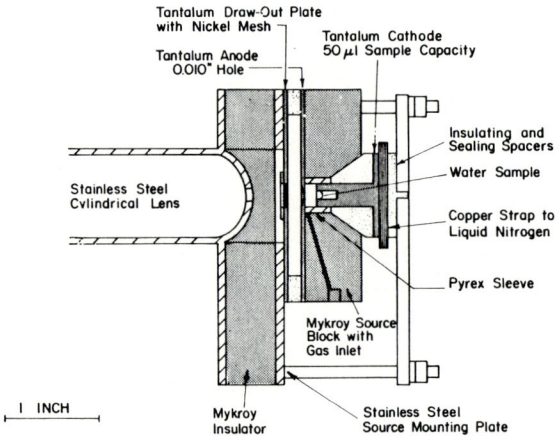

Fig. 12. Cross section of Cryogenic Hollow Cathode Ion Source (CHCIS).

Table 3. EPA standard TM575-3 water (Br as internal standard) 20 μL sample

Element	Determination Concentration ppma	ppmw[†]	EPA Value ppmw
Al	0.18	0.26	0.90
As	0.3	1.25	1.15
Be	<1.0	<0.5	0.40
Cd	<0.1	<0.6	0.07
Co	0.07	0.23	0.40
Cr	0.06	0.19	0.21
Cu	<0.1	<0.35	0.10
Fe	0.15	0.48	0.68
Hg	0.00076	0.0085	0.0094
Pb	<0.1	<1.0	0.35
V	0.05	0.1	0.16

$$^\dagger ppm(wt/wt)_{aqueous} = \frac{ppma \times At.Wt._{element}}{18}$$

per-million. In one of the test solutions 32 pg of Hg was detected in 20 µL of water containing Tl, Pb, Bi, Th, U, and Al. When the method was tested with a real sample from the Environmental Protection Agency the results given in Table 3 were obtained.

The feasibility of the method has been proven and a paper describing the results has recently appeared. More research is certainly warranted.

The laser mass spectrometric work was spearheaded by Bob Conzemius; the ICP/MS by R. Sam Houk and Alan Gray; and the cryogenic hollow cathode work by Gordon Foss. Most of the work I've alluded to was done in the Ames Laboratory operated for the U.S. Dept. of Energy by Iowa State University under contract No. W-7405-Eng-82. This research was supported by the Director for Energy Research, Office of Basic Energy Science.

BIBLIOGRAPHY

Laser Mass Spectrometry

R. J. Conzemius and J. Capellen, Int. J. Mass Spectrom. Ion Phys. 34, 197-271 (1980).

R. J. Conzemius, D. S. Simon, Shankai Zhao, G. D. Byrd, "Laser Mass Spectrometry of Solids: A Bibliography 1963-1982, Microbeam Analysis 1983, R. Gooley, Ed., San Francisco Press Inc., San Francisco, CA, 1983.

Y. A. Bykovskii, G. I. Zhuravlev, V. I. Belovsov, V. M. Gladskoi, V. G. Degtyarev, V. N. Novolin, Ind. Lab. USSR 44, 799-804 (1978).

F. Schmidt, R. J. Conzemius, O. N. Carlson, H. J. Svec, Anal. Chem. 46, 810-14 (1974).

R. J. Conzemius, H. J. Svec, Anal. Chem. 50, 1854 (1978).

R. J. Conzemius, F. Schmidt, H. J. Svec, Anal. Chem. 53, 1899 (1981).

Zhao Shankai, R. J. Conzemius, H. J. Svec, submitted to Anal. Chem., Oct. 1983.

ICP/MS

R. S. Houk, V. A. Fassel, G. D. Flesch, H. J. Svec, A. L. Gray, C. E. Taylor Anal. Chem. 52, 2283 (1980).

R. S. Houk, V. A. Fassel, H. J. Svec, Dyn. Mass Spectrom. 6, 234-51 (1981).

R. S. Houk, H. J. Svec, V. A. Fassel, Appl. Spectrosc. 35(4), 380 (1981).

R. S. Houk, V. A. Fassel, H. J. Svec, Org. Mass Spectrom. 17(5), 240 (1982).

R. S. Houk, J. J. Thompson, Am. Mineral. 67, 238 (1982).

R. S. Houk, J. J. Thompson, Biomed. Mass Spectrom. 10, 107 (1983).

R. S. Houk, A. Montaser, V. A. Fassel, Appl. Spectrosc. 37, 425 (1983).

A. L. Gray, A. R. Date, Dyn. Mass Spectrom. 6, 252-66 (1981).

A. R. Date, A. L. Gray, Analyst 106, 1255 (1981).

A. R. Date, A. L. Gray, Spectrochim. Acta B 38, 39 (1983).

A. L. Gray, A. R. Date, Int. J. Mass Spectrom. Ion Phys. 46, 7 (1983).

A. R. Date, A. L. Gray, Int. J. Mass Spectrom. Ion Phys. 48, 357 (1983).

A. L. Gray, A. R. Date, The Analyst 108, 159, 1083 (1983).

Cryogenic Hollow Cathode

G. O. Foss, H. J. Svec, R. J. Conzemius, Anal. Chim. Acta 147, 151 (1983).

W.S. Lyon (Editor), *Analytical Spectroscopy*
Elsevier Science Publishers B.V., Amsterdam — Printed in The Netherlands

A STATE-OF-THE-ART MASS SPECTROMETER SYSTEM FOR DETERMINATION OF URANIUM AND
PLUTONIUM ISOTOPIC DISTRIBUTIONS IN PROCESS SAMPLES*

Carol A. Polson

[1] E. I. du Pont de Nemours & Co., Savannah River Plant, Aiken,
South Carolina 29808

ABSTRACT

A Finnigan MAT 261 automated thermal ionization mass spectrometer system was purchased by the Savannah River Plant and recently installed by Finnigan factory representatives. This instrument is a refinement of the MAT 260 which has been used routinely for three years in the laboratory at SRP.

The MAT 261 is a highly precise, fully automated instrument. Many features make this instrument the state-of-the-art technology in precision isotopic composition measurements. A unique feature of the MAT 261 is the ion detection system which permits measurement of the three uranium or plutonium masses simultaneously. All Faraday cup measuring channels are of the same design and each is equipped with a dedicated amplifier. Each amplifier is connected to a linear voltage/frequency measuring system for ion current integration. These outputs are fed into a Hewlett-Packard 9845T desk-top computer.

The computer, and the Finnigan developed software package, control filament heating cycles, sample preconditioning, ion beam focusing, carrousel rotation, mass selection, and data collection and reduction.

Precision, accuracy, and linearity were determined under normal laboratory conditions using a NBS uranium suite of standards. These results along with other development in setting up the instrument will be presented.

INTRODUCTION

A need existed for a replacement thermal ionization mass spectrometer in the Savannah River Plant (SRP) Laboratories Department due to the increasing demand for fast, accurate, and precise isotopic analysis on process control and accountability samples. In 1980, the Laboratories Department selected and installed a MAT 260 automated mass spectrometer manufactured by Finnigan MAT Corp., Bremen, W. Germany (ref. 1). Based upon the performance of the MAT 260 under routine operation, the compatibility of electronics, and physical design, the decision was made to purchase the MAT 261, Finnigan's "state-of-the-art" thermal ionization mass spectrometer to further upgrade mass spectrometry capabilities (ref. 2).

---

*The information contained in this article was developed during the course of work under Contract No. DE-AC09-76SR00001 with the U.S. Department of Energy.

Performance specifications established at SRP for the MAT 261 were first demonstrated at the factory in Bremen, W. Germany (ref. 3) and again in the Building 772-F laboratory at SRP in August 1983.

This paper describes the MAT 261 mass spectrometer as it is installed in the laboratory and presents data obtained during and after specification testing on a suite of NBS uranium standards.

DISCUSSION

The MAT 261 met all performance requirements specified in the purchase agreement. Performance specifications required that the mass spectrometer be capable of analyzing a minimum of 12 uranium samples over a 12 hour period without operator attention. Specifications also required a relative precision (t·RSD) of 0.06% at the 95% C.L. for a single determination on twelve NBS U500 standards analyzed to determine the mass discrimination factor. In addition, an accuracy of 0.06% was required on prescribed groups of NBS uranium standards ranging from the NBS 005 to NBS 970. Minor ratio measurements were performed on the NBS 850 to determine the accuracy of the 234/236 ratio. The minor ratio was required to agree within 0.5% of the NBS value.

Most of the components of the MAT 261 are of the same design as the earlier Finnigan MAT 260. However, the ion source and magnetic control units were modified to improve the computer control of the mass spectrometer system. In addition, major improvements were made in the ion detection system to permit simultaneous measurement of 3 masses. Use of the new ion detection system improves the internal precision of the ratio measurements as well as decreases the analysis time.

Significant improvements were also made in the data acquisition system. The HP 9845 desktop computer and the vendor supplied software package provide total instrument control including filament heatup, sample carrousel focusing, lens focusing, peak jumping and data collection.

Ion Source

The MAT 261 ion source provides a high, stable ion yield with low mass discrimination. The source is designed for single or dual filament operation. At present, the SRP instrument utilizes the dual sample filament technique. Exact alignment of the filaments with respect to the axis of the ion-optical system is provided by a closed-loop regulator circuit.

A stepping motor controlled by the ion current signal, moves the filament into that position for which the maximum ion current signal is attained. Under computer control the filaments are heated to preselected outputs or pilot signals. The pilot signals are the rhenium ion current for the ionization filament and the ion current of the most abundant isotope for the evaporation filament. Adjustment of focus controls to optimize output is done automatically by computer control.

Generally, the MAT 261 ion source is the same as in the MAT 260. Modifications were made to the magazine (carrousel) drive unit which rotates and positions the sample magazine into place in the ion source. The design change eliminates slippage of the carrousel drive on the ceramic, feed-through drive shaft. This assures consistent and proper alignment of the filaments in the ion source. Other improvements were made in the filament electrical contacts which eliminate movement of the contact off the filament. Both of these items caused significant problems in the earlier MAT 260 models.

MAGNETIC SECTOR

The MAT 261 uses a 90° magnetic sector with a 23 cm. main radius. With a ion slit width of 0.2 mm and a collector slit width of 0.6 mm, the resolution is about 450 (at the 10% valley definition). The major improvement in the magnet system is the use of water-cooled coils to maintain a constant temperature. This prevents sudden shifts in the magnetic field as the ambient temperature changes.

The magnet is under computer control through the use of two digital-to-analog converters. One has a 16 bit resolution for mass setting and the other has a 10 bit resolution for peak centering and fine tuning in the magnetic field.

The computer can be set to look at any mass over the range of 5 to 350 amu. In addition, manual selectors are provided for mass setting and fine tuning of the peak.

Ion Collector System

The MAT 261 contains a triple Faraday cup collector system. The 3 collectors are physically separated so that the simultaneous measurement of 3 masses, 2 and 3 mass units apart, is achieved. The multi-collector system eliminates time variations that occur during the peak jumping sequence as required in a single collector system. The multi-collector system also speeds the measurement rate. Single collector mode of operation is available, but is normally not used because of the advantages of the multi-collector.

Vacuum System

The pumping system consists of a turbo molecular pump connected to the analyzer head and to ion getter pumps connected to the analyzer tube and collector flange. A cryopump is located on top of the analyzer head directly above the filaments. This arrangement allows the removal of vapors caused by filament heat-up, thus decreasing the chances of cross-contamination.

Pressure in the analyzer head is monitored by an ionization gauge, forevacuum is monitored by a thermal conductivity gauge, and the analyzer tube pressure is monitored by the ion pump control unit.

An isolation valve between the source and analyzer permits the analyzer section to remain at high vacuum during carrousel changeover. Pump down time of the source region typically is about 10-15 minutes. Pressure in the source region is maintained at about 10^{-7} torr and better than 10^{-8} torr in the analyzer tube. An automatic pneumatic valve is located between the source and the turbo pump. It closes upon loss of power or forevacuum to prevent flooding of the analyzer section.

Sample Carrousel and Sample Preparation

The sample carrousel holds up to thirteen dual filament assemblies. The disposal filament inserts have rhenium wire and are purchased from Finnigan. Blank filaments are degassed via a separate preconditioning unit to provide a contamination free environment for degassing filaments and to permit continued use of the ion source for sample analysis. Samples are loaded onto the evaporation filament using an Eppendorf pipette and a vendor supplied sample application device for exact positioning of the sample. The filaments, when in the loading position of the device, are connected to a heating unit. This unit provides program controlled heating and drying of 4 samples at a time. The sample drying procedure consists of heating via 0.75 amps of current for 3 minutes and then heating at an increasing rate of 0.2 amps per minute up to a maximum current of 2.1 amps where a red glow appears. After samples are dried, ionization filaments are placed onto the carrousel and then the shield plate is clipped on over the dual filament assembly to prevent filament to filament contamination. The carrousel is inserted into the source housing and the system is evacuated. Further conditioning of samples is done automatically during data acquisition.

Data Acquisition System

The HP 9845T desk top computer controls the MAT 261. The software package was developed by Finnigan and is stored on three cartridges. The Systems cartridge contains the various analysis programs and controls all of the mass spectrometer's operations. The Modules cartridge contains all parameters for instrument operation. The operator entered parameters are transferred to the Systems cartridge where the information is used in analysis. The Testhelp cartridge provides programs for bake out, preheat of filaments in the source, and additional support functions for trouble-shooting hardware and electronic malfunctions.

Prior to beginning analysis, the computer must have a reference point from which it is able to locate the masses. Reference masses are first defined by entering them into a mass-DAC reference table. Normally, mass numbers 185, 187, and the major isotope in the sample are used as reference masses. Along with the reference masses, corresponding DAC 16 (magnetic field coarse setting) and DAC 10 (high-voltage setting) values must be entered. Once the reference table is established, the reference mass input routine is used to store the exact positions of the reference masses which are set manually in the instrument.

The output of the voltage-to-frequency converter is compared with the stored values obtained in the mass calibration. The difference, if any, is corrected and the DAC values in the reference table are adjusted to reflect the correct mass location.

A peak centering routine is used to adjust the analyzing system to the center of the peak. By making small adjustments to the high voltage, the peak center is set as close to the center of the high voltage DAC range as possible.

A variety of programs are then used to define the sample sequence, establish parameters, rotate and focus the carrousel, analyze, and collect ratio data.

At SRP, results are printed out in terms of average isotope ratios. However, other print out formats can be selected if desired.

MAT 261 Performance

Performance tests were made at both the vendor's facility and at SRP. Accuracy runs were made by analyzing 4 groups of NBS uranium standards. Two groups were analyzed using the single collector and two groups using the multi-collector. Precision was determined on the NBS U500 standard using both the multi and single collectors. Abundance sensitivity calculations were performed on the NBS U005 standard. All analyses were made under routine operating conditions.

Accuracy

Accuracy of the 235/238 ratio over the range of NBS standards was found to be better than 0.06%. Tables 1 and 2 detail Groups I and II data respectively for accuracy runs using the multi-collector. All filament loadings were 1 µg quantities.

TABLE 1

Group I - Multi-collector

NBS Standard	Observed 5/8 Ratio	Theoretical 5/8 Ratio	% A
020	0.02080	0.02081	-0.05
050	0.05273	0.05278	-0.10
100	0.11355	0.11360	-0.04
350	0.5463	0.5465	-0.04
500	0.9992	0.9997	-0.05
850	6.148	6.148	-0.00
005	0.004922	0.004919	+0.05
970	186.75	186.78	-0.02
930	17.338	17.349	-0.06
005	0.004917	0.004919	-0.05

$A = 0.044$

Specification $A \leq 0.06\%$

Where:

A = Accuracy

TABLE 2

Group II - Multi-collector

NBS Standard	Observed 5/8 Ratio	Theoretical 5/8 Ratio	% A
005	0.004920	0.004919	0.01
005	0.004921	0.004919	0.03
005	0.004914	0.004919	-0.11
850	6.146	6.148	-0.03
850	6.149	6.148	0.02
020	0.02081	0.02081	0.00
100	0.11368	0.11360	0.07
970	186.82	186.78	0.02
100	0.11363	0.11360	0.03
350	0.5471	0.5465	0.11
005	0.004922	0.004919	0.05
930	17.357	17.349	0.05
005	0.004921	0.004919	0.03

$A = 0.055$

Where:

A = Accuracy

Precision

The precision of the measurement of major isotopes was determined by repeated measurements of the uranium 235/238 ratio on the NBS U500 standard. The precision of the 235/238 ratio was found to be better than the specification 0.06% (Table 3). The average 235/238 ratio is used to determine the mass discrimination factor which is applied to all 235/238 ratios on all standards.

TABLE 3
Precision Test - NBS U500 - Multi-collector

Observed 5/8 Ratio	
0.9997	X_{12} = 0.9998
0.9997	$S\bar{x}^2$ = 0.000157
0.9996	RSD = 0.016
1.0000	t · RSD = 0.035 (95% C.L.)
0.9998	% = 0.010
0.9998	
0.9998	
0.9999	
1.0001	
0.9995	
0.9999	
0.9997	

Mass Discrimination Factor Calculation:

$$B = \frac{1}{C}\left(\frac{R_{mean}}{R_{NBS}} - 1\right)$$

$$R_{corrected} = \frac{1}{1 + CB}$$

$$C = \frac{\Delta M}{M} = \frac{3}{238} = 0.0126$$

R_{mean} = 0.9998

R_{NBS} = 0.9997

R_{corr} = 0.9999

Minor Ratio Measurements

The 234/236 ratio was measured on the NBS 850 standard using the multi-collector and was found to be within 0.02% of the NBS value (Table 4). Specifications require that the 234/236 ratio agree within 0.5% of the NBS value. All filament loadings were 1 μg quantities.

TABLE 4
NBS U-850

234/236 Ratio - Multi-collector
1.7374
1.7368
1.7397
1.7374
1.7362
1.7345
1.7338
1.7397
1.7387
1.7363
1.7348
1.7444

X_{12} = 1.7375

$S_{\bar{x}}$ = 0.002896

RSD = 0.167

t · RSD = 0.367 (95% C.L.)

Δ% = -0.02

Theoretical = 1.7378

Abundance Sensitivity

The abundance sensitivity of the instrument is the ability to measure minute peaks adjacent to abundant ones. Specifications require that the contribution of the ^{238}U peak to mass 237 and 239 be less than 4 ppm on a NBS U 005 uranium standard. Measurements showed that the contribution of the 238 peak to the 237 and 239 peaks was significantly less than 4 ppm.

CONCLUSION

Tests have shown that the MAT 261 meets all specifications and can provide precise and accurate isotope analyses. Many significant improvements have been made in the collector and data acquisition systems. The MAT 261, along with the existing MAT 260, is expected to provide sufficient and timely results for process control and accountability at SRP.

REFERENCES

1. K. Habfast and D. Tuttas, *A System Calibration for Uranium*, Application Note No. 43, Finnigan MAT GmbH, Bremen, West Germany.
2. *MAT 261 Automatic Thermal Ionization Isotope Mass Spectrometer* (preliminary copy), Finnigan MAT GmbH, Bremen, West Germany.
3. *Mass Spectrometer MAT 261 Operating Manual*, Finnigan MAT GmbH, Breman, West Germany.

W.S. Lyon (Editor), *Analytical Spectroscopy*
Elsevier Science Publishers B.V., Amsterdam — Printed in The Netherlands

AMERICIUM DETERMINATION USING AN ISOTOPE DILUTION MASS SPECTROMETRY TECHNIQUE

DAVID W. CRAWFORD and RICHARD D. PEAVY
U. S. Department of Energy, New Brunswick Laboratory, 9800 S. Cass Avenue, Argonne, IL 60439

ABSTRACT

Americium-241, produced by the B-decay of plutonium-241, represents a significant impurity in plutonium materials, affecting the nuclear properties of the material, as well as measurements such as those for nuclear materials accountability. An isotope dilution mass spectrometry method offers some potential advantages over the current methods for the determination of americium including radiochemistry, nondestructive analysis, and wet chemical techniques.

INTRODUCTION

Precise and accurate americium measurements by isotope dilution mass spectrometry (IDMS) rely on the availability of (1) an americium tracer isotope that is well-characterized both isotopically and for total element content, and (2) a mass spectrometric technique for the precise ratio measurement of the americium sample and tracer isotopes. This paper addresses both of these conditions. The usefulness of the technique can be observed by comparing the IDMS result for americium content to the results of other analytical techniques for these samples, and comparing the precision and accuracy obtained by these different methods.

MASS SPECTROMETRIC EVALUATION

Oxides of americium-241 (^{241}Am) and americium-243 (^{243}Am) were obtained from the Oak Ridge National Laboratory. These materials were of high isotopic purity. In order to prepare the ^{241}Am/^{243}Am mixture used in the mass spectrometric precision evaluation study, each oxide was dissolved in 8N HNO_3 and portions of the respective solutions containing approximately equal amounts of ^{241}Am and ^{243}Am atoms were taken and mixed. This mixture was evaporated to dryness in the form of a nitrate and redissolved in ~5% HNO_3 for analysis by mass spectrometry.

Approximately 3 nanograms of americium were evaporated onto sample filaments and oxidized at a low temperature prior to loading into the mass spectrometer. A triple-filament assembly was used for the sample mounting. The mass spectrometer used for this work was a 30-cm radius of curvature 90° magnet sector thermal ionization instrument equipped with a Faraday cup and ion current detection system, and an HP programmable desktop computer for instrument control and data acquisition. Am^+ currents of 1-2 x 10^{-12} amp were obtained from the nanogram amounts of sample at sample filament temperatures of 1200°C and an ionization temperature of 2100°C. This level of signal allows for precise measurements of peak heights of near equal intensities (low dynamic range) representing a measured ratio of near unity.

Table I gives the data for the ^{241}Am/^{243}Am mixture analyzed under the aforementioned conditions. As is evidenced by the internal precision observed for several sets of ratios taken over a twenty to thirty-minute period of time, precisions of better than 0.2% can be obtained for the 3 nanogram sample size. Run-to-run reproducibility (i.e., external precision) for a single sample demonstrates similar precisions.

The greatest source of error in the mass spectrometry is probably the slight differences in the isotope fractionation between sample runs. Due to the two-mass separation between the isotopes being measured and the small sample sizes being analyzed, the rate of change in the measured ratio is somewhat more sensitive to and, consequently, may be affected by slight changes in temperature, pressure, and the point on the fractionation curve, with respect to time, at which the ratio measurements take place. Although the analyst can control these parameters, inherent inconsistencies and fluctuations do occur.

This study indicated that satisfactory ratio measurements could be made on small quantities of americium using ion current detection. This was expected since americium is a highly efficient ion producer in the thermal ionization process. This finding allowed for the handling of nanogram quantities of americium during the sample loading process. Precision obtained indicated that the maximum error contribution from the mass spectrometric aspect of the work would be on the order of several tenths of a percent for ratios near unity.

AMERICIUM SPIKE PREPARATION AND CALIBRATION

In order to perform IDMS measurements, a calibrated spike material must be prepared for use as a tracer. A suitable choice by convenience and availability was the ^{243}Am isotope obtained from the high purity stock used in the precision evaluation study. Table II gives the isotopic distribution of this material. The problem encountered at this point was that our laboratory was unable to perform conventional destructive assay of the ^{243}Am material, such as an EDTA titration, due to the small amount of americium available (~5 mg). An IDMS method for the calibration of the ^{243}Am was used which required the use of a known quantity of ^{241}Am as a spike.

A paper published by Kirby and Sheehan of the Monsanto Research Corporation (Mound Laboratory) in 1983 provided valuable information on the americium content of the three National Bureau of Standards (NBS) Plutonium Isotopic Standard Reference Materials (SRM).[1] The work described was the analysis of these materials by α-spectrometry over a period of nearly 6 years to determine the americium ingrowth. ^{241}Am/^{239}Pu ratios (atom ppm) were determined by α-spectrometry to a precision of several tenths of a percent. These results, furthermore, agreed reasonably well with the uncertified purification dates obtained from NBS. These materials, containing large quantities of americium relative to the ^{241}Pu content, provided a good ^{241}Am source with well-measured and documented americium content. NBS SRM 946 was chosen to be used as the ^{241}Am "spike" for the characterization of the ^{243}Am material. Table III gives the **^{241}Am** value for the SRM 946 determined (1) by Mound Laboratory using α-spectrometry, and (2) from purification dates for the materials (uncertified). The ^{241}Am value from the α-spectrometry measurement was used as the basis for the **^{243}Am spike calibration. Aliquots of a SRM 946** solution were taken to be used as spikes.

Throughout the aliquotting process, several samples were taken for plutonium assay since an assay value was necessary to determine the amount of americium present. Upon completion of the aliquotting, the solutions were dried at low heat to a nitrate and set aside for later use.

Controlled-potential coulometry was performed on three of the aliquots to determine the plutonium assay.[2] A value of 0.00038644 **g Pu/g solution was** obtained with a precision of <0.02%. Using the value, the americium content was determined based upon the ^{241}Am/^{239}Pu atom ratio value from Mound Laboratory, atomic weights of the ^{241}Am and SRM 946 (^{239}Pu) material, weight fraction of the ^{239}Pu in the SRM 946, and weight of each ^{241}Am spike aliquot. On the average each spike aliquot contained approximately 20 micrograms of ^{241}Am.

Aliquots of ^{243}Am were then prepared from the master ^{243}Am stock solution. These aliquots would be characterized and, later, would be used as spikes for samples analyses. Four of the ^{243}Am aliquots, containing approximately 20 micrograms of ^{243}Am, were mixed with four SRM 946 aliquots containing ^{241}Am for IDMS analyses. Chemical purification was necessary to separate the americium from impurities, most important, the ^{241}Pu isotope, which is isobaric with ^{241}Am and could create extreme measurement bias in the ratio determinations.[3] This was accomplished by valence adjusting all plutonium to the Pu(IV) valence state for the quantitative removal of the plutonium from the americium. Addition of ferrous sulfate to a weak HNO_3 solution containing the plutonium was done to convert the plutonium to Pu(III), and subsequent addition of strong HNO_3 solution to oxidize the plutonium to the Pu(IV) state. This solution was immediately added to a nitrate anion exchange column where the americium, as well as the iron and uranium in-growth products, were eluted. A second separation step was then performed to remove the uranium and iron from the eluted americium. This was done by converting the eluted solution containing americium to a chloride form and washing through a chloride anion exchange column where the americium was again eluted from the column which held these impurities on the resin surface. A purified americium fraction was obtained which was then reconverted to a nitrate and diluted in 5% HNO_3 to be analyzed mass spectrometrically under similar conditions as the precision evaluation study.

During the analyses, the $^{239}Pu^+$ peak was measured to insure that $^{241}Pu^+$ present would be subtracted from the total mass-241 peak, since in some cases, plutonium was observed to be present in the purified americium sample.

When Pu^+ was detected in the mass spectrum, the $^{241}Pu^+$ contribution to the mass-241 peak height was subtracted out by multiplying the $^{239}Pu^+$ intensity by the $^{241}Pu/^{239}Pu$ ratio of the plutonium sample. This was done automatically by the computer which was programmed to take alternate $^{241}(Am + Pu)/^{243}Am$ ratios and $^{239}Pu^+$ peak height measurements, and removing the $^{241}Pu^+$ contribution to yield a "plutonium-corrected" 241/243 ratio. Normally, this correction was less than 0.1% with respect to the americium ratios.

When calibrating a spike with an isotope which will later be measured using the spike, a correction for fractionation need not be applied to the measured ratios since any error in the calibration of the tracer isotope would be applied reciprocally to the unknown isotope in the sample. This applies to these measurements, since the unknown ^{241}Am in the sample is measured against the known ^{241}Am in the SRM 946, with the ^{243}Am being a common factor. As stated earlier, fractionation has to be controlled so that it can be reproduced between spike and sample analyses.

Table IV gives the results of the calibration of the ^{243}Am based upon the ^{241}Am α-spectrometry values for SRM 946.

SAMPLE ANALYSIS

Plutonium metal obtained from the Plutonium Metal Sample Exchange Program, a round-robin type analytical exchange program operated by Rockwell International Energy Systems Group (Rocky Flats), was analyzed by IDMS using this characterized ^{243}Am as a spike. These samples were high-enriched ^{239}Pu materials containing varying amounts of ^{241}Am, and are indicative of the types of materials normally analyzed by laboratories making nuclear measurements. Since these samples are distributed to various laboratories over a period of several years, a large amount of data on americium content by various methods is available for comparison purposes.

Three metals currently being analyzed, referred to by the Program and this paper as Metals A, B, and C, were dissolved, aliquotted, and dried to a sulfate. Several aliquots were obtained of each Metal and spiked with the ^{243}Am aliquots. Similar chemistry and chemical purification techniques as performed with the SRM 946/^{243}Am mixtures in the calibration phase of the project. These mixes were analyzed by thermal ionization mass spectrometry under similar conditions on the precision evaluation study and spike calibration. Tables V, VI, and VII, give the results of the americium content of Metals A, B, and C, respectively, using IDMS and comparisons between the IDMS results and other techniques used by other laboratories.

DISCUSSION

Due to the large spread of values for the americium content between all laboratories, only the precision of the IDMS method can be accurately evaluated. However, the values obtained by NBL using IDMS agreed with the Exchange average to within 2-3%. It is evident that the precision of the IDMS method greatly exceeds the precision of the other methods, in some cases by an order of magnitude. It should be noted that for Metal C, where the IDMS precision was 0.92% for 6 samples, if one sample was deleted from the set, the precision would have dropped to below 0.4%. All data, however, were kept for observation purposes since these data demonstrate a "worse case" situation.

In propagating the error around these values, the uncertainty associated with the calibration is on the order of 0.5%, combining the uncertainties in the half life of ^{241}Am and in the Mound Laboratory value for the ^{241}Am/^{239}Pu ratio of SRM 946. This uncertainty, coupled with the uncertainties around the values on the sample analyses makes the uncertainty of this IDMS method near 1%. A more accurately calibrated spike material with more precise values for the tracer isotope abundance can reduce this uncertainty by perhaps a factor of two.

SUMMARY

Precise americium measurements using IDMS can be made on nanogram quantities of samples using Faraday cup (ion-current) detection systems. Errors associated with the ratio measurements are a function of the isotopic fractionation control, sensitivities, and other thermal ionization characteristics. The uncertainties on the assay and isotopic values of the spike material influence the accuracy of the measurement. Data from this investigation demonstrate that the isotope dilution technique certainly provides a viable alternative to other, more conventional, non-destructive techniques for the determination of americium.

The determination of ^{241}Am by IDMS using ^{243}Am also can be used for the certification of ^{241}Am non-destructive assay calibration standards where gamma spectroscopy is used to estimate the ^{241}Am contribution in calorimetric measurements of ^{239}Pu.

TABLE I

AMERICIUM RATIO MEASUREMENTS - PRECISION EVALUATION

FILAMENT LOADING	MEASURED 241/243	RSD
1	0.58609	0.123% (n=4)
2	0.58388	0.205% (n=4)
3	0.58671	0.100% (n=4)
4	0.58494	0.157% (n=5)
5	0.58602	0.077% (n=3)
6	0.58615	0.112% (n=4)
7	0.58380	0.170% (n=3)
8	0.58470	0.070% (n=4)

$\bar{x}$ = 0.58529

RSD 0.19% (n = 8)

TABLE II

AMERICIUM-243 SPIKE

ISOTOPIC COMPOSITION

Am-241	0.012%
Am-242	<0.001%
Am-243	99.987%

TABLE III

NBS SRM 946 - AMERICIUM-241 CONTENT

	Am-241/Pu-239, (atom ppm)*	Am/Pu, (g/g)**
Mound (α-Spectrometry)	23145 ± 77 (0.33%)	0.020769 ± 0.000069
"Uncertified" Value (Purification date)	23317 ± 57 (0.24%)	0.020924 ± 0.000050

*(09/30/82)
**(09/01/83)

TABLE IV

CALIBRATION OF AMERCIUM-243 SPIKE
(Using NBS SRM 946)

MIX NO.	CONCENTRATION µg Am-243/g
1a	6.4218
b	6.4153
2a	6.4230
b	6.4270
3a	6.4287
b	6.4247
4a	6.4091
b	6.4272

$\bar{x}$ = 6.4221

95% Confidence Limit ± 0.09% (n = 8)

TABLE V

PU METALS EXCHANGE PROGRAM
"A" METAL

METHOD	LAB	Am (ppm)*	RSD
IDMS	NBL	501.0	0.43% (n=3)

EXCHANGE DATA

Gamma Spectroscopy	L	494.3	- (n=2)
Radiochemistry	M	452.3	6.0% (n=5)
	N	507.1	2.9% (n=5)
	P	516.0	2.2% (n=5)
	Q	558.7	1.9% (n=5)
Alpha Spectrometry	S	542.6	1.8% (n=3)

Exchange Average 511.8

*corrected for Am ingrowth and decay to 09/01/83

TABLE VI

PU METALS EXCHANGE PROGRAM
"B" METAL

METHOD	LAB	Am (ppm)*	RSD
IDMS	NBL	761.8	0.37% (n=3)

EXCHANGE DATA

Gamma Spectroscopy	L	780.0	7.9% (n=4)
Radiochemistry	M	711.9	5.1% (n=6)
	N	764.4	0.9% (n=6)
	P	785.6	3.3% (n=4)
	Q	823.5	1.7% (n=6)
Alpha Spectrometry	S	811.6	2.5% (n=4)

Exchange Average 779.5

*corrected for Am ingrowth and decay to 09/01/83

TABLE VII

PU METALS EXCHANGE PROGRAM
"C" METAL

METHOD	LAB	Am (ppm)*	RSD
IDMS	NBL	377.4	0.92% (n=6)

EXCHANGE DATA

Gamma Spectroscopy	L	374.5	- (n=1)
Radiochemistry	M	323.7	2.3% (n=4)
	N	366.6	5.8% (n=4)
	P	380.2	2.9% (n=3)
	Q	412.5	0.5% (n=4)
Alpha Spectrometry	S	403.9	- (n=2)

Exchange Average 376.9

*corrected for Am ingrowth and decay to 09/01/83

REFERENCES

1. H. W. Kirby, and W. E. Sheehan, "Determination of ^{238}Pu and ^{241}Am in ^{239}Pu by α-Spectrometry," Monsanto Research Corporation (Mound Laboratory), Miamisburg, Ohio (1983).
2. M. K. Holland, J. R. Weiss, and C. E. Pietri, NBL-299, (1981).
3. C. E. Pietri, B. P. Freeman, and J. R. Weiss, NBL-298 (1981).

W.S. Lyon (Editor), *Analytical Spectroscopy*
Elsevier Science Publishers B.V., Amsterdam — Printed in The Netherlands

AN IN-LINE ANALYZER FOR MONITORING GAS COMPOSITION IN TRITIUM PURIFICATION PROCESSES - DESIGN

Philippe Chastagner
E. I. du Pont de Nemours & Company, Savannah River Laboratory,
Aiken, South Carolina 29808

ABSTRACT

An in-line analyzer was developed to monitor impurity concentrations in tritium purification processes. Immediate detection of air in-leakage or impurity buildup is necessary to prevent interruption of the processes or damage to the equipment. The analyzer consists of a continuous-flow capillary sampling system, a multipoint sample selection manifold, and a quadrupole mass spectrometer that is optimized for the mass 2 to 44 range. The entire system is controlled by a dedicated computer. Up to twelve sample points can be monitored automatically for impurities at low ppm levels at a rate of one sample per 1.5 minutes. In practice, up to three of the sample points are used for calibrating and reference gases. Computer routines permit operation in three different modes:

- Continual sequential sampling of all points
- Intermittent sampling of less critical points while maintaining full surveillance of critical points
- Continuous analysis of critical points during process upsets

The computer selects sample points according to the desired routine, collects the data from the quadrupole mass spectrometer, and computes the sample composition. The nominal isotopic compositions of the gas streams of interest and their impurity concentrations are displayed on the computer terminal. They can also be printed out, if desired. The data are stored on a disk for later inspection. The computer triggers alarms if the impurity concentrations exceed the set points for any of the sample streams. These set points range from 10 to 100 ppm depending on the particular process that is being monitored.

INTRODUCTION AND SUMMARY

At the Savannah River Laboratory, we designed and contracted for the construction of an in-line analyzer to monitor impurity concentrations ≥ 10 ppm in tritium purification processes. We are particularly interested in the immediate detection of air in-leakage or impurity buildup in the feed lines to thermal diffusion columns and to cryogenic stills. Air in-leakage could cause formation of a mixture that would explode in the thermal diffusion columns, and air or

---

The information contained in this article was developed during the course of work under Contract No. DE-AC09-76SR00001 with the U.S. Department of Energy.

other gases would freeze in the cryogenic stills and their associated piping (Table 1). The analyzer, Figure 1, consists of a specially-designed continuous sampling system, a conventional quadrupole mass spectrometer, a sealed pumping system, and a computer. The analyzer is required to operate unattended 24 hrs per day at a rate of one sample point every 1.5 minutes. Complete analyses for each component in the mass 2 to 44 amu range are required at each sample point. The sample size is limited to 1 $cm^3 min^{-1}$ because tritium is a valuable material. Design constraints for the analysis system are given in Table 2.

TABLE 1
Gas Characteristics

Hydrogen	
Lower explosive limit in air	4.1%
Upper explosive limit in air	74.2%
Argon	
Melting point	-189.2°C
Nitrogen	
Melting point	-210°C
Oxygen	
Melting point	-218°C

TABLE 2
Design Constraints

Maximum sample flow	1 std $cm^3 min^{-1}$
Detection limit	$\leq$10 ppm
Dynamic range	$\geq 1 \times 10^4$
Sample rate	One sample/1.5 min.
Closed system - total gas recovery	
Remote analyzer for containment	

SAMPLE SYSTEM

The in-line analyzer is equipped with a 12 point sampling system (Figure 2). In the initial installation, six points will be connected to process lines, three will be connected to calibrating gas supplies, and the remaining three will be reserved for future applications. The system and its computer control are modular and can be expanded in blocks of six sample points. In this design, a set of continuous flow capillary lines connects process lines to a manifold system. The manifold consists of 12 pairs of pneumatic valves that operate under computer control. The computer directs connection of one sample at a time to the mass spectrometer inlet. The other samples are diverted to the recovery system. Each capillary is sized to deliver 1 std $cm^3 min^{-1}$ at the mean pressure in the process line to which it is connected. Typical capillary sizes are shown in Table 3. Fractionation of the samples in the manifold is prevented by maintaining viscous flow throughout the distribution system. Needle valves are provided in the diversion lines to adjust the nominal pressure in each sample line. The pressures in the manifold are balanced to provide samples at a nominal pressure of 100 microns to the molecular leak inlet of the mass spectrometer. A small portion of the gas in the molecular leak chamber passes into the mass spectrometer for analysis. The rest flows to the recovery system.

The volume of the sample capillaries ranges from 1 to 2 cm^3. In most cases, the sample gas arrives at the manifold within two minutes of the time it leaves the process line. Samples can be switched and be ready for analysis within 30 seconds.

The sampling system was designed to be consistent with present tritium practice at the Savannah River Plant (SRP). Stainless steel capillaries, valves, and fittings are used throughout. The pneumatic valves have Viton® (Du Pont) seats and the manual valves have metal seats. The molecular leak is gold and all the gaskets are metal.

MASS SPECTROMETER

The mass spectrometer specified is a conventional computer-controlled quadrupole mass spectrometer (QMS) that is designed for routine industrial process control. For our application, the spectrometer will be optimized for performance in the 2 to 50 amu range rather than the common 2 to 200 amu range. In order to better assure reliable long term operation, we have specified a dual filament electron impact ion source and Faraday cage ion detection. The required detection limit is $\leq$10 ppm for the gases of interest. The analyzer will be built into a frame separate from the electronic units so that the analyzer and the sample manifold can be put into a containment facility. The electronic units, computer, terminals, and printer will be in control rooms.

PUMPING SYSTEM

The pumping system (Figure 3) was designed to contain and recover all the tritium that is withdrawn from the process lines. The system has three sections: a mechanically pumped section for gas diverted in the manifold; a turbomechanically pumped section for gas that passes through the molecular leak chamber; and a second turbomechanical system for the QMS. The turbomechanical pumps will be

TABLE 3
Typical Capillary Sizes

Average Pressure, torr	Capillary Length, ft	Capillary ID, in.
900	100	0.011
1000	100	0.010
1300	100	0.009

tritium-hardened versions of Balzers model 170 TPU. The mechanical pumps will be Edwards E2M8HS hermetically sealed units. All three mechanical pumps will be equipped with molecular sieve foreline traps and with demisters in the exhaust lines. A type 13x molecular sieve filter will be installed between the mechanical pumps and the SRP recycle system to strip residual hydrocarbons from the gas. Both the Balzers turbomechanical pumps and the Edwards mechanical pumps are lubricated with hydrocarbon oils, not fluorocarbon oils. (The selection of lubricants for pumps that are in tritium service is the subject of a forthcoming article.) The vacuum system will be equipped with appropriate vacuum gages and interlocks to provide information about the system to protect the equipment.

COMPUTER SYSTEM

The in-line analyzer will operate under computer control. A variety of menu driven programs (Table 4) will be provided to:
- Calibrate the mass spectrometer
- Start and stop sample streams
- Change samples
- Analyze samples
- Display and print analyses
- Sound alarms
- Generate new analytical programs.

At the design rate of one sample every 1.5 minutes, samples will be switched and stablized in 30 seconds, and are analyzed in one minute. In practice, the various points will be sampled at various frequencies, depending on need. E.g., the feed lines to the thermal diffusion columns will be sampled at least every 15 minutes while the product lines may be sampled only at 2 to 4 hour intervals.

The computer control system requires a Digital Equipment Corporation (DEC) PDP 11/03 system with a one megabyte dual flexible disk system or its equivalent. The computer system will have two terminals: one with restricted capability, for routine operation by process operators; the other with full capability, for program development, maintenance, and modification. The computer will operate under control of a standard operating system like the DEC RT-11 system. The specific application programs will be written by the contractor in Fortran IV and will be modifications of existing process analysis software.

TABLE 4
Typical System Menu

(1) Run Process Analysis
(2) Create/Modify Internal Standard Method
(3) Create/Modify External Standard Method
(4) Create/Modify Analysis Method
(5) Create/Modify Calibration Method
(6) Create/Modify Sampling Sequence Method
    Select Option Number:

ANALYSIS SYSTEM CHARACTERISTICS

As shown in Table 5, this analyzer design meets the required criteria for an in-line analyzer to monitor gas streams in tritium purification processes. It is designed to operate in an unattended mode 24 hours per day. The design, though new, is completely compatible with existing tritium technology. All of the components are commercially available; no custom-made parts or materials are required.

TABLE 5
System Characteristics

Analysis rate (maximum)	One sample/1.5 min
Response time[a]	<4 min
Sensitivity	≤ 10 ppm
Dynamic range	$\geq 1 \times 10^5$
Calibration	Automatic
Sample consumption	1 std cm^3/min
Mass range	≥ 2 to 50 amu

[a] Time from the instant the sample enters the sample line until the analysis is complete.

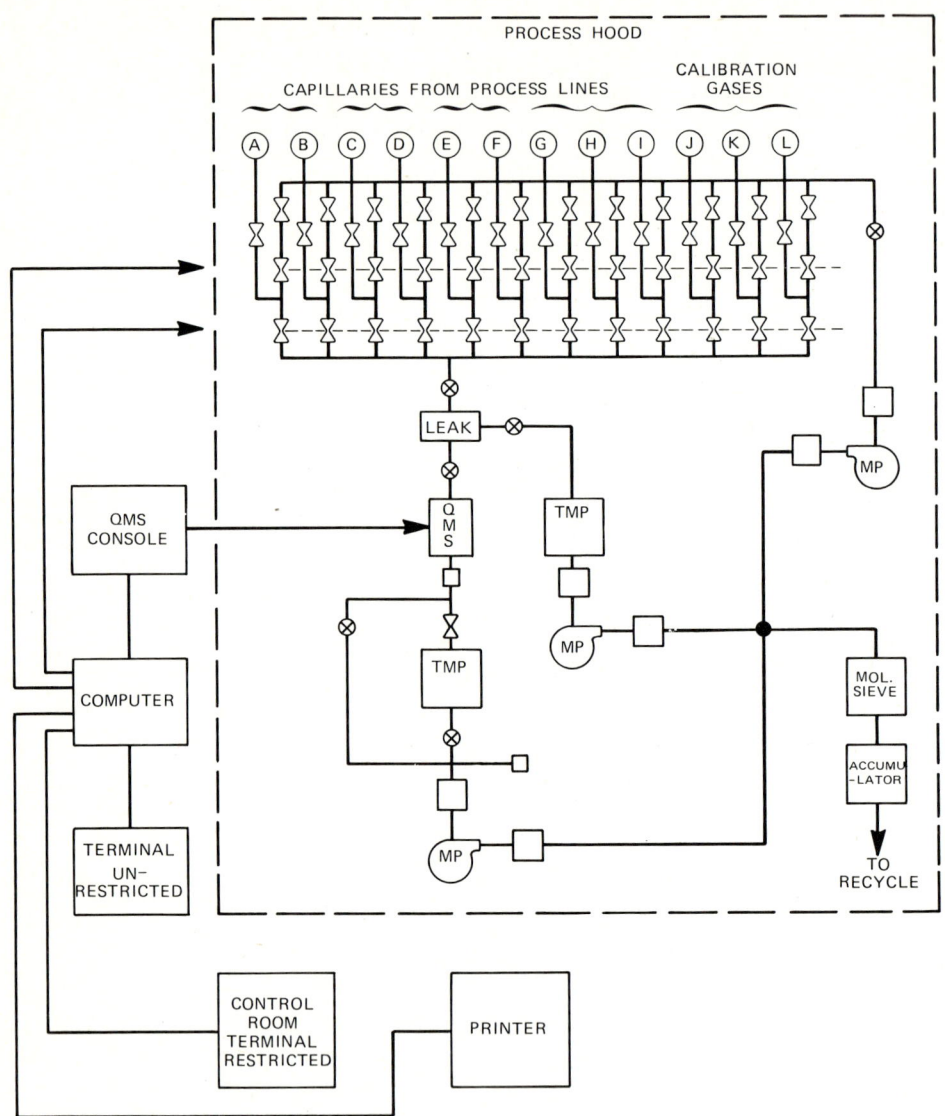

Figure 1. Analyzer Layout

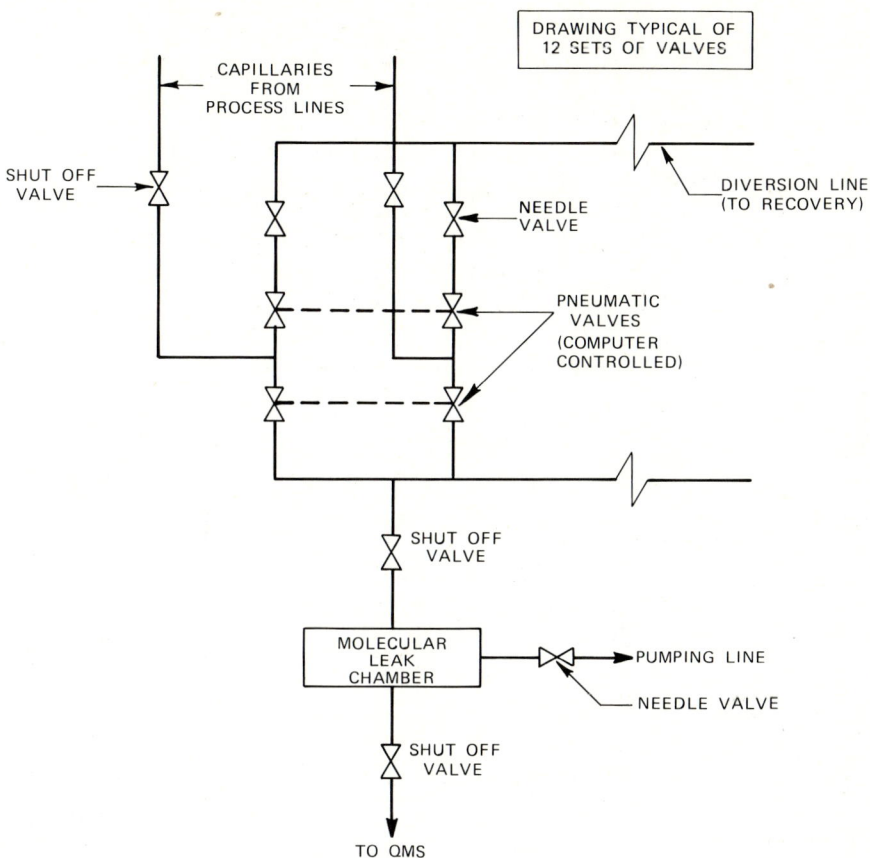

Figure 2. Sample System

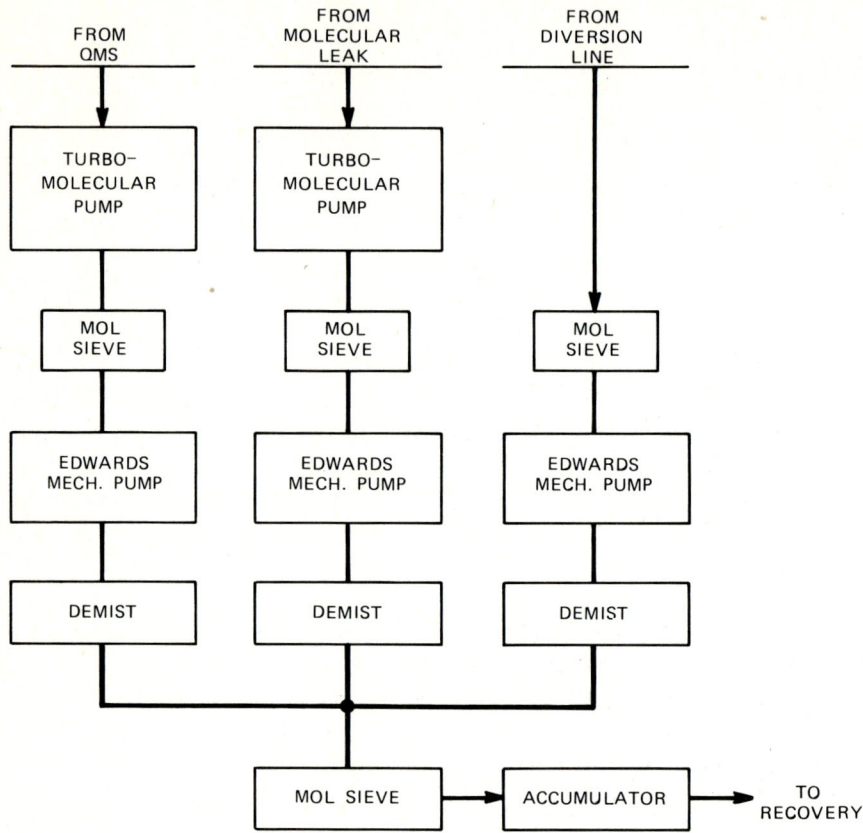

Figure 3. Pumping System

APPLICATION OF MASS SPECTROMETRY TO FUELS AND MATERIALS TESTING AT FFTF

C.E. PLUCINSKI, M.W. GOHEEN and J.J. McCOWN
Hanford Engineering Development Laboratory, P.O. Box 1970, Richland, WA, 99352

ABSTRACT

The Fast Flux Test Facility (FFTF) is a 400 MW(th) sodium cooled reactor and is the largest test reactor of its type in the world. It was designed and is being operated to serve two purposes: gaining liquid metal system experience and serving as a test bed for fuels and materials. During test operations it is possible that cladding breaches and escape of fission gas to the reactor cover gas region can occur. To identify the source of such a leak all 78 fuel pin assemblies contain "gas tag" with a unique "tag" mixture in each assembly. The mass spectrometric identification of tag isotope ratios makes possible rapid location and thus faster removal (if required) of breached test pins.

INTRODUCTION

The Fast Flux Test Facility (FFTF) is a sodium cooled reactor operated for the Department of Energy by Westinghouse Hanford Company. In FFTF the 78 fuel pin assemblies have been "tagged" with a unique combination of Kr and Xe isotopes. In addition to fuel pin assemblies, other test assemblies can also be similarly tagged as is the Materials Open Test Assembly (MOTA).

During power operations leaks are monitored by single channel gamma analyzers in the reactor cover gas system. When radioactive Kr and Xe isotopes are detected a gas tag sample trap (GTST) is used to take a sample of the reactor cover gas, the Kr and Xe are concentrated, and isotopic analysis is obtained by mass spectrometry.

DETECTION

The FFTF cover gas monitoring system includes a chromatographic column for concentration of the Kr and Xe isotopes and a Ge detector for gamma ray counting. The column contains several grams of activated charcoal through which cover gas flows continuously. As the Kr and Xe pass through the tube they are held up by the charcoal, thus concentrating the Kr and Xe to enhance the gamma count rates over those from interfering activated Ne and Ar isotopes.

Should tag gas be released into the cover gas system either because of a breach in a fuel pin or in a test assembly, the detection system measures γ-rays from gaseous radionuclides formed by activation and/or fission. A high Xe^{133} alarm along with a high Xe^{135} alarm signals that both fission gas Xe and gas tag Xe are present in the reactor cover gas. This combination of alarms will usually be indicative of a fuel pin breach. Test assemblies such as the MOTA are also tagged with Kr and Xe isotopes including Xe^{124} and Xe^{132} which neutron activate to Xe^{125} and Xe^{133}. A Xe^{125} + Xe^{133} alarm signals the release of activated gas tags, and indicates a release from a MOTA capsule.

LOCATION

Should all of the Kr and Xe tag gas from a single fuel pin escape into the reactor cover gas, the concentration of each would be only 100 ppb. Since this concentration is too low for accurate isotopic analysis, the sample must be separated from the bulk of the cover gas argon.

A large trap (GTST) is used to take a sample of the reactor cover gas. This trap contains a helical coil of stainless steel tubing which is partially filled with activated charcoal. The charcoal at cryogenic temperature is effective in trapping the Kr and Xe.

A sample of the Kr and Xe gas tag is obtained by flowing cover gas through the GTST. The inlet portion of the helical coil in the GTST is empty and merely acts as a prechiller in which the incoming gas is cooled prior to coming into contact with the charcoal. As the gas passes through the charcoal bed, the Kr and Xe are effectively adsorbed along with argon from the cover gas. Normally a sample consists of a total of 5 cubic feet of cover gas. When no gas tag is present the concentration of Kr and Xe is usually less than 0.1 ppb, and under ideal conditions is less than 0.02 ppb, which is our detection limit. When gas is present the concentration of Kr and Xe can range as high as 100 ppb Kr and Xe in the event of a fuel pin breach and up to several hundred ppb Kr and Xe should a test assembly be responsible for the tag release. The total amount of Kr and Xe is between 0.7×10^{-5} cc and 1.4×10^{-3} cc.

PROCESSING THE GAS

Following the sample step, which is performed by reactor operating personnel, the sample is sent to the analytical laboratory for further processing. The first phase is to desorb some of the argon, which is also trapped on the charcoal. This is accomplished by purging the trap with Ne gas. Because of the very low concentrations of Kr and Xe involved, it is essential that the purge gas not contain any contaminants which could interfere with subsequent Kr and Xe isotopic determinations. The purge gas must be relatively inert and not be significantly adsorbed on the charcoal bed. Several gases were tried to fulfill these requirements: N_2, He, and Ne. Nitrogen adsorbs significantly on the charcoal, and even ultra high purity nitrogen contains some objectionable impurities. Helium fulfilled the adsorption requirement but was not very effective as a purge gas in removing the adsorbed argon. Neon met the absorption requirements and also was much more effective in removing adsorbed argon from the charcoal. So even though high purity Ne is a relatively costly gas it was chosen as the purge gas for the first phase concentration of Kr and Xe from argon.

The second phase of Kr and Xe concentration consists of the transfer of Kr and Xe to a smaller charcoal bed at a controlled rate, again with the use of a purge gas. In this phase the GTST is heated to desorb the trapped Kr and Xe. Another smaller charcoal trap (SCT) containing only 3 grams of activated charcoal is cryogenically cooled and the Kr and Xe are purged by using He as the purge gas. In this phase He is effective in sweeping the Kr and Xe from the GTST into the SCT because of the higher GTST temperature.

The SCT is pumped at low temperature in the third phase of concentration in order to further remove argon, neon, and helium while still retaining the Kr and Xe on a cold charcoal bed. Argon, neon and helium are removed simply because they are adsorbed less strongly than the Kr and Xe. The SCT is pumped to a final pressure of about 5×10^{-7} torr. This results in the removal of most of the Ar, Ne and He while retaining nearly all of the Kr and Xe. This three phase concentration process results in a Kr and Xe recovery of nearly 85%. Assuming an original concentration of 1 ppb in the reactor cover gas, the final concentration of Kr and Xe is about 0.2%. The concentration factor is thus 2×10^6.

IDENTIFICATION

Identification by gas tagging requires the accurate determination of three isotopic ratios, $Kr^{78/80}$, $Kr^{82/80}$ and $Xe^{126/129}$. Although we routinely measure each of the six naturally occurring Kr isotopes, and each of the nine naturally occurring Xe isotopes only measurements of Kr^{78}, Kr^{80}, Kr^{82}, Xe^{126}, and Xe^{129} are essential. The remaining measurements simply act as additional data should some identification difficulties be encountered.

The gas tag design is defined by four Xe planes for fuel pin tags which are $Xe^{126/129}$ of 0.0100, 0.0121, 0.0149, and 0.0182. For the MOTA gas tags two additional Xe planes of 0.0350 and 0.0450 were chosen to differentiate from the fuel pin assemblies. The $Kr^{78/80}$ and $Kr^{82/80}$ values range from 0.1 to 4.5.

ANALYSIS OF KRYPTON AND XENON BY MASS SPECTROMETRY

When argon is ionized in a mass spectrometer mass 20 and mass 40 ions are produced. At high electron bombardment energy and high argon source pressure in the mass spectrometer some mass 80 ions are also produced. These mass 80 ions are the result of an ion molecule reaction between a neutral Ar^{40} atom and an Ar^{40} ion. At the normal electron energy of 70 volts a typical sample which contains over 99% argon and 0.2% Kr and Xe will produce enough Argon^{80} ions to result in a 50% interference with the Kr^{80} ions being measured. However, when this electron voltage is reduced from 70 volts to 30 volts, the ionization efficiency of both the Kr and xe is affected very little, whereas the production of Ar^{80} ions is very greatly reduced. The Ar^{80} interference is reduced from 50% or more down to less than 1%.

RESULTS OF ANALYSES

Many samples of reactor cover gas from the FFTF have been processed and analyzed for Kr and Xe isotopes. These samples range from the very lowest (background) of <0.02 ppb of total Kr and Xe to the highest (MOTA tag releases) which are several hundred ppb of total Kr and Xe.

In the Spring of 1982 gas was released from a fuel test assembly in the FFTF. At first only a very small amount of tag was presenat, too small to accurately identify. Eventually a larger release occurred which was easily identified.

Table 1 below summarizes the data from three gas tags and the tag which was released into the reactor.

TABLE 1

FUEL PIN FAILURE

Sample I.D.	$Kr^{78/80}$	$Kr^{82/80}$	$Xe^{126/129}$
Tag A	3.650	1.834	.0103
Tag B	1.828	1.714	.0101
Tag C	1.982	2.460	.0100
Tag Gas Release	1.806	1.720	.0100

Table 2 below shows the isotope ratios of 8 of the 40 unique tags used for the MOTA.

Table 3 shows data from four randomly selected releases of tag gas from the reactor which occurred between January and March of 1983. These are typical samples. As can be seen from the data three of the four results agree very closely with one of the original tags shown in Table 2. The fourth tag release is a combination of two tags released nearly simultaneously. The two tags were identified from the analytical result with the aid of a computer.

TABLE 2

MOTA TAG GAS RATIOS

Test I.D.	$Kr^{78/80}$	$Kr^{82/80}$	$Xe^{126/129}$
1	.1810	4.478	.0342
2	.1808	4.487	.0457
3	.4147	4.161	.0350
4	.4166	2.769	.0456
5	.7165	3.245	.0348
6	.5557	2.527	.0351
7	.9510	2.919	.0352
8	.4242	1.220	.0354

TABLE 3

MOTA TAG GAS RELEASES

Tag Release I.D.	$Kr^{78/80}$	$Kr^{82/80}$	$Xe^{126/129}$
A	.9429	2.929	.0349
B	.4234	4.168	.0359
C	.4220	1.240	.0350
D	.6220	2.840	.0363

CONCLUSIONS

Gas tagging has proven to be an effective method for positive identification of leaking assemblies, whether fuel pins or test capsules. Cryogenic adsorption of Kr and Xe on activated charcoal and subsequent mass spectrometric determination of Kr and Xe has proven to be an effective analytical method for the gas tag processing and analysis at the Hanford Engineering Development Laboratory (HEDL).

IMPROVED PRECISION AND SPEED OF THERMAL IONISATION MASS SPECTROMETRY WITH A MULTICOLLECTOR

P.J. TURNER, J.E. CANTLE, R.C. HAINES
VG Isotopes Limited, Road Three, Winsford, Cheshire, CW7 3BX, UK

The analytical accuracy of uranium analysis is limited by fractionation occuring before and during the data acquisition period. With peak jumping methods, increasing the data measuring period beyond 15 to 20 minutes can result in a deteriorating, internal precision because the ratio being measured is changing significantly. In addition the operator often has the choice of taking poor data when the beam is unstable or decaying rapidly or waiting for beam stability and measuring a more fractionated ratio to higher precision. Either way measurement accuracy is reduced.

Simultaneous measurement of the ion beams using a multi collector greatly eases these difficulties. Good internal precision is attained in a very few minutes and effectively allows the measurement to be taken at one point on the fractionation curve. The method is insensitive to fluctuations in ion beam intensity so that a decaying or unstable beam produces a minimal loss of internal precision.

Unlike peak jumping methods before such simultaneous measurements can be made it is necessary to know the relative gains of each collector accurately. In the 354 multicollector, prior to the analysis of each turret a current reference is switched in turn into each collector channel. Ratioing the integrated amplifier outputs provides the relative gain calibration. Analysis can then proceed in the same way as peak jumping methods up to the data acquisition period. The axial collector is used to measure and focus the 187 rhenium beam required to set the centre filament temperature. Note that an optical pyrometer can be used to set the centre filament temperature and can reduce the analysis time by about 10 minutes where analysis time is important.

The bead heating sequence is illustrated in Fig. 1.

A standard five collector system with all the separations adjustable can be set to measure 233, 234, 235, 236 and 238 isotopes simultaneously. The 233 isotope does not naturally occur and is sometimes added to a sample as a "spike" to measure the sample size. The Daly detector which uses the same collector slit as the axial Faraday bucket can be electronically selected by

134

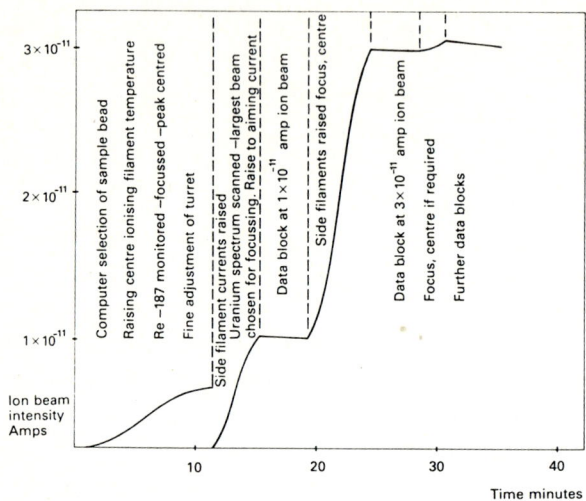

Fig. 1. Typical triple filament bead sequence

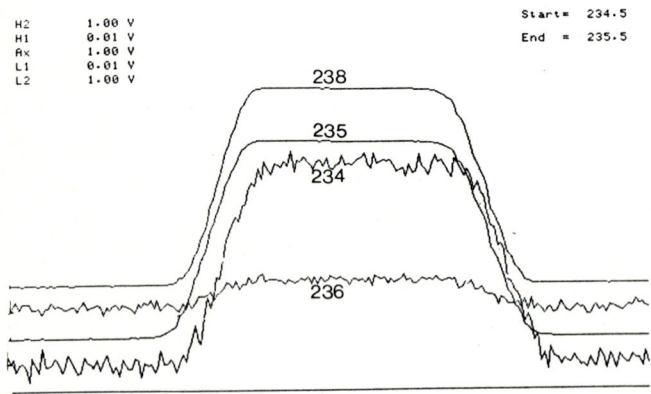

Fig. 2. A mass scan

simply turning on the high voltage to the Daly knob. This allows the peak jumping mode of analysis to be utilized, without any mechanical adjustment of the collector, for the measurement of isotopes at the parts per million level. The coincidence of the 234, 235, 236 and 238 isotopes are shown on a mass scan in Fig. 2. Changes in degree of coincidence of the buckets can be readily observed using the scan program. In the example shown the spectrum is scanned from 234.5 to 235.5 referred to the axial collector and the signal from all buckets monitored.

The results shown in Fig. 3 are from 21 1 microgram loadings of NBS U500 standard. The sample was loaded onto a tantalum side filament of a triple filament bead. The triple filament technique allows slow evaporation of the sample onto the hot rhenium centre filament used for ionization in the mass spectrometer. The samples were analysed in two batches, the first batch of 10 consecutive beads were analysed in about 45 minutes. In the second batch analysis was completed in under 35 minutes.

These 21 beads produced an external error (relative standard deviation) of .024 percent 1 sigma.

Fig. 4 shows a suite of 20 NBS standards in the range .5% 235 to 97% U235. It should be noted that the same "fractionation" factor has been applied to these ratios which range from 200:1 to 1:1 to 1:87 implying good detector linearity.

Note from the plot that the internal error bars on the measurements are insignificant compared to the external error even for high ratios. This indicates that higher ratios could be measured without loss of internal precision which is limited by uniformity of fractionation.

The relative standard deviation of these measurements is .032 percent 1 sigma.

The performance improvement using a multicollector is most easily noted by comparison with the previous uranium data published, 02.382, Uranium Systems calibration with NBS standards using the Isomass 54E. The main points are:-
1) Total analysis time reduced to less than one half.
2) Sample size is now 1 microgram whereas it was previously 2 micrograms.
3) The relative standard deviation for the measurement of the suite is reduced from .046 percent to .032 percent 1 sigma.
4) Data acquisition period is reduced to about 3 minutes from 15 minutes while internal precision (relative standard error) is improved by more than a factor of 3.

Summary

Multicollector analysis offers the following advantages for uranium analysis.

1. Greatly reduced analysis time.
2. Improved internal errors during data acquisition allowing:-
   (a) Smaller samples to be measured
   (b) Greater dynamic ratios to be measured before internal errors become comparable to the external errors.
3. Short data acquisition time giving better results on rapidly fractionating samples.
4. Great tolerance to unstable and rapidly decaying beams resulting in fewer "unsatisfactory" analyses.

N	MEASURED RAW V(N,3)	BIAS CORRECTED V(N,4)	NON-LIN CORRECTED V(N,2)	ERROR E(N)	TRUE V(N,1)	% FROM TRUE	T (MINS)	V
1	1.000829	.999400	.999400	.001	.999700	-.03	45	0.0
2	1.001320	.999890	.999890	.001	.999700	.02	45	0.0
3	1.000868	.999439	.999439	.001	.999700	-.03	47	0.0
4	1.001041	.999611	.999611	.001	.999700	-.01	47	0.0
5	1.001491	1.000061	1.000061	.001	.999700	.04	48	0.0
6	1.001093	.999663	.999663	.001	.999700	-.00	45	0.0
7	1.000864	.999435	.999435	.001	.999700	-.03	47	0.0
8	1.001463	1.000033	1.000033	.001	.999700	.03	49	0.0
9	1.000858	.999429	.999429	.001	.999700	-.03	46	0.0
10	1.001059	.999629	.999629	.001	.999700	-.01	46	0.0
11	1.001007	.999578	.999578	.003	.999700	-.01	33	0.0
12	1.000969	.999540	.999540	.001	.999700	-.02	33	0.0
13	1.001235	.999805	.999805	.002	.999700	.01	33	0.0
14	1.000844	.999415	.999415	.002	.999700	-.03	33	0.0
15	1.001448	1.000018	1.000018	.003	.999700	.03	33	0.0
16	1.001545	1.000115	1.000115	.002	.999700	.04	33	0.0
17	1.000751	.999322	.999322	.003	.999700	-.04	33	0.0
18	1.001171	.999741	.999741	.002	.999700	.00	33	0.0
19	1.001394	.999964	.999964	.001	.999700	.03	33	0.0
20	1.001266	.999836	.999836	.003	.999700	.01	33	0.0
21	1.001209	.999779	.999779	.003	.999700	.01	33	0.0

RELATIVE STANDARD DEVIATION 0.024 PERCENT 1 SIGMA

Fig. 3

N	MEASURED RAW V(N,3)	BIAS CORRECTED V(N,4)	NON-LIN CORRECTED V(N,2)	ERROR E(N)	TRUE V(N,1)	% FROM TRUE	T (MINS)	V
1	.004921	.004919	.004919	.014	.004919	.00	58	6.0
2	.010142	.010138	.010138	.007	.010140	-.02	59	5.8
3	.015562	.015556	.015556	.006	.015565	-.06	59	5.8
4	.052809	.052798	.052798	.002	.052780	.02	59	5.8
5	.113608	.113567	.113567	.0.2	.113600	-.03	58	5.9
6	.546878	.546680	.546680	.002	.546500	.03	59	5.9
7	1.000228	.999867	.999867	.001	.999700	.02	59	2.9
8	4.267173	4.265631	4.265631	.001	4.266000	-.01	58	1.6
9	186.824000	186.756488	186.756488	.014	186.780000	-.01	58	0.0
10	17.360300	17.354027	17.354027	.003	17.349000	.03	59	.3
11	.010147	.010143	.010143	.008	.010140	.03	59	.1
12	17.354240	17.347969	17.347969	.002	17.349000	-.01	43	0.0
13	.015582	.015576	.015576	.009	.015565	.07	44	0.0
14	.999687	.999326	.999326	.001	.999700	-.04	44	0.0
15	6.147758	6.145536	6.145536	.002	6.148000	-.04	44	0.0
16	17.361690	17.355416	17.355416	.003	17.349000	.04	42	0.0
17	6.149865	6.147643	6.147643	.003	6.148000	-.01	44	0.0
18	1.000030	.999669	.999669	.001	.999700	-.00	42	0.0
19	.113590	.113549	.113549	.002	.113600	-.04	41	0.0
20	.015573	.015567	.015567	.008	.015565	.02	45	0.0

RELATIVE STANDARD DEVIATION 0.032 PERCENT 1 SIGMA

Fig. 4

RESONANCE IONIZATION MASS SPECTROMETRY OF IRON--QUANTITATIVE ASPECTS

J.D. FASSETT, L.J. MOORE, and J.C. TRAVIS
Center for Analytical Chemistry, National Bureau of Standards, Washington, D.C. 20234

ABSTRACT

Resonance ionization mass spectrometry has been used to study the formation of ions of iron. Iron was thermally vaporized from a filament at 1250 K. A one-wavelength, two-photon ionization scheme was employed which utilized the tunable UV light provided by a Nd:YAG pumped dye laser with frequency doubling. The systematics of the resonance ionization process has been studied and compared with thermal ionization.

INTRODUCTION

Resonance ionization mass spectrometry (RIMS) is a relatively new technique that combines optical spectroscopy with mass spectrometry (1-3). The technique exploits lasers to selectively ionize gas phase atomic species and achieve high sensitivity. By tuning the laser to discrete, resonant electronic transitions of an element, the ionization probability increases manyfold due to the stepwise absorption of photons. Most of the initial applications of RIMS have been made with modified thermal ionization mass spectrometers (TIMS). RIMS has the potential to be applied to TIMS measurement problems where lack of sensitivity (failure to produce ions) and/or non-selectivity (isobaric interferences) result in fundamental limitations. This study was intended to provide a basis for isotope dilution mass spectrometry of iron where thermal ionization is particularly insensitive. Systematics of the measurement process including atomization, ionization, and mass filtering is discussed.

EXPERIMENTAL

The instrument consists of three basic components: a laser system capable of producing tunable UV radiation; a magnetic sector mass spectrometer with a suitably modified source; and, a detector system capable of measuring the pulsed ion currents produced in the experiment. The instrument has been described in detail previously (3,4). The laser consists of a Molectron Corp.* MY-34 Nd:YAG laser with 2nd and 3rd harmonic generation, capable of producing 250 mJ pulses at 532 nm, which is used to pump a Molectron Corp. DL-18 tunable dye laser. This laser produces 4-ns pulses at 10 Hz. The dye Rhodamine 6G was used in this study to access the wavelength region 558-578 nm, frequency doubled to 279-289 nm. The laser is automated to provide wavelength scanning with a step size of 0.5 pm in the UV. The mass spectrometer is a $60°$, 15 cm magnetic sector thermal ionization instrument whose source was modified to allow for the entry and exit of laser radiation, parallel to the thermal filament. The ions were detected using an electron multiplier operated at low gain whose output was directed into a preamp and a boxcar averager. The output of the boxcar averager was connected to a digital voltmeter interfaced to a HP9845A calculator to control signal processing and graphics.

Nanogram and microgram sized samples of Fe were prepared by drying solutions of ferric nitrate onto the Re filaments with a heat lamp and heating under a hydrogen atmosphere.

DISCUSSION
Atomization

Resonance ionization requires an efficient mechanism to produce a gas phase reservoir of atoms for ionization, and thermal atomization has been shown to be broadly applicable (4). Both experience and theory of thermal ionization suggest that thermal sources can be tailored to produce atoms for a majority of elements in the periodic table. In thermal ionization mass spectrometry, a highly purified sample is dried on a filament substrate. Ions are formed by directly heating the sample. The Saha-Langmuir equation is usually invoked to describe the efficiency of the ionization process:

$$n^+/n^0 \propto \mathrm{Exp}[(W-IP)/kT] \qquad (1)$$

---

*Certain commercial equipment, instruments, or materials are identified in this paper to specify adequately the experimental procedure. Such identification does not imply recommendation or endorsement by the National Bureau of Standards, nor does it imply that the materials or equipment identified are necessarily the best available for the purpose.

where W is the work function of the surface (5 eV for Re), IP is the ionization potential of the element, k is the Boltzmann constant and T is the temperature in K. It can be seen that for elements with high ionization potentials thermal ionization is extremely inefficient. For iron with an IP of 7.9 eV at 1250 K the Saha-Langmuir factor is 2×10^{-12}; at 2250 K it is 3×10^{-7}.

The Saha-Langmuir relationship is not applicable to systems in which there is a significant loss of sample due to the non-atomic vaporization of the element. Molecular species may form from chemical interaction of the element with the filament, from the chemical species loaded with the sample, or from residual gases in the mass spectrometer. It is now clear that an example of this type of problem was encountered in the original RIMS experiments done in this laboratory (3). Vanadium vaporized principally as VO and not only was the efficiency of atomization reduced, but also the RIMS process was compromised by the multi-photon ionization (and dissociation) of the VO. The reduction procedure used for loading the iron samples produced stable, long-lived atomic beams with no evidence for vaporization of molecular iron species.

The RIMS signal versus temperature was measured for iron. The log(I) versus 1/T linear behavior, predicted by the Clausius-Clapeyron relationship, was observed between 1180 and 1370 K. The ionization signal, and thus the gas phase concentration of iron, increased by a factor of approximately 500 in this range. Since the slope of the line relating log(I) to 1/T is equal to the heat of vaporization of the element, this observation strongly suggests that controlled atomic vaporization of iron occurs.

All of the atoms of an element thermally vaporized may not be in the ground state. The intensities and positions of observed spectral lines can be qualitatively explained by assuming a Boltzmann distribution of excited state populations leaving the filament (5). These excited state species will be long-lived (metastable) if the transition to ground state is not radiatively allowed. Iron is an example of an element with low level electronic states which will be populated in the thermal vaporization process. Table 1 summarizes the spectral information collected for iron in the wavelength range 279-289 nm. The lowest electronic levels and their Boltzmann fractional population are also listed. Observed spectral lines that are not listed in reference

6 are also noted in this table. The obvious result of metastable excited atom formation is that the absolute sensitivity must be reduced by a factor associated with the fractional population of the level from which excitation occurs. For iron this fraction is 0.533 for resonance ionization from the ground state.

TABLE 1

Iron Spectral Information, 279-289 nm

Level	Theoretical Population	Lines[a]	Experimental Intensities
0 cm.$^{-1}$	.533	4	1.00-0.24
416	.255	2	0.36-0.24
704	.130	0	
888	.069	0	
978	.019	0	
6928	1.9×10^{-4}	0	
Unknown[b]		8	0.36-0.01

[a] Transitions from the energy level in the experimental wavelength region, tabulated from reference 6.
[b] Untabulated lines in spectrum.

Ionization

The RIMS system in this laboratory was developed around an ionization scheme in which two photons of equal wavelength are absorbed. This scheme requires that the resonant level be more than halfway to the ionization continuum; it is generally applicable to roughly 50 elements with resonant transitions in the range 260-355 nm. Our calculations indicated that the present laser system with 2-3 mJ/pulse would be capable of saturating the ionization process; that is, at the intersection of the atom and laser beams, within the acceptance volume of the mass spectrometer, all atoms resonantly excited should be ionized and detected. The limiting factor in ionization using this scheme is the probability of absorption of the second photon since photoionization cross sections are typically 3 orders of magnitude less than the cross sections of bound-bound transitions.

The relationship between laser power and ionization signal was experimentally checked for the resonant transition at 283.6 nm with and without a 30 cm focusing lens. Saturation was observed with the lens in position and was not observed without the lens. The effect of the differing geometrical overlap of the laser and atom beams with and without the lens in position was also demonstrated. Although saturation did not occur without the lens, the ionization signal was approximately twice as intense as the ionization signal obtained with the lens. This increase can only be explained by an increased geometrical overlap of the laser and atom beams without the lens. An estimate of the absolute efficiency of the ionization process would require a mapping of the plume emitted from the filament and a knowledge of the overlap of the laser and this plume. These experiments suggest that the majority of the atoms in the atom plume are ionized and that with careful engineering of the geometry of the system, 100% ionization per laser pulse of a thermally produced atom reservoir is feasible.

For the experimental RIMS system used here, the 10 Hz repetition rate of the laser provides a significant limitation to the sensitivity. The duty cycle of a laser producing short light pulses and a thermal source producing a continuous reservoir of neutral atoms can be defined as that fraction of atoms traversing the laser focal volume which are irradiated. This duty cycle will be determined by the diffusion time of the atoms, the size of the laser focal volume, and the repetition rate of the laser. Iron atoms at 1250 K take 4.1 us to diffuse 2.5 mm, approximately the diameter of the laser beam. A laser pulsed at 10 Hz capable of ionizing all the atoms in the laser focal volume in 4 ns will have a duty cycle of atomization/ionization of 4.1×10^{-5}. Given the optimistic assumptions of 100 % ionization and atomization efficiencies, one can combine the duty cycle of the process, the Saha-Langmuir factor, and the population of iron in the ground state to estimate that RIMS sensitivity will be about $(4 \times 10^{-5}) \times .53/(2 \times 10^{-12})$ or 10^7 more sensitive than thermal ionization at 1250 K, the temperature of stable Fe vaporization.

A pulsed thermal source has been developed and evaluated in our laboratory in order to better match the atomization and ionization duty cycles (6). With a miniaturized thermal filament, a 30-fold improvement in sample utilization, defined as the ratio of atoms detected to atoms leaving the filament, was demonstrated relative to a continuous thermal source. Thus, the overall duty cycle as defined above would be improved for iron to 1×10^{-3}.

The bandwidth of the laser is specified as 0.3 cm^{-1}. Because this bandwidth is broad, energy level shifts caused by isotope effects are insignificant. The possibility of an isotope effect between ^{57}Fe and ^{56}Fe was checked by scanning the laser wavelength over the resonance line with the highest step resolution for both masses. When the results were normalized and compared, no measurable shift in the wavelength position could be detected. The half-height band width of the resonance line was 0.02 nm.

Mass Filtering

The pulsed ion beam produced by the laser has a time of flight, from formation to detection in the mass spectrometer, proportional to the square root of the ratio of mass to accelerating potential. Thus, the instrument has the potential for dual mass selectivity by the use of both magnetic and time filtering. The time-of-flight filters out the DC background, or thermally produced ions, at any given mass channel. It was observed that detectable thermal peaks existed in the iron mass range, probably caused by hydrocarbon species. By using a boxcar averager with a window of 500 ns, the contribution of these peaks in the time discriminated mass channels will be 5×10^{-6} of the signal for laser ionization. However, since the position of the iron isotopes differs by roughly 40 ns per mass under the experimental conditions here, it is obvious that care must be taken to provide complete or equal integration of all mass channels for quantitative isotope ratio measurements.

REFERENCES

1. D.L. Donohue, J.P. Young and D.H. Smith, Int. J. Mass Spectrom. Ion Phys., 43 (1982) 293-307.
2. C.M. Miller, N.S. Nogar, A.J. Gancarz and W.R. Shields, Anal. Chem., 54 (1982) 2277-2378.
3. J.D. Fassett, J.C. Travis, L.J. Moore and F.E. Lytle, Anal. Chem., 55 (1983) 765-770.
4. J.D. Fassett, L.J. Moore, J.C. Travis and F.E. Lytle, Int. J. Mass Spectrom. Ion Phys., in press.
5. L.J. Moore, J.D. Fassett and J.C. Travis, in preparation.
6. C.E. Moore, Atomic Energy Levels, Nat. Stand. Ref. Data Ser., Nat. Bur. Stand. (U.S.), 35 (1971).
7. J.D. Fassett, L.J. Moore, R.W. Shideler and J.C. Travis, in preparation.

ISOTOPIC MEASUREMENTS OF URANIUM AND PLUTONIUM BY RESONANCE IONIZATION MASS SPECTROMETRY*

D. L. DONOHUE, J. P. YOUNG, D. H. SMITH

Analytical Chemistry Division, Oak Ridge National Laboratory, P. O. Box Y, Oak Ridge, TN 37831

ABSTRACT

Isotopic measurements of the elements uranium and plutonium have been carried out using laser-excited resonance ionization mass spectrometry (RIMS). This technique makes use of the high elemental selectivity inherent in the resonance ionization process which results in complete removal of isobaric (same-mass) interferences. The requirements of the laser system and atomization source will be discussed with emphasis on maximizing the efficiency of temporal overlap, thus increasing the sensitivity of the technique. Results will be shown for resin-bead loaded samples containing a mixture of NBS standard reference materials of U and Pu at the nanogram level. Data will be presented which demonstrate the selectivity, accuracy, and precision of RIMS compared to the conventional thermal ionization technique.

INTRODUCTION

The technique of resonance ionization mass spectrometry (RIMS) has been extensively studied in recent years (refs. 1-9). Work in our laboratory has centered on understanding the physical processes involved in ionizing rare earth (refs. 2,3) and actinide (ref. 4) elements using thermal filament vaporization and pulsed dye lasers. This paper will describe the most recent results for the elements uranium and plutonium, demonstrating the precision, accuracy, and sensitivity achieved to date. The goal has been to make isotope ratio measurements on these elements comparable to those obtained by the thermal ionization method, incorporating the unique advantage of RIMS, namely its high elemental selectivity which can reduce or eliminate isobaric interferences.

---

*Work sponsored by the U. S. Department of Energy, Office of Basic Energy Sciences, under Contract W-7405-eng-26 with the Union Carbide Corporation; J. P. Young received partial support from the Office of Health and Environmental Research.

DISCUSSION

Resonance ionization techniques are based on the photo-excitation of electrons in free (gas-phase) atoms. Various schemes or excitation ladders, leading to ionization have been reported for nearly all the elements proceeding through one or more intermediate state and involving one or more wavelengths of excitation energy (ref. 10). The simplest scheme (Scheme 1) involves absorbing a photon of one wavelength, raising the electron from a low-lying energy level (usually the ground state) to an allowed intermediate state. From this excited state the atom can then be ionized by absorption of a second photon of the same wavelength. This scheme can be applied to most of the lanthanide and actinide elements.

In classical resonance ionization spectroscopy (RIS), the free electron which is produced in the photo-excitation process can be multiplied and detected in a gas proportional counter. However, more information can be obtained by accelerating and mass analyzing the positive ion produced. The hybrid technique of RIMS was developed to take advantage of the elemental selectivity inherent in the resonance ionization process. This results in less interference between elements having overlapping isotopes or isobars. An additional advantage is in the reduction of molecular ion interferences.

The isotopic analysis of U and Pu is important for fissile materials accountancy and safeguards. These two elements have an isobaric overlap at mass 238 which degrades the accuracy of thermal ionization measurements. In addition, ^{241}Pu suffers from overlap with its decay product ^{241}Am, which grows in rapidly ($t_{1/2}$ = 14.3 years). These two interferences occur when using the "resin-bead method" (refs. 11,12) of sample preparation, in which both U and Pu are adsorbed onto anion resin beads which are then used for the sequential mass spectrometric analysis of both elements. Americium is produced in situ and retained by the resin bead matrix. Delays of only a few weeks between resin bead preparation and mass spectrometric analysis can result in a significant bias in the ^{241}Pu assay.

Our experimental set-up for RIMS consists of a flashlamp-pumped dye laser (Chromatix Model CMX-4, LCD/Milton Roy Co., Sunnyvale, CA) interfaced to a 2-stage magnetic mass spectrometer of ORNL design (ref. 13). The laser produces 5-10 mJ pulses of tunable radiation, 1 μsec in duration, with a 0.1 nm (3 cm^{-1}) bandwidth. The pulses are directed into the ion source region of the mass spectrometer by mirrors and a silica lens. The focused laser beam interacts with a cloud of sample atoms produced by a heated metal filament in a manner similar to triple filament thermal ionization mass spectrometry. The laser generated ions are accelerated and passed through two 90° magnetic sectors in tandem (for abundance sensitivity of 1 x 10^6) and are collected by an electron multiplier detector operated in the current integration mode. Each bundle of ions at a given mass produced by a single laser pulse is digitized and accumulated in our computer (MINC-11, Digital Equipment Corp., Maynard, MA). Typically, 7,000-10,000 pulses are accumulated during an analysis for a major ion species, with up to 30,000 pulses collected for a minor isotope. The total number of ions collected for each isotope is approximately 5 x 10^5, giving a theoretical precision of $\pm 0.14\%$.

One problem associated with pulsed laser ionization is the poor temporal efficiency. The laser is "on" for only 30 μsec/sec, for an efficiency of 3 x 10^{-5}. If the sample atoms are being produced continuously, there will be a significant loss of sample during the "off" time of the laser. Therefore, we have implemented a pulser circuit to heat the sample filament briefly before each laser pulse (ref. 14). It has also been found that the efficiency of this pulser depends on the conduction of heat away from the filament between pulses. Using heavy gauge Pt filament legs, it was found that sample utilization efficiency was improved by a factor of 10 compared to continuous heating.

This pulser, along with optimization of the laser beam optics, has allowed us to make high quality isotope ratio measurements with samples as small as 10 ng of U or Pu. The ratios shown in Tables 1 and 2 were obtained with samples containing 10 ng of each element adsorbed on anion resin beads. The Pu and U were analyzed sequentially from the same resin bead, with each analysis requiring 15 minutes. The precision obtained for $^{240}Pu/^{239}Pu$ is excellent, $\pm 0.24\%$, and compares favorably with thermal ionization results for the same size sample. Precision for the minor isotope ^{241}Pu is not as good, reflecting the smaller number of ions collected in the analysis. Results for $^{235}U/^{238}U$ from the NBS-500 standard are also quite good and compare well with the $\pm 0.3-0.5\%$ precision obtained by conventional single filament mass spectrometry.

The analytical bias for these measurements appears to be a characteristic of RIMS, at least at the wavelengths used in this study. The U wavelength, 591.5 nm, is known to be a three-photon scheme 1 process. This means that two transitions must match in energy closely enough to allow ionization with photons of a single color (within the bandwidth of the laser). These three-photon transitions seem to occur for many actinide elements due to their densely packed upper energy levels. Since these processes originate at or near the ground state, they are usually the most intense features in the ionization spectrum in the wavelength range we used (R6G dye at 580-605 nm). The Pu wavelength used was 588.1 nm and is also believed to be a three-photon transition.

TABLE 1
RIMS - plutonium measurements

Analysis	$^{240}Pu/^{239}Pu$	$^{241}Pu/^{239}Pu$
1	0.2456	0.0353
2	0.2471	0.0362
3	0.2467	0.0361
4	0.2460	0.0351
5	0.2464	0.0354
Avg.	0.2464	0.0356
SD	0.0006	0.0005
RSD	0.24%	1.4%
NBS Cert.	0.2414	0.0341

TABLE 2
RIMS - uranium measurements

Analysis	$^{235}U/^{238}U$
1	1.025
2	1.034
3	1.036
4	1.022
5	1.024
Avg.	1.028
SD	0.0067
RSD	0.65%
NBS Cert.	0.9997

Apart from the expected hyperfine splittings and isotope shifts of the transitions involved, the three-photon excitation mechanism requires an exact match of two transitions. The degree to which they match and the fine structure of each transition will result in more or less bias as a function of wavelength for a given isotope ratio. The important consideration is, therefore, how well can the wavelength (and thus the bias) be controlled from one analysis to another. We have used a computer routine to locate the centroid of the U line at 591.54 nm and that of a Ne line at 591.38 nm using the optogalvanic effect in a U hollow cathode lamp. The computer automatically scans the Ne line at the beginning of each analysis and sets the laser wavelength to the same position on the U line with a precision which is better than the short-term drift of the laser itself. The resulting precision in the isotope ratio measurements can be seen in Table 2.

The selectivity obtained for mixed U-Pu samples was determined from the $^{239}Pu/^{235}U$ ratio at both RIMS wavelengths and using thermal ionization. At the Pu wavelength, 588.1 nm, a selectivity ratio of 200 was obtained for Pu with respect to U, normalized to equal thermal ion signals. The selectivity for U compared to Pu at 591.5 nm was found to be 3400. These values are not as high as those reported elsewhere (refs. 5,7) for classical scheme 1 RIMS transitions, but can be explained by non-resonant ionization processes, especially for U, producing a "background" which is independent of wavelength.

Recent studies have sought to measure the ionization spectra of the actinide elements Th, U, Np, Pu, Am, and Cm. Such studies provide information about ionization routes (three- vs two-photon schemes) as well as help in the selection of analytical wavelengths for highest sensitivity and selectivity. It has been found, for instance, that most actinide elements can be ionized via three-photon processes originating near the ground state, and that these lines can be the most intense in the spectrum. Other lines have been attributed to a two-photon (scheme 1) mechanism originating from high-lying excited levels. These excited states are probably populated by hybrid resonances occurring when small molecular species such as dimers are photo-dissociated.

It was found that the analytical line of Pu, 588.1 nm, was close to a relatively strong line of Am at 588.2 nm, thus potentially eliminating any selectivity for Pu with respect to Am. However, in measurements of samples containing equal amounts of Pu and Am (verified by thermal ion signals), a selectivity ratio for Pu of 15,000 was obtained at 588.1 nm. It appears that the Pu line is a three-photon excitation process originating from the ground state, while the Am line originates from a highly excited level and proceeds to ionization via a two-photon process. The slight mismatch in wavelength between the Pu and Am lines was responsible for the high selectivity observed.

SUMMARY

Laser-induced resonance ionization has been found to be a powerful tool when coupled with isotope ratio mass spectrometry. Our studies have shown that the ionization process is frequently more complex than originally predicted. This, however, provides the analyst with many possible routes, expanding the wavelength range over which ionization can be accomplished. In addition, RIMS represents a unique tool for studying the processes of thermal atomization and excitation of neutral atoms. In the future, we will continue to develop RIMS into a viable analytical method, and to use it to probe physical and chemical systems of interest.

REFERENCES

1. D. W. Beekman, T. A. Callcott, S. D. Kramer, E. G. Arakawa, G. S. Hurst, and E. Nussbaum, Int. J. Mass Spectrom. Ion Phys., 34 (1980) 89.

2. D. L. Donohue, J. P. Young and D. H. Smith, Int. J. Mass Spectrom. Ion Phys., 43 (1982) 293-307.

3. J. P. Young and D. L. Donohue, Anal. Chem., 55 (1983) 88-91.

4. D. L. Donohue and J. P. Young, Anal. Chem., 55 (1983) 378-379.

5. C. M. Miller, N. S. Nogar, A. J. Gancarz, and W. R. Shields, Anal. Chem., 54 (1982) 2377-2378.

6. C. M. Miller, J. B. Cross, and N. S. Nogar, Optics Communications, 40 (1982) 271-276.

7. C. M. Miller and N. S. Nogar, Anal. Chem., 55 (1983) 1606-1608.

8. J. D. Fassett, J. C. Travis, L. J. Moore, and F. E. Lytle, Anal. Chem., 55 (1983) 765-770.

9. J. D. Fassett, L. J. Moore, J. C. Travis, and F. E. Lytle, Int. J. Mass Spectrom. Ion Phys., 1983, in press.

10. G. S. Hurst, M. G. Payne, S. D. Kramer, and J. P. Young, Rev. Mod. Phys., 51 (1979) 767.

11. R. L. Walker, R. E. Eby, C. A. Pritchard, and J. A. Carter, Anal. Lett., 7 (1974) 563.

12. D. H. Smith, R. L. Walker, and J. A. Carter, Anal. Chem., 54 (1982) 827A.

13. D. H. Smith, W. H. Christie, H. S. McKown, R. L. Walker, and G. Hertel, Int. J. Mass. Spectrom. Ion Phys., 10 (1972/73) 343.

14. J. D. Fassett, NBS, personal communication, May, 1983.

SPUTTER INITIATED RESONANCE IONIZATION SPECTROSCOPY
FOR TRACE ELEMENT ANALYSIS

J. E. Parks[1], H. W. Schmitt[1], G. S. Hurst[2], and W. M. Fairbank, Jr.[3]

[1]Atom Sciences, Inc., 114 Ridgeway Center, Oak Ridge, Tennessee 37830
[2]Consultant from the Oak Ridge National Laboratory, Oak Ridge, Tennessee 37830
[3]Visiting Scientist on leave from Colorado State University, Ft. Collins, CO

ABSTRACT

This paper describes a new technique, Sputter-Initiated Resonance Ionization Spectroscopy (SIRIS)*, a technique being developed commercially by Atom Sciences, Inc., for ultrasensitive elemental analysis of solid samples. SIRIS combines resonance ionization spectroscopy and ion beam sputtering to provide analyses for all the elements except helium and neon with predicted sensitivities down to 1 part in 10^{12} in routine analysis, and greater for special cases. Recent results in the development of the new SIRIS process and apparatus are presented.

INTRODUCTION

Sputter-Initiated Resonance Ionization Spectroscopy (SIRIS) is an ultrasensitive analytical procedure being developed commercially by Atom Sciences. The principal technology in SIRIS is Resonance Ionization Spectroscopy (RIS) and is now a well documented (refs. 1-2) and proven technology. The uniqueness of RIS is that it is an extremely selective and sensitive process. It is selective in that only atoms of a given element are ionized, since the intermediate excited states through which the ionization proceeds may be chosen such that they are uniquely characteristic of that element. RIS is extremely efficient in that all atoms of the selected element which are in the laser beam are ionized, provided only that the laser power is sufficient.

An inherent condition of almost all RIS techniques is that the sample must be in the gas phase. In a few cases, the sample is initially a gas, but most samples of interest are either solid or liquid. SIRIS uses sputtering to accomplish this.

---

*Patent pending

In SIRIS a pulsed ion beam (usually argon) is focused onto a solid sample, thus producing a cloud of vapor immediately above the target. RIS lasers then selectively ionize atoms in the vapor cloud of the chosen element which are subsequently accelerated through a mass filter for detection at an electron multiplier. Secondary ions, produced by the impact of the ion beam, can be rejected by electrostatic fields, electrostatic energy analysis, the relative timing between the ion beam pulse and RIS laser pulse, or combinations of these.

SIRIS provides ultrasensitive analysis of solid samples for all the elements except helium and neon. Sensitivities down to 1 part in 10^{12} will be possible in routine SIRIS analysis, and greater sensitivities will be possible for special cases.

THE SIRIS APPARATUS

The SIRIS apparatus, designed and constructed by Atom Sciences, has been described previously (ref. 3). The apparatus has a duoplasmatron ion source which supplies energetic ions, generally argon. The ion beam is passed through an analyzing magnet and is pulsed by deflection across a pair of chopping slits. The beam is then focused onto a sample in the target chamber. The target chamber allows precision sample manipulation with X-Y-Z rastering and depth profiling.

The RIS process is carried out with neodymium YAG pumped tunable dye lasers. The ions of the selected element produced by the RIS process are electrostatically accelerated. They are then analyzed with an electrostatic sector for energy analysis and a magnetic sector for mass analysis. The ions are detected with an electron multiplier equipped with a conversion dynode.

The SIRIS apparatus is microprocessor controlled and has automated multiple sample handling capability. The apparatus uses cryogenic, sorption, and ion pumping to achieve an ultrahigh, ultraclean vacuum system. A vacuum lock with bakeout capability is used for sample introduction; thus the main system vacuum need never be broken, the ultrahigh vacuum is maintained, and chamber cleanliness is assured. Data are collected with a computer-based data acquisition system.

DATA AND RESULTS

The development of the SIRIS technique and apparatus has continued since the initial construction phase was completed in late 1982. Recent results of trace element analysis of gallium, indium, and aluminum in samples of silicon and stainless steel are now presented.

Data were taken to demonstrate the validity of the SIRIS technique by scanning the magnetic field of the mass analyzer over a range that would include several isotopes of the impurity atoms of interest and other adjacent masses. The computer based data acquisition system was programmed to set the magnetic field at a specified value while SIRIS signals were measured for a specified number of laser pulses, typically 300 (at 30 hertz, for a time of 10 seconds). Two modes of signal processing were used, single ion counting with pulse height discrimination and analog pulse averaging. In the single ion counting mode, count rates were less than one ion per laser pulse. Ideally, the rate should be kept at less than one out of ten, but in order to extend the dynamic counting range higher rates (but less than one out of one) were allowed. The data was then corrected using Poisson statistics and the corrected data was used as the representative measurement of SIRIS ions arriving at the detector. Single ions produced a voltage pulse of about one volt and a discrimination level of 0.3 volts was set to reject noise pulses. The single ion counting mode was used for impurity levels of a few ppm and lower.

In the analog pulse averaging mode, more than one ion arrives at the detector for each laser pulse and the pulse height of the signal is proportional to the number of ions and the electronic gain. The pulse height of each laser generated event was measured and the average taken for the number of laser pulses used. Due to a dc voltage offset in the electronics, a small background noise level occurs in the spectra.

Well characterized silicon samples doped with gallium were supplied by Hughes Research Laboratories of Malibu, California and were used for the first demonstration experiments. A typical spectrum is shown in Figure 1. The spectrum shows two isotopes of gallium, gallium-69 and gallium-71, for a sample of silicon doped at the 58 ppm level. The relative natural abundance of gallium 69 and 71 is 60.5% and 39.5% respectively and the data agrees with that fact. Data points are seen as inflection points on the graph where the points have been connected with straight lines. The data was collected using the analog pulse averaging mode. A similar sample of silicon containing 0.5 ppm gallium was measured using the single ion counting mode. Those results are shown in

Figure 2. It is clear that the single ion counting mode has essentially no background noise. The data indicates that 26 gallium-69 ions were counted in 10 seconds, 300 laser pulses at 30 hertz. In a 5 minute counting time more than 780 ions would have been counted. If the background could be kept relatively small, 100 ions could be counted in 5 minutes for a 0.06 ppm sample.

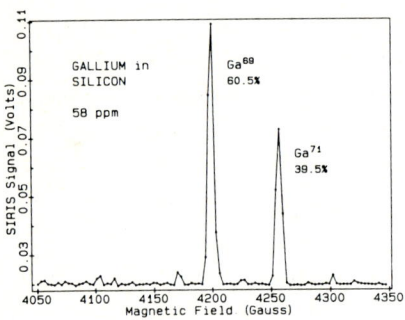

Figure 1. Mass spectrum of 58 ppm gallium in silicon

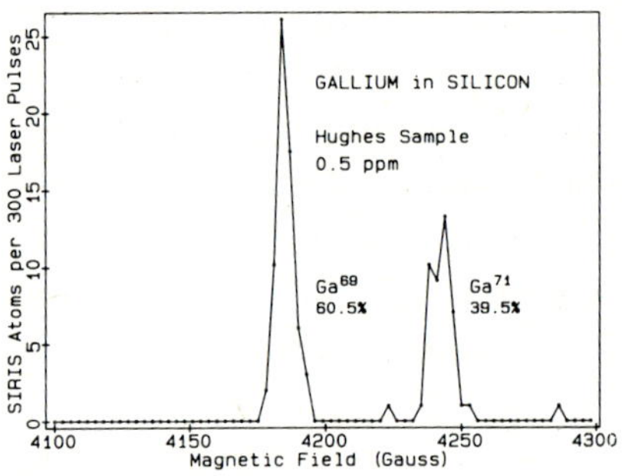

Figure 2. Mass spectrum of 0.5 ppm gallium in silicon

A series of gallium in silicon samples containing 58, 25, 2, 0.5, and 0.05 ppm were measured for their gallium-69 content. A composite plot of those results are shown in Figure 3. The gallium-69 peaks were integrated and background was subtracted for each of the samples and the measured values were correlated with the values supplied with the Hughes samples. The results are shown as points in the log-log correlation plot in Figure 4. The straight line indicates the one to one correlation. Two different samples at the 0.05 ppm level were measured repeatedly and yielded anomolous results where the SIRIS measured value is approximately 100 times higher than that specified by Hughes.

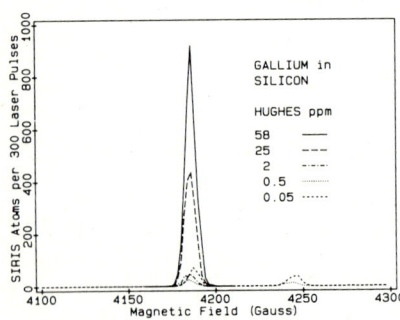 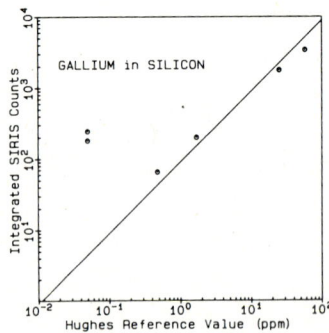

Figure 3. Composite of SIRIS spectra of gallium in silicon samples

Figure 4. Comparison of SIRIS measurements with Hughes reference values

A set of five stainless steel samples with certified analyses of various impurity elements were obtained from the National Bureau of Standards. These samples were analyzed by SIRIS to determine the aluminum concentration. The spectra obtained in those measurements are shown in Figure 5 and a log-log plot of the SIRIS results versus the NBS values is shown in Figure 6. The measurements were made using the SIRIS apparatus in the analog pulse averaging mode since the concentrations of aluminum were well above the 1 ppm level. The data were analyzed as with the gallium in silicon data, by integrating the peaks and subtracting the background. The measurements were repeated a second day and were found to be reproducible. The SIRIS value for NBS sample #1265 was found to be slightly higher than the NBS reference value. By sputtering the surface, the samples were cleaned, and the repeated results of the second day were in good agreement with the NBS values. Surface contamination was particularly noticeable with the 7 ppm aluminum in stainless steel sample, as indicated by the somewhat high first-day run. Note that in Figure 6 the first day's results are shown as o's and the second day's results are shown as x's. NBS samples #'s 1264 and 1265 were not certified as to their aluminum content, and the values were obtained from direct measurements by NBS.

The same samples were measured for their vanadium content and were compared with the vanadium concentrations as measured and certified by NBS. The results of those measurements were similar and gave good agreement in a correlation plot.

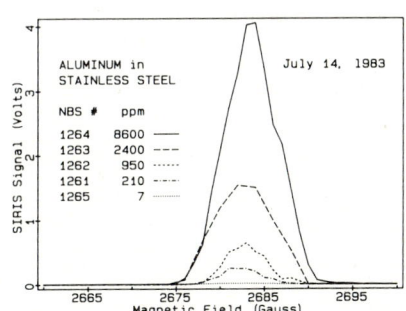 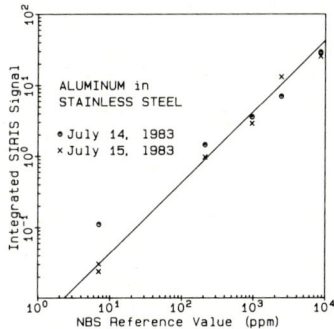

Figure 5. Composite of SIRIS spectra of aluminum in stainless steel

Figure 6. Comparison of SIRIS measurements of aluminum with NBS

A set of indium doped silicon samples were obtained from Hughes Research Laboratories and measured with the SIRIS technique. The results of these measurements are shown in the composite graph in Figure 7. These samples were characterized by Hughes and their indium content had been determined to be 7.8, 4.1, 0.40, and 0.036 ppm. The SIRIS results were analyzed as the previously described samples and compared with the Hughes values. That comparison, shown in the correlation plot of Figure 8, shows good agreement, however, the uncertainty of the results of the 0.036 ppm sample are quite large since the signal was at the background noise level.

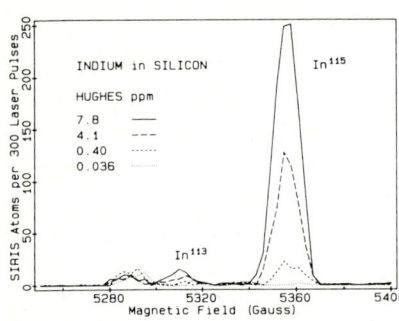

Figure 7. Composite SIRIS spectra of indium in silicon

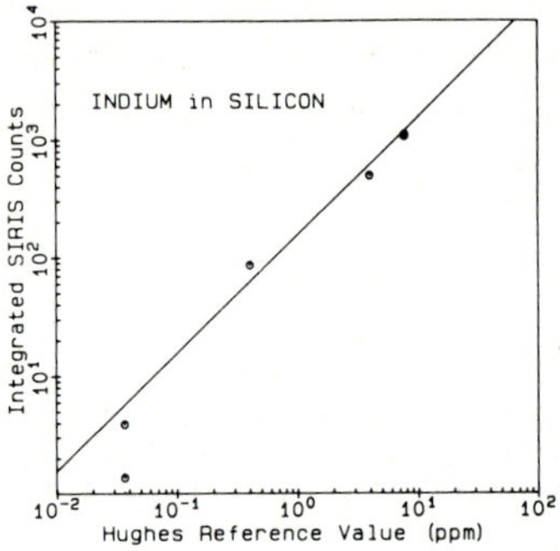

Figure 8. Comparison of SIRIS measurements with Hughes reference values

SUMMARY

The results presented here are the recent results of the Sputter-Initiated Resonance Ionization Spectroscopy technique determined with the apparatus in its present stage of development. Diagnostic tests of the instrument components have been undertaken to determine the nature of any losses in design performance. Improvements in the ion optics of both the ion beam and mass analyzer sections of the apparatus are planned. The desired sensitivity of 1 part in 10^{10} for a five minute counting period (with statistical uncertainty of ±10%) should be reached in the near future.

ACKNOWLEDGEMENTS

The authors gratefully acknowledge discussions with, and preparation of samples by, O. J. Marsh and J. P. Baukus, Hughes Research Laboratories. The partial support of this work by the Defense Advanced Research Projects Agency is gratefully acknowledged.

REFERENCES

1 G. S. Hurst, M. G. Payne, S. D. Kramer, and J. P. Young, Rev. Mod. Phys. 51, (1979) 767.
2 G. S. Hurst, M. G. Payne, S. D. Kramer, and C. H. Chen, Phys. Today 33, (1980) 24.
3 J. E. Parks, H. W. Schmitt, G. S. Hurst, and W. M. Fairbank, Jr., Thin Solid Films 8 (1983) 69.

RESONANCE IONIZATION MASS SPECTROMETRY AT LOS ALAMOS NATIONAL LABORATORY

N. S. NOGAR,[1] S. W. DOWNEY,[1] R. A. KELLER,[1] and C. M. MILLER[2]

[1]Chemistry Division, Los Alamos National Laboratory, Los Alamos, NM 87545
[2]Isotope and Nuclear Chemistry Division, Los Alamos National Laboratory, Los Alamos, NM 87545

ABSTRACT

We present recent results on the application of resonance ionization mass spectrometry to the precision measurement of isotope ratios, particularly in the presence of isobaric interferences. Emphasis is placed on hardware developments having the potential to increase ionization efficiency.

INTRODUCTION

The use of lasers as a selective ionization source for elemental and isotopic analysis is becoming a reasonably commonplace technique (refs.1-12). Resonance ionization mass spectrometry (RIMS) offers well established advantages over conventional ionization in terms of both sensitivity and selectivity. The selectivity inherent in the RIMS process can be particularly useful in reducing or eliminating isobaric interferences in isotopic analysis (refs.3-6,9).

While most RIMS work has utilized pulsed lasers, we have recently been concentrating (ref.13) on the application of cw lasers to the resonance ionization process. There are several reasons for this choice. First, most RIMS work to date has utilized thermal atom sources (refs.3-9), which produce gas-phase atoms continuously. Ionization with pulsed lasers typically leads to duty cycles $\lesssim 10^{-4}$. Continuous lasers obviously will exhibit better temporal overlap with such a continuous source. Secondly, cw lasers are better suited for use with pulse-counting electronics, since pulsed laser ionization typically produces ions in such short time intervals that the counter dead time and pulse pile-up can be serious problems. Lastly, cw lasers produce ions at the highest rate when the laser focal volume is very small ($\lesssim 10^{-4}$ cm^3), in contrast to pulsed laser ionization where ionization efficiency is greatest when a relatively large volume (≥ 1 cm^3) is saturated. This smaller ion source volume should lead to better extraction efficiency and resolution in the mass spectrometer with current ion source optics. These considerations suggest (ref.14) that cw lasers may be more effective than pulsed lasers in producing ions for mass spectral analysis. In fact recent experiments at this laboratory (ref.9) have shown cw ionization rates to be >10^3 times that observed with a flashlamp pumped dye laser.

Ionization with cw lasers is not without drawbacks however. The range of accessible wavelengths is more limited than with pulsed lasers. The rather low powers available dictate (ref.14) that the laser must be focused (10^{-3} cm $\leq$ beam waist $\leq 10^{-2}$ cm) in order to have an appreciable probability of ionizing an analyte atom. The volume sampled is thus quite small, with a consequent geometrical loss of sample. In the remainder of this paper we discuss methods for increasing the effective photon intensity in the ionization region, in order to increase ionization efficiency and concentration of the sample in order to improve the volumetric overlap of laser and sample.

Intracavity RIMS

With relatively low peak power flashlamp pumped- and cw-dye lasers, the laser power circulating within the cavity is usually much larger than the power coupled out. To a first approximation, $I_{intra}/I_{extra} = \frac{1}{1-R}$ where R is the reflectivity of the output coupler. This power enhancement has been used to good effect in photoacoustic (ref.15), photothermal (ref.16), and fluorescence (ref.17) spectroscopy, and for photochemical (ref.18) experiments. To our knowledge, there has been no previous report of RIMS being performed intracavity.

The cavity of a flashlamp-pumped dye laser is extended to encompass the source region of a time-of-flight mass spectrometer. A periscope containing two dielectric mirrors is used for beam steering, the windows on the mass spectrometer are AR coated, and the normal output coupler of the laser is used to couple light out for diagnostics. The total length of the extended cavity is 3 meters. It was found that insertion of a 500 mm focal length positive lens near the center of the cavity was necessary (ref.19) to ensure lasing. Ions were detected with a channel electron multiplier, and after passing through a preamplifier and amplifier, the signal was processed with a boxcar integrator operated in a swept-gate mode. Signals were monitored in real time with an oscilloscope. While these experiments were performed with a flashlamp-pumped dye laser, we expect the results to be applicable to cw laser excitation.

Initial experiments were carried out on uranium, using the doubly resonant, three-photon ionization (ref.20) at 591.5 nm. After adjusting the filament current to provide an adequate supply of uranium atoms, the laser discharge circuitry was adjusted to fire the lamp at a chosen voltage. Measurements of ion current were then made with the mass spectrometer both intra- and extra-cavity. In both cases, the laser output was maximized by fine adjustment of the cavity, and the wavelength optimized by tuning the birefringent filter to produce the maximum optogalvanic signal (ref.21) in a uranium hollow cathode lamp. Sample time-of-flight spectra are shown below.

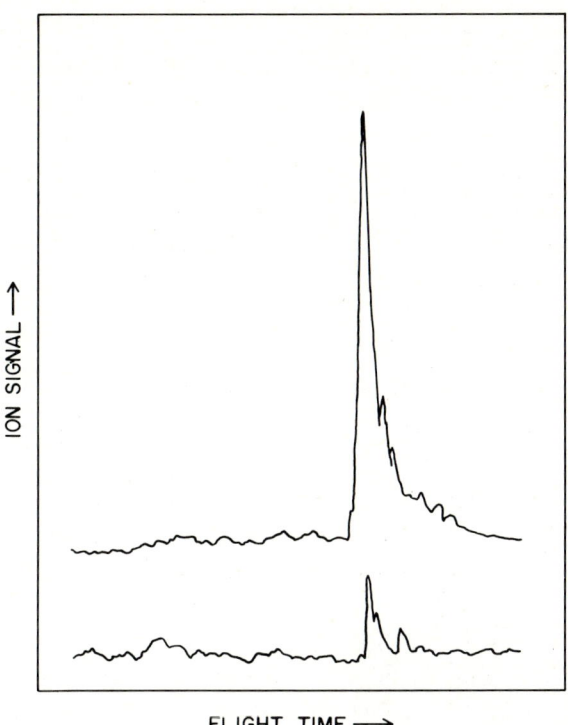

Fig. 1. Time-of-flight spectra for intracavity (upper) and extracavity (lower) excitation. Pulse energies were 0.6 mJ and 1.2 mJ, respectively.

The laser output with the extended cavity was somewhat lower, for a particular discharge voltage, than with the normal cavity. This is presumably due to reflective losses at the intracavity windows and lens, and mirror imperfections in the periscope.

The intracavity pulse energy for the example shown in Fig. 1 is expected to be ~0.6 mJ × (100/15) = 4 mJ for the 85% output coupler used in these experiments (ref.22). If the ionization probability is linearly dependent on fluence, as it may be if the cross section for the bound-free transition is much smaller than for the resonant transitions, then one would predict: Signal (intracavity)/ signal (extra-cavity) $\simeq$ 4/1.2 $\simeq$ 3.3, in rough agreement with experimental observation. This advantage may be even greater for cw lasers, where higher reflectivity output couplers are often used. In summary, we expect intracavity excitation to offer a significant improvement in ion yield in cases where a single wavelength is sufficient to affect ionization.

Gasdynamic Focusing for Sample Concentration.

Thermal filaments typically produce a cosine squared distribution of atoms above the flat filament surface. In our apparatus, a focused cw laser beam, passing 1 mm above the filament surface, will intercept only ~3% of the evaporated atoms.

However, if the atoms are evaporated into a flowing (inert) gas stream, which then passes through a converging nozzle, it is possible to significantly reduce the cross sectional area of the sample stream. This principle of hydrodynamic focusing has been successfully applied to the analysis of liquid phase samples (ref.23), and to the production of aerosol beams (ref.24). We hope to apply this same principle to gasdynamic focusing of samples for resonance ionization.

In preliminary experiments, we have used laser-induced fluorescence, of I_2, seeded in Ar to characterize the spatial distribution and demonstrate the principle of focusing. Figure 2 shows the results of an early experiment carried out in this laboratory.

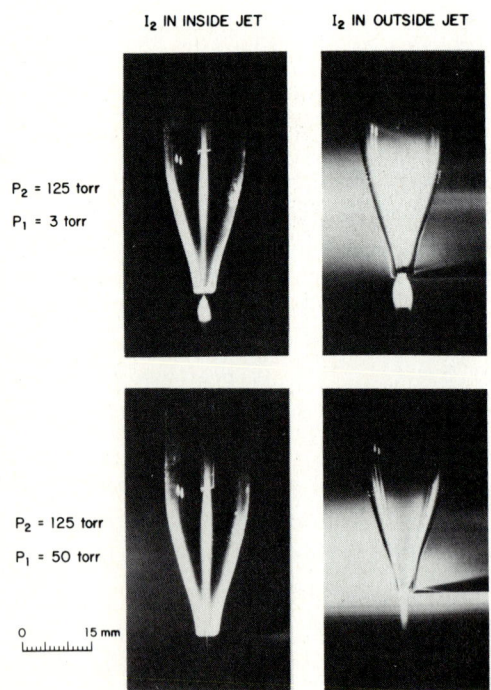

Fig. 2. Gasdynamic focusing. P_2 is the reservoir pressure, P_1 is backing pressure.

In the photographs on the left, the I_2 seeded argon stream is flowing from top to bottom through a glass capillary in the center of the glass nozzle. The carrier stream of pure argon is flowing around the capillary, and down the glass nozzle. The area outside the glass nozzle is evacuated by a mechanical pump. Laser excitation is provided by a multimode Ar^+ laser beam (514 nm) which enters from the left. Fluorescence was photographed through a long pass filter with a standard 35-mm camera. The photographs on the right were taken with iodine seeded argon flowing through both the capillary and the outer nozzle.

Under proper conditions (lower left photograph) it is obvious that substantial concentration of the sample is taking place over a considerable linear distance. In this particular case, we have determined the diameter of the sample stream to be ~250 μm. In other cases we have attained sample streams ≤100 μm in diameter. Even when expanded into rather low pressure, as in the upper left photograph, the sheath flow ensures that the sample volume is very small immediately after emerging from the nozzle. This low pressure operation may allow for more expeditious interfacing with a mass spectrometer. A secondary advantage is that the flow stream lines in the nozzle prevent the sample from coming into contact with the nozzle walls. In addition, solid samples should be cooled from their evaporation temperature (~1000°C) to at least 50°C, which should reduce the electronic partition function and increase both selectivity and sensitivity.

We feel that gasdynamic focusing has considerable potential to increase the sensitivity of various laser-based detection schemes. Experiments have been initiated at Los Alamos to observe solid samples evaporated from a hot wire in the sample flow. An extension of this work to detection by resonance ionization is anticipated.

ACKNOWLEDGMENTS

Financial support of the Department of Energy through Los Alamos National Laboratory is gratefully acknowledged, as is the excellent technical assistance of F. Archuleta and A. Garcia.

REFERENCES

1 S. Mayo, T. B. Lucatorto and G. G. Luther, Anal. Chem., 54 (1983) 553-556.
2 D. J. Beekman, T. A. Callcott, S. D. Kramer, E. T. Arakawa and G. S. Hurst, Int. J. Mass. Spec. Ion Phys., 34 (1980) 89-97.
3 C. M. Miller, N. S. Nogar, A. J. Gancarz and W. R. Shields, Anal. Chem. 54 (1982) 2377-2378.
4 D. L. Donahue, J. P. Young and D. H. Smith, Int. J. Mass. Spec. Ion Phys., 43 (1982) 293-307.
5 J. P. Young and D. L. Donahue, Anal. Chem., 55 (1983) 88-91.
6 D. L. Donahue and J. P. Young, Anal. Chem., 55 (1983) 378-379.

7   J. D. Fassett, J. C. Travis, L. J. Moore and F. E. Lytle, Anal. Chem., 55 (1983) 765-770.
8   N. S. Nogar, R. K. Sander, S. W. Downey and C. M. Miller, Proc. SPIE, Los Alamos Conf. on Optics, 1983, in press.
9   C. M. Miller and N. S. Nogar, Anal. Chem. 55 (1983) 1606-1608.
10 F. M. Kimock, J. P. Baxter and N. Winograd, Surf. Sci., 124 (1983) L42-L48.
11 N. Winograd, J. P. Baxter and F. M. Kimock, Chem. Phys. Lett., 88 (1982) 581-584.
12 G. S. Hurst, M. G. Payne, S. D. Kramer and J. P. Young, Rev. Mod. Phys., 51 (1979) 767-819.
13 C. M. Miller, N. S. Nogar, and S. W. Downey, Proc. SPIE, 27th Int. Tech. Symp. of Internat. Soc. Opt. Eng., San Diego, CA, 1983.
14 C. M. Miller and N. S. Nogar, Anal. Chem., 55 (1983) 481-488.
15 R. G. Bray and M. J. Berry, J. Chem. Phys. 71 (1979) 4909-4922.
16 K. V. Reddy, Rev. Sci. Inst., 54 (1983) 422-424.
17 W. R. Anderson, J. A. Vanderhoff, A. J. Kotlar, M. A. Dewilde and R. A. Beyer, J. Chem. Phys., 77 (1982) 1677-1685.
18 K. V. Reddy and M. J. Berry, Chem. Phys. Lett., 66 (1979) 223-229.
19 S. W. Downey and N. S. Nogar, Submitted to Appl. Spec.
20 H. L. Chen and C. Borzileri, J. Chem. Phys., 74 (1981) 6063-6069.
21 R. A. Keller, R. Engleman, Jr., and E. F. Zalewski, J. Opt. Soc. Am., 69 (1979) 738-742.
22 This is the normal output coupler supplied by the manufacturer, Chromatix, Sunnyvale, CA.
23 M. Melamed, P. F. Mullaney, M. L. Mendelsohn, Flow Cytometry and Sorting, Wiley, New York, 1979.
24 B. Dahneke and H. Flachsbart, Aerosol Science, 3 (1972) 345-349.
25 R. A. Keller and N. S. Nogar, Appl. Opt., submitted.

THE DETERMINATION OF URANIUM IN URINE BY ISOTOPE DILUTION MASS SPECTROMETRY USING RESIN BEAD LOADING

M. P. May, R. L. Walker, T. G. Scott, F. F. Dyer, and J. R. Stokely
Analytical Chemistry Division, Oak Ridge National Laboratory,
Oak Ridge, Tennessee 37831

ABSTRACT

Ion exchange separations are coupled with isotope dilution mass spectrometry to provide a highly sensitive method for the determination of uranium in urine. This method was examined for its sensitivity, practicality, reproducibility, and accuracy. Our goal, as suggested by the Nuclear Regulatory Commission, was to measure uranium at the 0.1 ng/ml $\pm$ 0.01 ng/ml level for 10 mL aliquots of urine. The ability to measure this low concentration of uranium can be limited by both instrumental background and contamination from working environment, reagents, and labware in the procedure (i.e., procedural blank). Initially, high blank levels of uranium were observed so much emphasis was placed upon lowering the procedural blank. To achieve a 10% uncertainty, this blank must remain low and be reproducible throughout all determinations. The validity of the method was tested by analyzing an artificial urine spiked with a known concentration of uranium and several human urine samples analyzed by an independent method. Details and results of the method plus its possible bioassay applications are presented.

INTRODUCTION

The urine of workers at nuclear facilities is sampled periodically and analyzed for uranium in compliance with federal regulations. The Nuclear Regulatory Commission (NRC) is searching for cost effective, practical techniques that have greater sensitivity than the presently used methods for bioassay analyses. In this work, ion exchange separations are coupled with the resin bead isotope dilution mass spectrometry (IDMS) technique. The method is examined for its sensitivity, accuracy, precision, and practicality for the analysis of uranium in urine.

METHOD

The resin bead method is divided into two stages. First, the uranium isotopes are isolated from the urine by ion exchange separations. Second, they are loaded onto two or three resin beads which are analyzed by thermal ionization IDMS for uranium concentration and isotopic ratios. Efforts in this work have focused on the methodology for the separation and isolation of uranium prior to mass spectrometry measurements. Based upon information from the literature and experimental work, the following procedure was adapted to urine matrices. Ten milliliters of urine were adjusted to 8M HCl with 20 mL of concentrated Ultrex HCl (high purity acid by Baker). Then 6.6 ng of ^{233}U (with known isotopic composition) was added as a tracer for the mass spectrometry measurements. This solution was allowed to digest overnight and then passed through a Dowex 1x8 anion exchange resin column that had been preequilibrated with 8M HCl. Under these conditions, uranium is retained on the resin. A wash with 8M HCl was used to remove interfering radionuclides plus salts and organics present in the urine that remained on the column, and uranium was then eluted with 0.1N HCl. Eluates were taken to dryness and dissolved in 0.2-0.4 mL of 8M HCl. Two or three beads of the same anion exchange resin were added and allowed to equilibrate with the solution for about two hours by shaking, during which time the uranium loaded onto the three beads. Chemical yield measurements showed that about 7.4% of the ^{233}U loaded onto three resin beads or 0.1 ng ^{233}U loads per bead. A single bead was crimped into a rhenium filament and analyzed by thermal ionization mass spectrometry. Isotope dilution calculations based upon the added ^{233}U tracer provided uranium isotopic ratios and concentrations.

One of the goals of the study was to measure uranium at the 0.1 $\pm$ 0.01 ng/mL level for 10-mL aliquots of urine. Initial "blank" measurements, in which 10 mL of deionized triply-distilled water was substituted for the urine and analyzed, showed uranium levels of 0.5 ng/mL. This amount of uranium exceeds not only the above goal but also the uranium concentration found in normal urine. Because it was not known whether this contamination came from reagents, labware, or the working environment, it was necessary to make changes in the entire procedure to minimize contamination from all possible sources. Project operations were moved to a special low-level laboratory with restricted entry regulation and with better control of air flow to minimize particulate contamination. Reagents were limited to Baker's Ultrex HCl and HNO_3, and deionized triply-distilled water produced in a quartz still. An extensive cleanup procedure was adopted for all labware and the ion exchange resin which included washes with 4M HNO_3, deionized H_2O, and deionized triply-distilled H_2O. By taking these actions the procedural blank was successfully lowered to 0.30 ng U total which corresponds to 0.030 $\pm$ 0.0047 ng/mL for 10 mL of water or sample plus 20 mL HCl.

RESULTS

Normally, a procedure is verified by analyzing standard reference materials containing certified amounts of the elements in question; but no urine standards certified for uranium are known to exist. As a substitute, a spiked synthetic urine consisting of an artificial urine matrix spiked with a known concentration of natural uranium (7.024 μg/kg or 7.234 ng/mL) as well as a blank synthetic urine (matrix with no added U) were obtained from Battelle-PNL and analyzed. Five analyses of the synthetic blank gave a mean value of 0.022 ± 0.002 μg/kg, which lies at or below the procedural blank. The spiked synthetic urine was analyzed in 1-mL aliquots and after subtracting blank values, 7.19 ± 0.06 ng/mL U was found. The results showed that the spiked synthetic urine could be analyzed with both accuracy and precision. As a second verification method, several human urine samples that had been measured by independent analytical methods were obtained and analyzed for uranium. The results were precise (see Table 1) but ran consistently lower than the independently determined values. This poor agreement was thought to have two possible causes: (1) large errors associated with the independent methods when measurements at these low U concentrations are made or (2) an isotopic exchange problem between the mass tracer and uranium in the urine. Therefore to test the adequacy of isotopic exchange several aliquots of the same urine were subjected to extensive digestions with nitric acid and perchloric acids to destroy the organic matrices present. The uranium concentrations observed in these samples agreed well with those obtained by the normal procedure. (See Table 1.) Thus the poor agreement was attributed to errors associated with the independent methods.

The accuracy and precision of the proposed method was determined from a calibration curve obtained by spiking human urine samples of unknown (but low) uranium concentrations with increasing known quantities of ^{238}U tracer (0.873 ng, 4.36 ng and 34.9 ng respectively). The blank human urine and the spiked urine samples were all analyzed in quadruplicate using 10-mL aliquots. By plotting observed uranium concentration as determined by IDMS vs spiked or added uranium concentration, a calibration curve was obtained which showed the mean of the observed values for the blank human urine to be 0.062 ± 0.005 ng/mL. After subtracting the 0.030 ng/mL procedural blank, the net concentration of 0.025 ng/mL U was obtained.

As a shorter alternative to the normal procedure, both the spiked synthetic urine and human urines were analyzed by directly loading uranium from the urine onto the beads, thus eliminating the ion exchange separations. In this variation, 0.5-mL and 1.0-mL aliquots of urine were adjusted to 8M HCl with Ultrex HCl, and ^{233}U tracer was added. Three resin beads were then added and the mixture was equilibrated by shaking for two hours. As previously described, the beads were analyzed by mass spectrometry. As seen in Table 1, the results obtained by direct loading agree very well with both our normal procedure and the extensive chemical digestion procedure.

TABLE 1: URANIUM CONCENTRATION IN HUMAN AND SYNTHETIC URINE BY THREE DIFFERENT IDMS METHODS

URINE

Normal Method[1]		Destructive Method[2]		Direct Bead Loading[3]	
Sample volume (mL)	U concentration (ng/mL)*	Sample volume (mL)	U concentration (ng/mL)*	Sample volume (mL)	U concentration (ng/mL)
Human Urine A					
1.0	2.86	1.0	2.81		
1.0	2.81	1.0	2.94	0.5	2.96
1.0	2.78	1.0	2.88	0.5	2.86
Mean($\bar{x}$)	2.82		2.88		2.91
SD[4]	0.04		0.07		0.07
Human Urine B					
1.0	8.25				
1.0	8.31			0.5	8.28
1.0	8.30			1.0	8.24
Mean($\bar{x}$)	8.29				8.26
SD	0.03				0.03
Spiked Synthetic Urine (7.234 ng/mL)[5]					
1.0	7.25				
1.0	7.24				
1.0	7.20				
1.0	7.17			0.5	7.19
1.0	7.10			0.5	7.28
Mean($\bar{x}$)	7.19				7.24
SD	0.06				0.06

*Corrected for blank values.
[1] The method as described here.
[2] The urine matrix is completely destroyed with $HNO_3/HClO_4$ and heat.
[3] Uranium is loaded directly from the acid-adjusted urine onto 2 or 3 resin beads and analyzed.
[4] SD=Standard deviation.
[5] 7.024 µg/kg or 7.234 ng/mL based upon measured density of 1.0299 g/mL.

CONCLUSIONS

Several conclusions have been made from this work. The blank comes primarily from uranium contamination in the procedure (i.e, reagents, glassware and airbone particulates) rather than from the instrument background. Both the synthetic and human urine can be analyzed by resin bead IDMS with <u>accuracy</u> and <u>precision</u>. The sensitivity (ability to determine 0.1 $\pm$ 0.01 ng/mL) is dependent upon: 1) adequate chemical yields, 2) isotopic exchange between the ^{233}U mass tracer and uranium in the urine, and 3) low and reproducible blanks. The direct loading of uranium from urine onto two or three resin beads provides a possible time-saving alternative to the ion exchange separations utilized here, but it should be noted that this technique was not tested at concentrations below about 3 ng/mL. Urine standards that are certified for uranium at several different concentrations are needed to verify procedures for determining low concentrations of uranium in urine. Finally, the simplicity, time-efficiency and moderate cost of this method make it practical for routine analysis.

W.S. Lyon (Editor), *Analytical Spectroscopy*
Elsevier Science Publishers B.V., Amsterdam — Printed in The Netherlands

SIMS STUDY OF COMPOSITIONAL CHANGES OBSERVED IN A PuO_2 HEAT SOURCE CLADDING ALLOY

W. H. Christie,[1] D. H. Taylor,[2] R. E. Eby,[3] and D. Pavone[4]

[1,3] Analytical Chemistry Division, ORNL, Oak Ridge, TN 37831
[2] Savannah River Laboratory, Aiken, SC 29801
[4] Los Alamos National Laboratory, Los Alamos, NM 87545

ABSTRACT

Secondary ion mass spectrometry (SIMS) has been used to investigate changes that occur in an advanced Ir-0.3W alloy during high temperature aging. This alloy is used to clad $^{238}PuO_2$ heat sources used in thermoelectric generators for deep space reconnaissance satellites. Long-term direct contact with PuO_2 at 1400°C leads to physical and chemical changes within the cladding alloy that affect its metallurgical properties. SIMS was used to show that Cr, Fe, Ni, and in some cases O, diffuse from the PuO_2 into the alloy. Thorium and aluminum diffuse out of the alloy in these same regions. This SIMS study suggests that inward O diffusion and subsequent formation of ThO_2 on grain boundaries may stabilize the alloy against enhanced grain growth.

INTRODUCTION

Pu-238 oxide is used as a general purpose heat source (GPHS) to power radioisotopic thermoelectric generators for NASA deep space probes, such as the upcoming Galileo mission to Jupiter (launch 1985). After being fabricated into ceramic pellets approximately 2.5 cm long by 2.5 cm in diameter, the PuO_2 is encapsulated in a 700 μm thick iridium shell which is vented on one end to permit escape of the decay helium. The major purpose of the iridium shell, or clad vent set (CVS), is to provide containment of the PuO_2 in the event of a mission failure involving a high velocity impact of the GPHS capsules with the ground. An iridium alloy (DOP 26) consisting of 0.3% tungsten, and nominally 50 wt ppm aluminum and 60 wt ppm thorium was chosen as the cladding material because of its high temperature strength and ductility and chemical inertness, especially to PuO_2. The alloying elements enhance the impact strength and ductility by increasing the coherent strength of the grain boundaries (ref. 1). Perhaps even more importantly these elements, in particular thorium, help maintain ductility during high temperature aging of the iridium by preventing grain growth (ref. 1). Thorium is effective as a grain growth inhibitor because it exists primarily as discreet particles on the grain boundaries.

Although the microstructure and grain boundary chemistry of the iridium have been well characterized (ref. 2) for the case where the iridium has been heat treated alone, the effects of high temperature aging in the presence of PuO_2 are unknown. This paper describes the results of using SIMS to investigate some of the effects on iridium DOP 26 alloy of high temperature aging under service conditions in the presence of PuO_2. Optical microscopy of polished specimens, prepared from aged sections of this cladding alloy (Fig. 1), showed a significant increase in average grain size in the near-surface regions exposed to PuO_2. One sample exhibited less grain growth in this region and SIMS was used in an attempt to determine what chemical differences exist between samples exhibiting large grain growth and those exhibiting lesser growth.

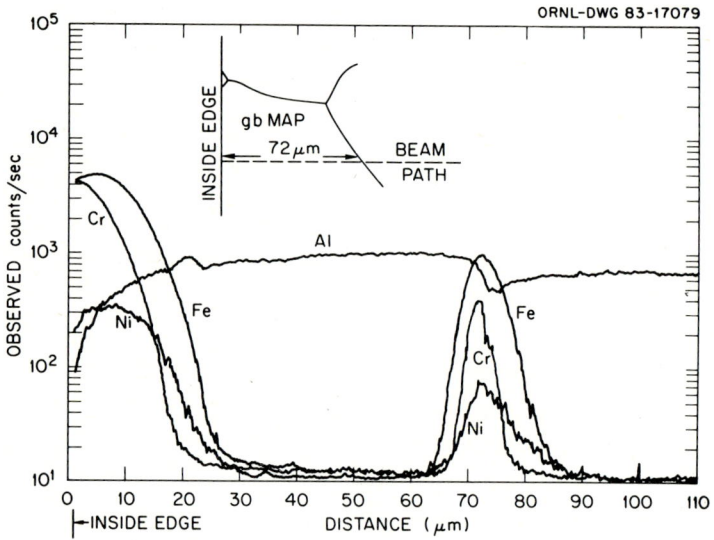

Fig. 1. Aged specimen of Ir-0.3W showing enhanced grain growth at inside surface (top of micrograph).

METHODS

Sections were cut from three CVS's (A, B, and C) that had been aged for 6 months in vacuum at 1440°C (fuel temperature). These specimens were mechanically abraded and then chemically etched (HF/HNO_3) to remove adhering PuO_2.

Transverse cross sections of these decontaminated specimens were then mounted in epoxy and metallurgically polished until a smooth mirror-like finish was obtained. The mounted specimens, which showed no smear removable alpha activity, were then overcoated with 20-30 nm of spectrographic grade carbon, prior to SIMS analysis, so as to render the entire surface conducting.

Secondary ion mass spectrometry was used to study the movement of impurity elements from the PuO_2 fuel into the cladding alloy as well as changes in the cladding dopant elements Al and Th. Several modes of data acquisition were used. Mass spectra were taken from regions representing the inside edge (close to the PuO_2 fuel), the center and outside edge of each specimen. These spectra were evaluated to determine which elements needed further study. Line scans were made by traversing a finely focused (2-5 μm) O_2^+ or N_2^+ ion beam across the 700 μm width of the alloy. Elements identified as being of interest in the mass spectra were monitored in this fashion. Elements found via the line scans to have undergone diffusion or precipitation were then examined in detail via ion imaging and depth profiling techniques.

RESULTS AND DISCUSSION

Prior to fabrication of the cladded heat sources, certain dopants were added to the PuO_2 fuel. Major dopants, determined by spark source mass spectrometry (SSMS), are listed, along with sample pretreatment steps in Table 1.

TABLE 1
SSMS Analysis of PuO_2 Fuel (Wt. ppm)

Element	Sample			Reference Specification
	A[2]	B[1]	C[2]	
Al	30	60	900	<150
Ca	2	3	600	<300
Cr	-	-	2000	<250
Fe	900	2000	10000	<800
Ni	-	-	2000	<150
Si	40	100	1000	<200
P	4	4	40	<15

1. PuO_2 heat treated for 6 hours at 1525°C in oxygen.
2. PuO_2 heat treated as in 1, followed by vacuum outgassing for 1 hour at 1500°C.

Dopant levels above the reference specification levels in Table 1 were added to determine their effect on the cladding alloy during the aging studies.

Mass spectra showed all three samples of the aged cladding alloy (A, B, and C) to have significant Cr, Fe, and to a lesser extent Ni contamination in the inside surface region. The deepest penetration (up to 150 μm) and highest levels were generally observed in sample C. Line scans across the 700 μm thickness also showed little Ca and Si and no detectable Pu to have diffused into any of the cladding samples. It was also observed that Al, added as a dopant in the cladding alloy, underwent some depletion from the alloy in regions where Cr and Fe had diffused in. Thorium, also a cladding dopant, was less abundant in the inside surface region as compared to the center of the sample. Other workers have reported Th loss associated with grain growth in thermally aged samples (ref. 2). Line scans across the specimens demonstrated some variability in amount and depth of penetration of Cr and Fe into the alloy. Ion images were then made to further understand these observed variations. These studies showed that Cr and Fe diffused into the alloy primarily along grain boundaries. The amount of Cr or Fe observed in any particular line scan and the depth of penetration depended to some extent on how that line scan intersected grain boundaries; but in general it was shown that for sample C the concentration maxima at the inside surface was approximately 0.5 to 1.0 atom % and decreased to background (5-10 ppm atomic) at a depth of about 150 μm. For sample A the maximum concentration of Cr and Fe was about the same as sample C at the surface, but the depth of penetration was significantly less being only 30-50 μm. The concentration of Cr and Fe at the surface of sample B was also approximately the same as for C and the depth of penetration was midway between A and C.

Figure 2 presents data obtained by scanning a 1 nA, O_2^+ primary beam across sample C in a direction perpendicular to the inside surface. The scan region chosen was such that no grain boundaries were within 50 μm of the point where the scan started on the inside edge of the sample (see grain boundary map in Fig. 2). Presumably, the Cr, Fe, and Ni that penetrated to a depth of 20-30 μm arrived via bulk diffusion. The peaks seen at 70-80 μm depth clearly represent diffusion along grain boundaries. SIMS showed that the concentration of Fe and Cr was maximum at the iridium grain boundaries and fell off with depth into the grains perpendicular to the grain boundaries. For grains near the surface of the iridium the concentration fell off very little from the maximum at the grain boundary. However, for grain boundaries away from the iridium surface, the fall off became sharper with depth into the iridium until at the maximum depth of penetration, the Fe and Cr were located only at the grain boundaries. Iron was also shown to diffuse more rapidly along the grain boundaries and into the bulk iridium. The SIMS data graphically demonstrate that Fe and Cr diffuse rapidly into Ir along grain boundaries and from the grain boundaries into the bulk iridium.

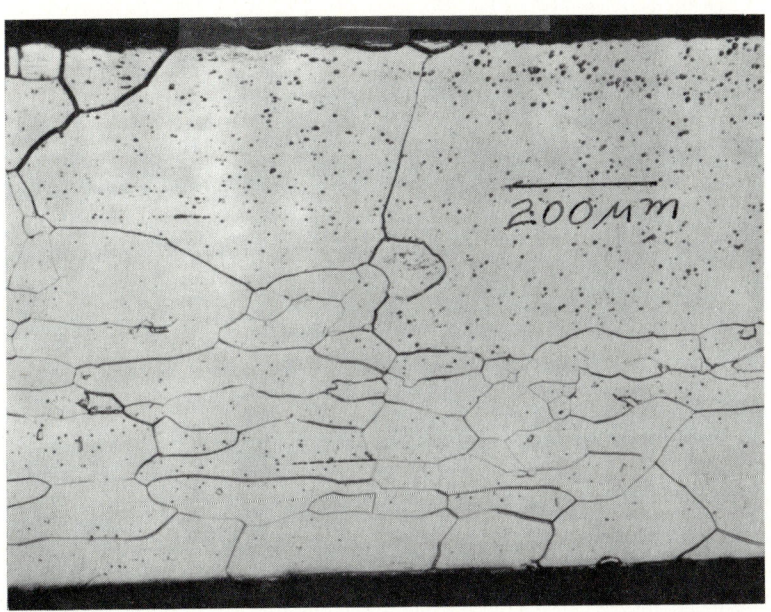

Fig. 2. SIMS line scans (1 nA, 2 μm O_2^+ primary beam) showing diffusion of Cr, Fe, and Ni into sample C. Grain boundary diffusion accounts for peaks at 70-80 μm depth from inside edge. Aluminum loss is also noted in regions where Cr and Fe have intruded.

Ion imaging was used to show that in the interior of all samples Th was segregated to grain boundaries and also existed as discrete precipitates in the interior of grains and on grain boundaries. Earlier studies (ref. 2) showed that near surface grain growth is facilitated by loss of Th from the alloy by diffusion along grain boundaries until it reaches the outer surface where it is presumably oxidized to ThO_2. For the present samples, loss of Th on the grain boundaries was confirmed for near surface grains by Auger electron spectroscopy (AES). However, in sample C where growth of near surface grains was least pronounced, SIMS revealed a striking correlation between reduced grain growth and the oxidation of Th on grain boundaries by the inward diffusion of oxygen. Two separate regions on the interior edge were studied, both showed ThO_2 precipitates to extend along grain boundaries over 100 μm in depth from the surface exposed to PuO_2. The SIMS study suggests that if the oxygen partial pressure is high enough (as it must have been in capsule C), oxygen can diffuse inward along grain boundaries and oxidize Th in situ. In the locations where Th was seen to exist as ThO_2, minimum grain growth was observed. This suggests that excessive grain growth may be inhibited by oxidation of Th on grain boundaries interior to the sample. The formation of ThO_2 below 10-15 μm in depth was not observed in samples A and B.

CONCLUSIONS

Chromium, Fe, and Ni migrate from the PuO_2 fuel to the Ir cladding and alter the chemistry of the near surface grains and grain boundaries. Although the effect of these impurities is not now known, they should probably be excluded from the Ir until their effect is understood. The observation that in situ oxidation of Th on grain boundaries seems to reduce grain growth should be investigated further. Currently no information is available concerning the effect of grain boundary ThO_2 on high temperature impact ductility. This study does suggest that this oxidation phenomena may be a way to stabilize the alloy against exaggerated grain growth and thereby improve its impact resistance.

REFERENCES

1. C. T. Liu and H. Inouye, "Development and Characterization of an Improved Ir-0.3%W Alloy for Space Radioisotopic Heat Sources," ORNL-5290, Oct., 1977.
2. C. L. White and C. T. Liu, Acta Metall., 29(2) (1981) 301-310.

ACKNOWLEDGMENT

Research sponsored by the U. S. Department of Energy, Division of Basic Energy Sciences, under Contract W-7405-eng-26 with the Union Carbide Corporation.

LASER ENHANCED IONIZATION IN GLOW DISCHARGE MASS SPECTROMETRY

W.W. HARRISON, P.J. SAVICKAS, R.K. MARCUS, and K.R. HESS
Department of Chemistry, University of Virginia
Charlottesville, Virginia 22901

ABSTRACT

The output of a tunable dye laser was used to ionize selectively the atomic population produced in a glow discharge. Potassium was atomized by both a Knudsen cell and the glow discharge for comparison. Wavelength ionization spectra and mass spectra showed the ionization created by the laser radiation.

INTRODUCTION

Lasers offer the analytical chemist a powerful tool based on two major characteristics: (1) a highly intense, coherent beam of radiation, and (2) the ability to select or tune this radiation to specific desired frequencies. The former suggests the availability of localized energy to create analytically advantageous situations, while the latter promises the opportunity to allow only certain desired energy transfer to occur, selectively excluding other competing steps. Laser applications in mass spectrometry (refs.1,2) have not been nearly as extensive as in fluorescence (refs.3,4), which is to say that the laser has been considered more of an excitation than an ionization source. However, lasers have been used as combination atomization/ionization sources for elemental analysis (ref.5) using high power thermal methods. Requiring a laser to serve as the atomization source for low volatility samples (e.g., metals and alloys) imposes a severe energy requirement with attendant cost and complexity. If however a laser is used for ionization alone, in conjunction with a complementary atomization source, more flexibility is afforded. The fact that such atomization/ionization combinations are not yet found in broad general use in analytical chemistry suggests that a better atomization/laser ionization system is needed. This has led us to couple a versatile glow discharge (as the atomization source) with a tunable laser (for selective ionization), using a simple quadrupole mass spectrometer as the measurement system.

The glow discharge has long been used as a sputter source to obtain thin film depositions. This efficiency of creating an atomic population has led to its application as a solids source for atomic emission, atomic absorption, and atomic mass spectrometry. Use of the glow discharge for trace element analysis by mass spectrometry has been of particular interest in our laboratory (refs.6-8).

The discharge plasma has been estimated to ionize less than 1% of the sputtered atoms in a general, nonspecific mode. Lasers, however, are reputed to ionize up to 100% of a given microvolume and offer the promise of great selectivity. We have coupled an eximer pumped dye laser to a glow discharge, achieving selective ionization of atoms sputtered from an analytical sample.

METHODS

Figure 1 is a diagrammatic representation of the instrumentation involved. The eximer and dye lasers (Lumonics) produce a tuned frequency which is directed just in front of the ion exit orifice of the glow discharge ion source, thus interacting with a population of neutral atoms sputtered from the adjacent sample cathode. An argon discharge at one torr, 3 mA, and 800 v serves as the atomization means. For potassium, a small amount of potassium chloride (approx. 10%) is mixed with high purity graphite and an electrode (1.5mm x 10mm) is formed with a hydraulic press. Conducting metals can be cut to produce directly a pin electrode.

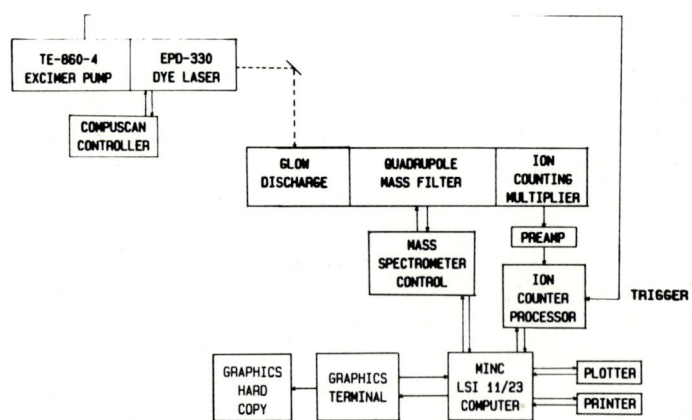

Figure 1. Diagram of instrumentation.

A quadrupole spectrometer (Extranuclear) employs ion counting detection and an ion counter/processor (Princeton Applied Research); the system is controlled by a DEC MINC 11 computer. The laser pulse creates a packet of ions which appears at the detector approximately 60 µsec later. Use of a delayed data gate in the counter/processor permits detection of these ions while a background gate set to open between laser pulses provides a stripping mode to subtract ions formed from modes other than the laser pulse.

RESULTS AND DISCUSSION

To determine the feasibility of using the glow discharge as an atomization source for <u>r</u>esonance <u>i</u>onization <u>m</u>ass <u>s</u>pectrometry (RIMS) (ref.9), we selected an element, potassium, which has a relatively simple optical spectrum and which has been shown to be successful in previous RIMS experiments (ref.10). A scheme 1 designation has been assigned in that by using either 404.4 nm or 404.7 nm, the absorption of two equal energy photons in a resonant mode can promote an electron into the ionization continuum.

Initially, a Knudsen cell was used to produce an effusive beam of potassium atoms. By directing potassium hollow cathode radiation through this beam, atomic absorption measurements were made, confirming the presence of ground state potassium atoms. The hollow cathode lamp also allowed accurate wavelength calibration by use of the <u>o</u>pto<u>g</u>alvanic <u>e</u>ffect (OGE) (ref.11). When the laser radiation was directed into the cathode cavity, a voltage change was produced when the laser frequency was coincident with the intrinsic potassium resonance steps. The tuned laser output could thus be calibrated to within $\pm$ 0.05 nm. For the Knudsen cell potassium beam, a wavelength ionization spectrum taken between 400-410 nm showed two peaks (representing sharp increases in $^{39}K^+$) at 404.4 nm and 404.7 nm, consistent with previous observations (ref.10).

The Knudsen cell was replaced with a glow discharge employing a graphite/KCl cathode. Figure 2 shows the wavelength ionization spectrum for potassium over the 403-406 nm range. The ion signal peaks for $^{39}K^+$ occurred at the anticipated wavelengths, indicating a 2-photon absorption to cause ionization. No other peaks were observed over the broader range scanned of 400-410 nm. Compared to the spectrum obtained using the Knudsen cell, the glow discharge peaks were broader and not entirely resolved. These effects were attributed to atoms of greater energy spread and perhaps too high power levels from the laser.

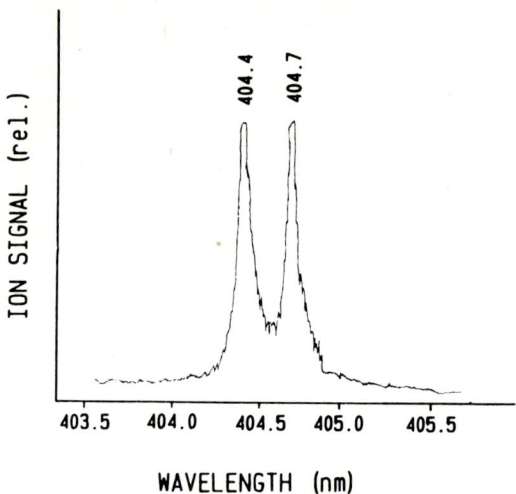

Figure 2. Wavelength ionization spectrum of potassium in a glow discharge.

Selecting the 404.4 nm radiation for ionization, a mass spectrum of the glow discharge ion output was taken. Because the discharge source also forms ions by electron impact and by collisions with metastable atoms, the ion counter/processor was set to the difference mode, so that only those ions produced by the laser pulse were displayed in the resultant spectrum. Figure 3 shows the potassium ions, $^{39}K^+$ and $^{41}K^+$, which were the only difference ion peaks evident over the 1-70 mass range. The choppy ion signal results at least in part from the stripping of large peaks at masses 40 (Ar^+) and 41 (ArH^+). This illustrates the potential advantage of removing mass interferents by the use of RIMS coupled with spectral stripping.

These preliminary experiments show that the glow discharge is an effective atomization source which, at least for potassium, creates ground state atoms capable of being ionized by multiphoton absorption. Whether many other elements lend themselves to comparable studies awaits the results of glow discharge sputter experiments to determine atomic energy distributions. The potential exists, however, for the use of the glow discharge as a simple and efficient source to couple with a tuned laser for direct multiphoton ionization of solids.

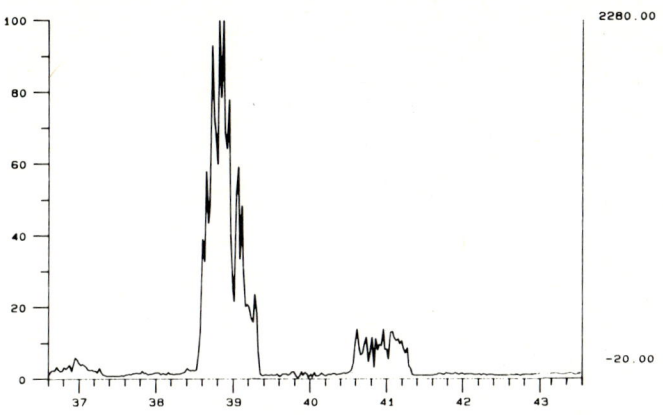

Figure 3. Resonance ionization mass spectrum of potassium in a glow discharge.

ACKNOWLEDGMENT

This work was supported by the Department of Energy, Division of Chemical Sciences, award no. DEAS0580ER10635.

REFERENCES

1. D.L. Donohue, J.P. Young and D.H. Smith, Int. J. Mass Spectrom. Phys., 43 (1982) 293-307.
2. V.S. Antonov and V.S. Letokhov, Appl. Phys., 24 (1981) 89-106.
3. L.M. Fraser and J.D. Winefordner, Anal. Chem., 43 (1972) 1444.
4. J.C. Wright and M.V. Johnston, Anal. Chem., 53 (1981) 1050-1054.
5. R.J. Conzemius and J.M. Capellen, Int. J. Mass Spectrom. Phys., 34 (1980) 197-271.
6. W.W. Harrison and C.W. Magee, Anal. Chem., 46 (1974) 461.
7. C.G. Bruhn, B.L. Bentz and W.W. Harrison, Anal. Chem., 50 (1978) 373.
8. T.J. Loving and W.W. Harrison, Anal. Chem., 55 (1983) 1526.
9. J.P. Young and D.L. Donohue, Anal. Chem., 55 (1983) 88-91.
10. D.W. Beekman, T.A. Calcott, S.D. Kramer, E.T. Arakawa and G.S. Hurst, Int. J. Mass Spectrom. Phys., 34 (1980) 89-97.
11. D.S. King, P.K. Schenck, K.C. Smyth and J.C. Travis, Applied Optics, 16 (1977) 2617-2619.

EXAMINATION OF EXCITED STATE POPULATIONS IN SPUTTERING USING MULTIPHOTON RESONANCE IONIZATION

F. M. KIMOCK, J. P. BAXTER, D. L. PAPPAS, P. H. KOBRIN and N. WINOGRAD
Department of Chemistry, The Pennsylvania State University, 152 Davey Lab., University Park, PA 16802

ABSTRACT

Multiphoton Resonance Ionization has been employed to study the populations of excited state atoms ejected from ion bombarded metal surfaces. Preliminary investigations have focused on three model systems: aluminum, indium and cobalt. In this paper we examine the effect of primary ion energy (2 to 12 keV Ar^+) on excited state yields for these three systems. The influence of the sample matrix on excited state populations of sputtered atoms is also discussed.

INTRODUCTION

In many cases, neutral species comprise the vast majority of all particles which eject from an ion bombarded surface. Recently we have demonstrated that sputtered atoms can be ionized with high selectivity and efficiency using Multiphoton Resonance Ionization (MPRI) (ref.1). It has been shown that this selectivity will be useful for the analysis of mixtures or compounds in which isobaric interferences make conventional analysis difficult (ref.2). The high efficiency (nearly 100% in some cases (ref.3)) has several ramifications for the use of MPRI. First, MPRI promises to yield a great deal of information which will be important in elucidating the mechanisms which govern the sputtering process. Secondly, MPRI should improve the use of ion beam methods for trace analysis of solids. Detection of dopants or impurities at the ppb level should be feasible (refs.3,4).

For fundamental investigations, understanding the formation of excited state neutrals is significant. Recently, laser induced fluorescence techniques have been used to measure the velocity distribution of sputtered metastable calcium atoms (ref.5) and to characterize excited state populations of sputtered uranium atoms (ref.6).

Equally important is the effect of excited state populations on trace analysis studies. For an experiment with microprobe capabilities, or when the laser must be focused, the ion beam may have to be less than 100μ in diameter to insure efficient sampling of the sputtered particles. In order to achieve sufficient ion beam current with such a small spot, high keV ion beam energies may be required to overcome space-charge effects. (Generally, the minimum ion beam spot size varies as $I/V^{3/2}$, where I is the ion beam current, and V is the beam energy.) When high ion beam current needs to be delivered into a small spot (consequently with high beam energy), the population of excited state atoms may become significant. It is especially important to know which atomic state(s) will give the maximum ion yield by MPRI, since only one state can be probed at a time.

In this work we present some preliminary results on the population of excited state atoms sputtered from polycrystalline aluminum, indium and cobalt. For several reasons, these three systems provide excellent models for studying excited state populations by MPRI. Both aluminum and indium have only one low-lying excited state, $^2P_{3/2}$. For Al, this state is only 112 cm^{-1} (0.014 ev) above the ground state, while for In, the first excited state lies 2212 cm^{-1} (0.27 eV) above the ground state. For each atom, the ground state is $^2P_{1/2}$, so transitions of the type $^2P_{3/2} \rightarrow {}^2P_{1/2}$ are dipole forbidden. The next nearest excited state in each atom is greater than 3 eV from the ground state, so that all of the cascading from higher states should have occured before the MPRI ionization. In the case of Co, the first 10 excited states are located less than 1.1 eV above the ground state. The ground state and all of these excited states have even parity, so cascading between them is forbidden. The first state of odd parity is located 2.9 eV above the ground state.

Schematic energy level diagrams of Al, In and Co are presented in Fig. 1 along with the MPRI schemes used to examine the states of interest.

EXPERIMENTAL

Both the instrument and data acquisition system have been previously described (refs.1,3). The primary beam of Ar ions is generated by a Danfysik 911A Sidenius-type hollow cathode source which is capable of producing ions of up to 30 keV energy. For these particular experiments, the ion gun was operated in the d.c. mode and typically 0.5 µA of ion beam current was delivered to the sample in a defocused condition. The beam energy was varied from 2 to 12 keV. These operating conditions are required to reproducibly vary the ion beam energy without changing the ion beam spot size and position on the sample. All experiments were conducted in an ultra-high vacuum system with a base pressure of $\sim 10^{-9}$ torr. All of the samples were foils and were cleaned by acid etching in a solution of 30% HNO_3, 10% H_3PO_4, 10% H_2SO_4 and 50% glacial acetic acid, followed by ion beam sputtering in vacuum. Al and Co were sputtered by 5µA/cm^2 of 7 keV Ar$^+$ for about 16 hours over two days immediately before the analysis. The In sample was also sputtered and was determined to be clean when the In$_2$/In ratio as measured by MPRI (with 4102 Å light) reached a maximum value (ref.1). The direct dye output and frequency doubled dye output of Exciton DCM (in methanol) were used to ionize Al, In and Co (see Fig. 1). For ionization of Al and In, the laser was defocused (6mm diameter). In order to increase the MPRI signal for Co, the laser was tightly focused using a 30 cm lens. All ions were detected using a Cu-Be electron multiplier.

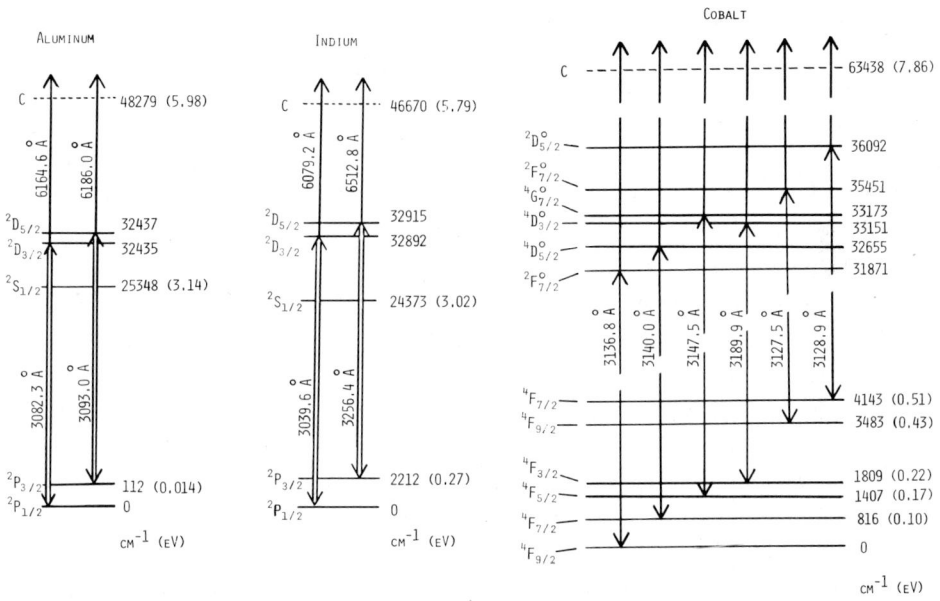

Fig. 1. Schematic energy level diagrams of Al, In and Co (not to scale.) Arrows indicate the MPRI schemes employed to study both ground and excited state atoms. Wavelengths (Å) are experimentally observed. Energies in cm^{-1} and eV are from (ref.7).

RESULTS AND DISCUSSION

The variation of the $^2P_{3/2}$ to $^2P_{1/2}$ ratio as a function of primary ion beam energy for sputtered Al and In is presented in Fig. 2. Each plot is normalized to the maximum value, and the error bars give a measure of the magnitude of scatter in the data. For sputtered Al atoms (Fig. 2a), there is almost no variation in the ratio of atoms in the first excited state to the ground state as a function of primary ion energy from 1.5 to 11 keV. A similar plot for sputtered In atoms (Fig. 2b) shows a dramatic increase up to 4 keV, followed by a plateau region. The absolute $^2P_{3/2}$: $^2P_{1/2}$ ratio for In sputtered by 5 keV Ar$^+$ has previously been measured to be ~0.10 (ref.1). From this value, we calculate an effective Boltzmann temperature of ~1180°K for In bombarded by 5 keV Ar+. Similarly, using the data from Fig. 2b, we calculate an effective temperature of ~725°K for In bombarded by 2 keV Ar$^+$. These temperatures are in reasonable agreement with those reported by Wright, et al. (920 ± 100°K) for uranium bombarded by 500 to 3000 eV Ar$^+$ (ref.6). At present, we have no reliable measurement of the $^2P_{3/2}$: $^2P_{1/2}$ ratio for Al, although a value of unity for ion beam energies from 1.5 to 11 keV is consistent with our ion intensity and saturation measurements.

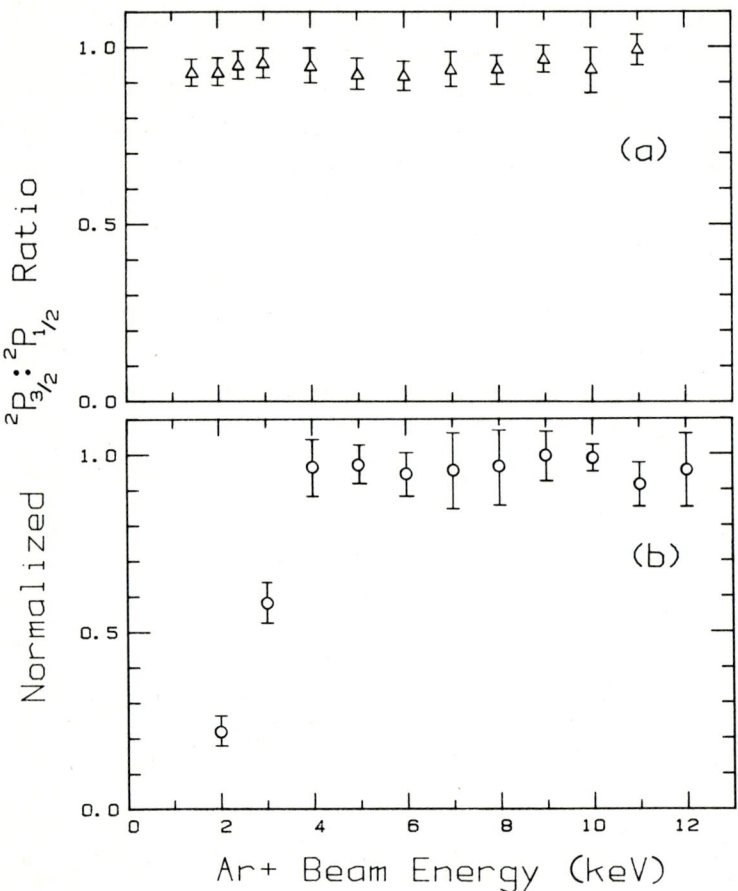

Fig. 2. $^2P_{3/2}$: $^2P_{1/2}$ ratio as a function of primary ion beam energy for (a) aluminum and (b) indium.

There are several factors which may influence the shapes of these curves. The $^2P_{3/2}$ state of In is a factor of 20 higher in energy than the same state of Al. It is possible that the dissipation of 1.5 keV Ar$^+$ energy is sufficient to saturate the excited state population of Al, whereas 4 keV is required to saturate the $^2P_{3/2}$ population of In. Secondly, there may be strong matrix effects in these systems which can give rise to variations in the curve shapes. We are confident that the data for In was taken from a clean sample. However,

Al is notoriously difficult to clean and to keep clean, even under ultra-high vacuum conditions, due to the stability of aluminum oxide. We have recently removed the quadrupole mass spectrometer from our analysis chamber, so surface cleanliness could not be established by SIMS.

The effect of surface oxidation on the $^2P_{3/2} : {}^2P_{1/2}$ ratio is revealed in Fig. 3. For In, one monolayer of In_2O_3 is known to form after exposure to ~800 L of oxygen (ref.8). This monolayer formation leads to an increase in the ratio from 0.1 to ~0.2; it subsequently increases to 0.36 after the oxidation has been saturated. Thus, it is clear that alteration of the sample matrix can have a profound effect on the excited state population. We have not yet examined the effect of chemisorbed oxygen on the shape of the curve in Fig. 2b. Such information will be useful in the interpretation of the Al curve (Fig. 2a).

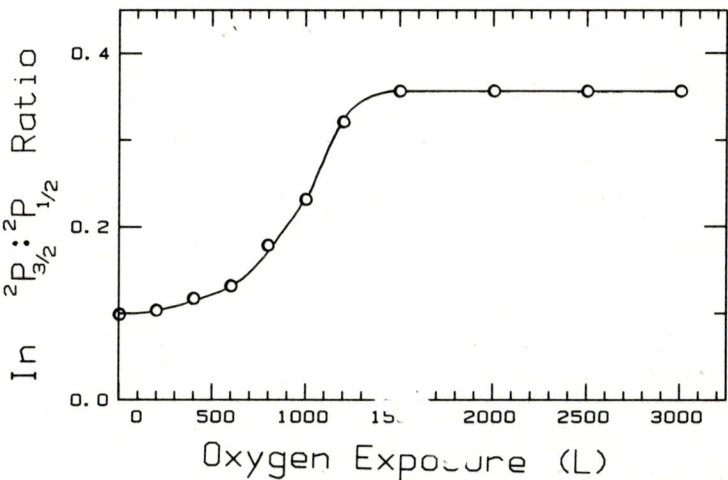

Fig. 3. Variation of absolute $^2P_{3/2} : {}^2P_{1/2}$ ratio of sputtered In atoms with oxygen exposure. Data was collected using 5 keV Ar^+. (1 L = 10^{-6} torr sec.)

The influence of primary ion beam energy on the excited state to ground state ratio for 4 low-lying excited states of Co is illustrated in Fig. 4. Significant ion intensities could not be accurately measured below 3 keV beam energy due to the weak signal level of excited state Co relative to Al and In, and the need to employ a defocused ion beam. The signal from the $^4F_{3/2}$ state (1809 cm^{-1}) was also quite weak at 3189.9 Å. The data suggest that much like In, excited state populations may rise to a plateau regime for some states; although more measurements are clearly needed. At this date, we have no estimate of the influence of the sample matrix on excited state formation in sputtered Co.

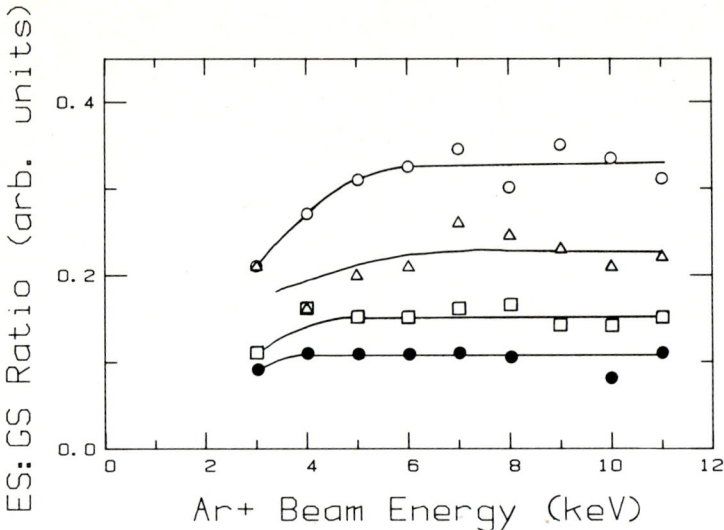

Fig. 4. Influence of primary ion beam energy on excited state (ES) to ground state (GS) ratio for 4 excited states of Co. (O) $^4F_{7/2}$, 816 cm^{-1}; (△) $^4F_{5/2}$, 1407 cm^{-1}; (□) $^4F_{9/2}$, 3483 cm^{-1}; (●) $^4F_{7/2}$ 4143 cm^{-1}.

In summary, we have begun to collect data which indicates that the primary ion beam energy and sample matrix can influence excited state formation of sputtered atoms. This type of data is important for understanding the process of energy dissipation in the solid, as well as the mechanism of secondary ion formation, and for the development of MPRI as an analytical tool. The effective sputtering temperatures that we have calculated are on the order of the temperatures employed in hot filament evaporation techniques.

ACKNOWLEDGEMENTS

The authors are grateful for the financial support of the National Science Foundation, the Office of Naval Research, the Air Force Office of Scientific Research and the Petroleum Research Foundation.

REFERENCES

1. F. M. Kimock, J. P. Baxter and N. Winograd, Surf. Sci., 124 (1983) L41.
2. J. P. Young and D. L. Donahue, Anal. Chem., 55 (1983) 88.
3. F. M. Kimock, J. P. Baxter, P. H. Kobrin, D. L. Pappas and N. Winograd, Proc. SPIE 27th Int. Tech. Symp., San Diego, Aug. 21-26, 1983, in press.
4. J. E. Parks, H. W. Schmitt, G. S. Hurst and W. M. Fairbank, Jr., Proc. SPIE 27th Int. Tech. Symp., San Diego, Aug. 21-26, 1983, in press.
5. W. Husinsky, G. Betz and J. Girgis, Phys. Rev. Lett., 50 (1983) 1689.
6. R. B. Wright, M. J. Pellin, D. M. Gruen and C. E. Young, Nucl. Instrum. Meth., 170 (1980) 295.
7. C. E. Moore (Ed.), Atomic Energy Levels, NBS Circular 467, Vol. 1-3, U. S. Gov't. Printing Office, Washington, D. C., 1971.
8. R. W. Hewitt and N. Winograd, J. Appl. Phys., 51 (1980) 2620.

PLASMA

A PRELIMINARY STUDY ON THE DETERMINATION OF BORON AND CADMIUM IMPURITIES IN URANIUM OXIDE AND URANIUM METALS BY INDUCTIVELY COUPLED PLASMA-ATOMIC EMISSION SPECTROSCOPY

B. R. BEAR[1], M. C. EDELSON[1], B. GOPALAN[2], and V. A. FASSEL[1]

[1] Ames Laboratory-USDOE, Iowa State University, Ames, Iowa 50011
[2] Scientific Office QC(F), Control Laboratory, Nuclear Fuel Complex, Hyderabad-500762, India

ABSTRACT

Boron and cadmium impurities in U metal and U_3O_8 may be determined by inductively coupled plasma-atomic emission spectroscopy (ICP-AES) after their separation from the U matrix by multiple extractions with tri-n-octyl phosphine oxide (TOPO). Useful precision and accuracy were obtained at concentrations of ~0.2 µg/g U.

INTRODUCTION

The purity requirements for nuclear grade uranium are very stringent. Impurities not only affect the metallurgical properties of the uranium, but some impurities, such as B, Cd, Li and the rare earths, also affect the chain reaction efficiency, even at concentrations as low as the fractional ppm range (1).

For approximately 40 years, the d-c carbon arc, carrier-distillation, atomic emission spectroscopic approach has been used to determine many impurities in uranium (1-3). Although this procedure has made it possible to determine many impurities satisfactorily, the determination of those elements that form relatively involatile oxides have presented difficulties.

Procedures based on dissolution of the sample prior to the separation of impurities from the U matrix via liquid-liquid extraction, followed by inductively coupled plasma-atomic emission spectroscopic (ICP-AES) analysis have been described (4-6). Although these extractions were more time consuming than the carrier-distillation approach, they were applicable to the determination of most of the impurities with acceptable accuracy and precision.

The extractants used in the above studies were tributylphosphate (TBP) dissolved in CCl_4 and triethylhexylphosphate (TEHP) dissolved in hexane or heptane. Tri-n-octyl phosphine oxide (TOPO) has also been reported to be an excellent extractant for the quantitative removal of uranium in nitric acid solution (7,8). In previous studies in our laboratory, Gopalan et al. (9) found that ~99.9% of a 1 g U sample present in a 2M nitric acid solution was removed from the aqueous phase by three extractions with a 0.2M TOPO solution dissolved in CCl_4. The aqueous phase was then analyzed via ICP-AES to determine Al, Co, Cu, Fe, Mn, Ni, Si and V impurities in a well-characterized U metal. The results were in excellent agreement with previous analyses of this material performed with d-c arc, carrier-distillation spectroscopy and by ICP-AES analysis using a TEHP extractant.

In this paper we discuss the extension of this earlier work to include the elements B and Cd. To avoid exposure to a known carcinogen, cyclohexane was used in preference to CCl_4 to dilute the TOPO.

EXPERIMENTAL

Sample Preparation

Uranium metal and uranium oxide samples were obtained from NLO, Inc., Cincinnati, OH, and from the New Brunswick Laboratory (NBL), U. S. Dept. of Energy, Argonne, IL. The uranium metal samples were washed with 1:1 HNO_3 and then dried with acetone before weighing. One or two gram samples were weighed into previously leached Teflon beakers and dissolved with 1:1 HNO_3 by gently heating the solutions to near dryness. The HNO_3 used was reagent grade and was then further purified by distillation in an all-quartz apparatus. Boron and cadmium were not detectable by ICP-AES in the HNO_3 or in the diluted TOPO used in this study.

Reference Solutions and Blank Preparation

All solutions used to prepare the analytical calibration curves initially contained the equivalent of 1 g U (as uranyl nitrate) in 2M HNO_3 and were subjected to the same liquid-liquid extraction procedure used for the samples to be analyzed. A set of "synthetic" standards (i.e., standards that contained known concentrations of analyte but no added U) were similarly treated. Single element reference stock solutions of B and Cd (Spex Industries, Inc., Metuchen, NJ) were used to provide the desired analyte concentrations.

Extraction Procedure

Each dissolved sample was diluted to ~5 mL with 2M HNO_3 and then transferred with ~5 mL to a previously leached 125 mL Teflon separatory funnel. Twenty mL of a 0.2M solution of TOPO (Eastman Kodak) in reagent grade cyclohexane was then added to the funnel. This organic solution had been previously prewashed with 2M HNO_3 to remove water miscible contaminants. Each sample was agitated by hand for approximately 2 minutes and the phases were then allowed to separate for 3 minutes before the aqueous phase was withdrawn and the process repeated with a fresh aliquot of extractant. The 1 g samples were typically extracted 3 times and the 2 g samples extracted 4 or 5 times before the final aqueous layer was removed for analysis. This procedure was adapted from the work of Gopalan (9) with only minor modifications. If fine particulates were visibly evident in the aqueous solutions, the solutions were filtered before analysis. The reference calibration solutions were extracted in a similar manner.

Operating Conditions

The plasma and nebulizer operating conditions and the spectral lines used for analysis are given in Table I. Ames Laboratory ICP software was used to control the stepper-motor driven monochromator and to acquire the data. The B and Cd analyte lines and nearby background structure were measured over an ~0.25 nm spectral range with intensities measured every 0.005 nm. These spectra were stored on disk for later analysis. The resident software package allowed spectral subtraction ("stripping") to be performed on the stored analyte spectra. The spectral structure near the analyte line was markedly simplified by stripping the contribution of the nitric acid reference blank from the total spectrum.

TABLE I. Operating Conditions and Analysis Lines

Plasma Generator Power	
Forward	1100 Watts
Reflected	20 Watts
Plasma Gas Flows	
Outer	14 L/min
Intermediate	0 L/min
Inner	1.2 L/min
Nebulizer	Perkin Elmer Corrosion Resistant Pneumatic
Sample uptake	1.5 mL/min (with pump)
Torch	Perkin Elmer demountable
Injector tube	Al_2O_3
Monochromator	McPherson Model 2051, 1 meter
Grating	1800 g/mm, holographic
Data Acquisition System	As described by Floyd, et al. (5)
Spectral Lines	Cd (II) 214.44 nm
	B (I) 249.68 nm
	U (II) 263.55 nm

Boron Contamination and Memory Effects

To preclude B contamination from a borosilicate glass nebulization system, a plastic Scott spray chamber and nebulizer were used (Table I). The spray chamber was etched by the manufacturer to promote efficient drainage of coalesced analyte solution from the chamber walls. The cleanout time (i.e., time required for the analyte signal intensity to decrease to 1% of its maximum value upon removal of the analyte solution) for B was measured to be approximately one-half that of an Ames Laboratory crossflow pneumatic nebulizer equipped with a borosilicate Scott spray chamber. The problems associated with B contamination and prolonged memory effects have been previously noted (10-12).

RESULTS

Linear analytical calibration curves for B and Cd were obtained over the concentration range of 0.1 to 1.6 µg/gU. The detection limits (13) for B and Cd were calculated to be 0.05 and 0.02 µg/gU for synthetic standards based on a 2σ criterion.

Two certified uranium oxide samples (U_3O_8) (NBL 98-3 and 98-6) and three well-characterized U and U_3O_8 sample pairs (NLO, Inc.) were analyzed. The results obtained for the NBL samples are given in Table II and for the NLO samples in Figure 1. Only B results are presented for the latter group of samples because the Cd content was below our limit of quantitative determination. In terms of present capabilities and difficulty of these determinations, the level of agreement shown by the data presented in Table II and Figure 1 is good. Further refinements and optimization should improve the correlations with other analytical data.

TABLE II. Analysis Results Obtained for NBL Reference Samples (µg/gU)[a]

	B		Cd	
	Measured	Certified	Measured	Certified
98-3[b]	1.10 ± 0.14	1.2 ± 0.2	1.12 ± 0.08	1.4 ± 0.1
98-6[c]	0.45	0.2 ± 0.0	0.43 ± 0.15	0.3 ± 0.0

[a]) Average of 3 separate determinations
[b]) 1 g samples extracted 3 times
[c]) 2 g samples extracted 4 times
[d]) 1 determination

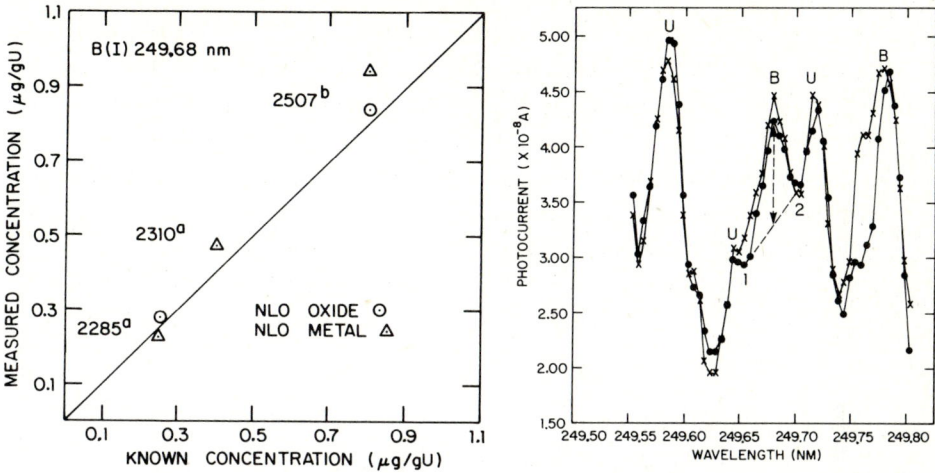

Fig. 1. Correlation of results obtained for NLO samples.

Fig. 2. Spectra of two uranium oxide samples (NLO 2285) known to contain 0.25 µg B/g U.

DISCUSSION

The TOPO extractions exhibited variable degrees of U removal; the 2 g U samples contained as much as 5000 µg/mL U in the aqueous phase after four extractions (97.5% extraction efficiency). The high residual U content resulted in complex background structure near the B line that complicated the B determination. Moderate U concentrations (<1000 µg/mL) could be tolerated if the spectral background of the reference blank solution was stripped from the total profile and if judicious background corrections were chosen to compensate for U spectral structure close to the B line. These background corrections were based on a linear interpolation of readings made at positions 1 and 2 as labeled in Figure 2. The interpolation position was chosen to coincide with the wavelength of the B 249.68 nm analyte line. The background structure due to 100 µg/mL U at the B line is shown in Figure 3(A). A graphic illustration of the effects of relatively large quantities of unextracted U on the B spectra of three NBL 98-6 samples is shown in Figure 3(B). It is readily apparent that increasing U concentrations will cause a significant positive bias due to a weak on-line interference that is not explicitly compensated by the simple background correction method employed. For this reason, analytical results on solutions containing relatively high U concentration (~1000 µg/mL) are subject to a considerable positive bias.

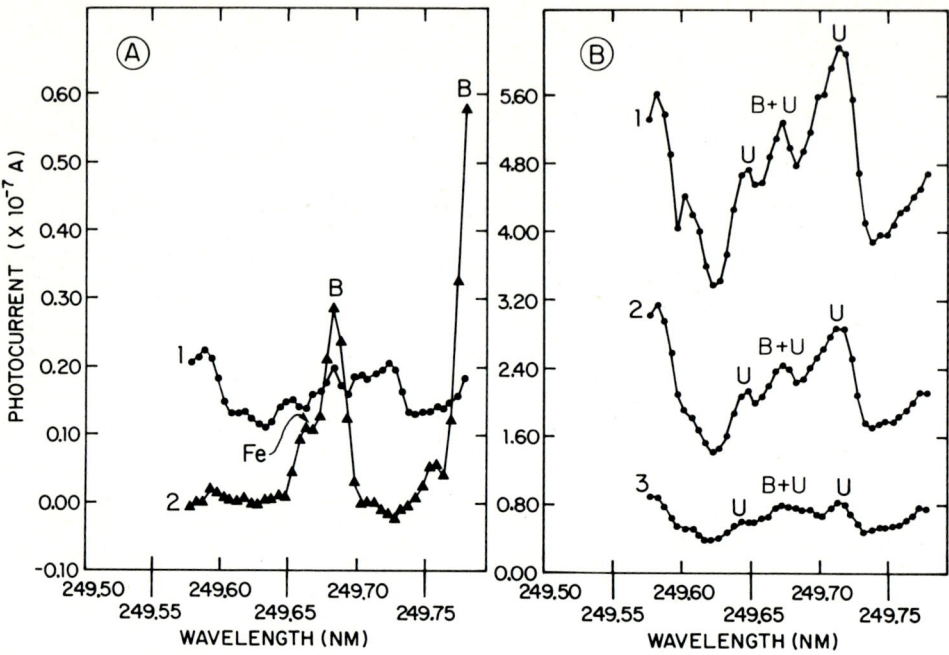

Fig. 3. A) Comparison of spectra in the vicinity of the B line at 249.68 nm.
1) 100 µg/mL U
2) NBL 98-1 (B=5.5 µg/g U, Fe≈500 µg/g U)
B) Spectra of the three NBL 98-6 samples containing varying amounts of residual U after four TOPO extractions.
1) ~5000 µg/mL U
2) ~3000 µg/mL U
3) ~1000 µg/mL U

Two avenues for improvement of the ICP-TOPO method may be advanced for future consideration. Improved computer software can be developed to mitigate the influence of the U background structure at the B analysis line. This can be done by monitoring the U intensity, as was done in this work, and constructing a normalized U blank for stripping from the spectrum of the analyte. The TOPO extraction conditions, using cyclohexane as a diluent, should also be more carefully optimized in future studies.

REFERENCES

1. B. F. Scribner and H. R. Mullin, J. Res. Natl. Bur. Stnd., 37 (1946) 379-389.
2. A. G. I. Dalvi, C. S. Deodhar, T. K. Sheshagiri, M. S. Khalap, and B. D. Joshi, Talanta, 25 (1978) 665-668.
3. C. E. Pepper, NLO, Inc., Cincinnati, Ohio, private communication.
4. J. P. Maney, V. Luciano, and A. F. Ward, Jarrell-Ash Plasma Newsl., 2 (1979) 11-13.
5. M. A. Floyd, R. W. Morrow, and R. B. Farrar, Spectrochimica Acta, 38B (1983) 303-308.
6. B. W. Short, H. S. Spring, and R. L. Grant, "Determination of Trace Impurities in Uranium Hexafluoride by an Inductively Coupled Argon Plasma Spectrometer", Goodyear Atomic Corporation, Piketon, Ohio, CAT-T-3184, 1983.
7. J. C. White, "The Use of Trialkyl Phosphine Oxides as Extractants in the Fluorometric Determination of Uranium", Oak Ridge National Laboratory, Oak Ridge, Tennessee, ORNL-2161, 1956.
8. W. D. Shults and L. B. Dunlap, Anal. Chim. Acta, 29 (1963) 254-260.
9. B. Gopalan and V. A. Fassel, unpublished data.
10. B. R. Bear, MS Thesis, Iowa State University, Ames, Iowa, 1983.
11. G. F. Larson, R. T. Goodpasture, and R. W. Morrow, in R. M. Barnes (Ed.), Application of Plasma Emission Spectrochemistry, Heydon and Son Inc., Philadelphia, Pennsylvania, 1979, 53.
12. R. M. Barnes and H. S. Mahanti, Appl. Spectrosc., 37 (1983), 17-19.
13. G. L. Long and J. D. Winefordner, Anal. Chem., 55 (1983), 712A-724A.

ANALYSES OF NUCLEAR FUEL AND HIGH-LEVEL WASTE
BY INDUCTIVELY COUPLED PLASMA-EMISSION SPECTROSCOPY*

CHARLES J. COLEMAN
E. I. du Pont de Nemours & Company
Savannah River Laboratory, Aiken, South Carolina 29808

ABSTRACT

Containment of the inductively coupled plasma (ICP) source has increased the application of ICP emission spectroscopy at the Savannah River Laboratory (SRL). Nuclear materials are analyzed for up to 41 elements simultaneously with an ICP polychromator detection system. This technique is used for determining trace impurities in uranium that can adversely affect nuclear fuel processing. Uranium samples, after solvent extraction removal of uranium to reduce spectral interferences, are satisfactorily analyzed for impurity elements. ICP emission spectroscopy is also effective for analyzing sludge from the Savannah River Plant (SRP) high-level waste tanks. Methods were developed for analyzing a wide range of sludge compositions for major and minor elements. ICP analyses of waste tank sludge are required for SRP defense waste solidification programs. Method development and analyses of typical samples will be presented.

INTRODUCTION

ICP emission spectroscopy (ICP-ES) has been for some time the principal method of elemental analysis for virtually all nonradioactive samples at SRL. Recently, the ICP emission source was contained to permit analysis of low-to-intermediate level radioactive samples. ICP-ES methods were needed to analyze two important types of radioactive samples at SRL: (1) uranium metal and uranium oxides for trace-level impurities, and (2) SRP waste tank sludge for major and minor elements. Uranium impurity determinations are required to support fuel reprocessing research programs. In addition, uranium used in nuclear fuel processing must be analyzed for quality control purposes. The latter samples are now analyzed by carrier distillation-dc arc spectroscopy, but a method with greater precision is desired. SRP waste tank sludge historically has been analyzed by atomic absorption spectrophotometry. However, the simultaneous multielement capability of ICP-ES is clearly an advantage for analyzing samples for a large number of elements.

---

*The information contained in this article was developed during the course of work under Contract No. DE-AC09-76SR00001 with the U.S. Department of Energy.

Uranium and SRP waste tank sludge present a challenge for ICP-ES analysis. Uranium emission lines interfere with the analytical lines of many elements, resulting in poor ICP-ES detection limits and biased determinations. Solvent extraction separation of uranium from the impurities prior to analysis is the most effective way to eliminate uranium spectral interference. Several uranium extracting reagents have been used successfully to achieve this separation. Selection of an extraction reagent most favorable to SRL's analytical requirements was a major goal of this work. SRP waste tank sludge analysis is challenging because the wide variation in sludge compositions makes it impractical to optimize ICP-ES conditions for every sludge matrix. The approach taken here is to prepare synthetic standards that simulate some of the extreme compositions and demonstrate that a single ICP-ES method will provide reliable sludge analyses.

METHODS

ICP-ES Determination of Trace Impurities in Uranium

Three uranium extracting reagents were evaluated for use in the ICP-ES method: tris(2-ethylhexyl)phosphate (ref.1,2,3) (TEHP) in hexane (1:1,V/V); tri-n-octyl phosphine oxide (ref.4,5) (TOPO) in cyclohexane (0.2 M); and neat tri-n-butyl phosphate (ref.6,7) (TBP). To compare the relative selectivity and efficiency of each reagent, synthetic standards containing 250 g/L uranium and 10 mg/L of 31 added impurity elements were extracted and then analyzed by ICP-ES.

TEHP, in our hands, exhibits slightly better selectivity and extracting efficiency than TOPO or TBP and is used at SRP for uranium extractions in which the aqueous phase is retained for analysis. Recovery data (Table 1) show that 90 to 100% of most impurity elements are recovered in the aqueous phase after

TABLE 1
Recovery of impurity elements

The following elements were added to a 250 g/L uranium solution to make the concentration of added impurity elements 10 mg/L: Al, As, B, Ba, Be, Bi, Ca, Cd, Co, Cr, Cu, Fe, La, Li, Mg, Mn, Mo, Na, Ni, P, Pb, Ru, Si, Sn, Sr, Ti, V, Y, Yb, Zn, Zr. Recoveries of the elements in the aqueous phase after the solvent extraction step were quantitative (90-100%) except for the elements listed below:

TEHP	TOPO	TBP
Si 50%	Bi 20%	Mo 30%
Zr 50%	Mo 50%	Si <10%
	Ti 50%	Sn 40%
	Y 70%	Zr <10%
	Zr <10%	

the extractions with TEHP. Three TEHP extractions quantitatively remove uranium from HNO_3 solutions containing up to 250 g/L uranium when a 3:1 ratio of TEHP-hexane solution to uranium solution is used. Under similar conditions, five TOPO extractions are required to remove the uranium from solution. Neat TBP forms a poorly separated interface with a 250 g/L uranium solution. To effect adequate phase separation, the uranium concentration is reduced to about 125 g/L. Two or three TBP extractions remove the uranium from this solution.

Solid uranium samples can be analyzed with good accuracy and precision for impurities by dissolving the sample in concentrated HNO_3, extracting the uranium three times with TEHP, and then analyzing the aqueous layer by ICP-ES. Provisional U_3O_8 standards supplied by the New Brunswick Laboratory (NBL) were treated in this manner to test the effectiveness of this method. ICP-ES analyses of three standards are compared with the NBL values in Table 2. The average difference in the values is about 15%. The precision, based on the

TABLE 2
ICP-ES analyses of U_3O_8 standards

Element	Std 123-1 (µg/gU) SRP Mean		NBL Mean	Std 123-2 (µg/gU) SRP Mean		NBL Mean	Std 123-3 (µg/gU) SRP Mean		NBL Mean
Al	214	±15[b]	194	95	±7[b]	97	51.4	±4[b]	56
B	5.0	±0.3	5.6	2.2	±0.2	2.6	0.8	±0.1	1.0
Ca	213	±12	198	96	±12	118	39	±5	51
Cd	4.4	±0.2	5.3	2.2	±0.2	2.8	0.8	±0.1	1.0
Cr	97	±4	108	47	±5	58	19	±2	22
Cu	46	±4	45	22	±4	28	8	±1	10
Fe	198	±9	202	97	±4	99	51	±3	60
Mg	104	±9	110	46	±6	54	18	±2	21
Mn	46	±4	49	22	±2	28	8	±1	11
Mo	97	±6	103	44	±4	55	17	±2	20
Ni	170	±28	233	84	±10	105	38	±4	52
Pb	43	±8	47	21	±3	24	16	±4	11
Si	a		190	a		95	a		100
Na	371	±33	405	171	±10	205	59	±8	52
Sn	39	±8	45	26	±5	23	6	±2	11
V	47	±5	50	22	±3	25	8	±1	10
Zn	221	±12	218	102	±5	100	54	±2	52
Zr	a		207	a		99	a		50
	Average Difference 8%			Average Difference 15%			Average Difference 20%		

[a] Si and Zr are not reported due to poor recoveries.
[b] Ranges are based on the standard deviation of 5 determinations.

standard deviations of five determinations, ranges from 10 to 15% for most elements that have final concentrations several times above the detection limit. SRL determinations exhibit a small negative bias compared with the NBL values, probably because of small analyte losses in the dissolution and extraction steps. However, the accuracy and precision of this method is well within SRP requirements and it is now the standard method for analyzing uranium samples for research programs. Efforts are also underway to establish this method for quality and process control analyses of uranium.

ICP-ES Analysis of SRP Waste Tank Sludge

Three compositional types of SRP waste tank sludge were prepared as synthetic standards for ICP-ES method development. Two of the standards simulate extreme sludge compositions: high-iron sludge (HIS) containing about 25% iron, and high-aluminum sludge (HAS) containing about 23% aluminum. The third sludge (BLEND) is a nominal blend of sludges that is suitable for incorporation into glass in the sludge processing facility planned for SRP. Standards were prepared by mixing reagent grade solids of the major sludge components with pure solutions of the minor components. This mixture was dried and ground into a fine powder for dissolution.

Total sludge dissolution is critical for obtaining accurate determinations of all sludge components. Two dissolution methods are used in tandem at SRL to achieve optimum results. Na_2O_2 fusion of the sludge followed by uptake of the residue with HCl completely dissolves the sludge. A drawback to this method is the quantity of dissolved solids can affect nebulizer performance. Treating the sludge with a mixture of HF and HCl at elevated temperature and pressure produces a solution with low dissolved solids content. However, some metal fluorides are too insoluble in this solution to be determined accurately. The best dissolution procedure for determining each sludge element was found experimentally by analyzing the synthetic standards and is denoted in Table 3.

Comparison of the estimated concentration of each element in the three standards with the ICP-ES determinations (Table 3) shows good agreement of the values. Differences between the values for both major and minor elements is generally less than 15%. Close agreement of the values for the minor elements is due to careful corrections for background and spectral interferences. The precision of the method, obtained by taking the standard deviation of ten determinations, is estimated at better than 10% for major elements and better than 20% for most minor elements. Actual sludge samples with a wide diversity of compositions have been analyzed with similar accuracy and precision. These results indicate that ICP-ES methods will provide reliable elemental analyses in support of waste tank sludge encapsulation programs.

TABLE 3
ICP-ES analyses of synthetic sludge standards

Element	Dissolution	HIS Wt %		HAS Wt %		BLEND Wt %	
		Est.	ICP-ES	Est.	ICP-ES	Est.	ICP-ES
Fe	HF-HCl	25.4	24.8 ±2.7[a]	5.3	5.5 ±0.4[a]	20.9	21.1 ±0.9[a]
Mn	HF-HCl	7.6	7.2 ±0.3	1.6	1.6 ±0.1	4.3	4.3 ±0.2
U	Na_2O_2-HCl	8.6	8.6 ±0.8	1.0	1.2 ±0.3	5.2	5.5 ±0.7
Al	Na_2O_2-HCl	2.3	2.3 ±0.2	23.2	25.6 ±1.8	5.7	6.3 ±0.3
Ca	Na_2O_2-HCl	1.2	1.2 ±0.1	0.9	1.0 ±0.1	2.0	2.2 ±0.1
Na	HF-HCl	3.7	4.4 ±0.6	2.6	1.8 ±0.5	3.6	4.2 ±0.5
Ni	HF-HCl	3.7	3.8 ±0.1	0.6	0.6 ±0.1	2.0	2.1 ±0.1
Si	HF-HCl	1.0	1.1 ±0.1	2.2	1.9 ±0.3	1.4	1.3 ±0.2
Ru	Na_2O_2-HCl	0.30	0.33 ±0.02	0.10	0.11 ±0.01	0.30	0.34 ±0.03
Th	Na_2O_2-HCl	0.10	0.09 ±0.02	1.2	1.0 ±0.1	0.50	0.52 ±0.07
Ba	Na_2O_2-HCl	0.40	0.42 ±0.02	0.10	0.09 ±0.01	0.20	0.21 ±0.01
Ce	Na_2O_2-HCl	0.20	0.23 ±0.04	0.10	0.13 ±0.03	0.30	0.31 ±0.04
Pb	HF-HCl	0.10	0.12 ±0.05	0.10	0.12 ±0.04	0.20	0.25 ±0.04
La	Na_2O_2-HCl	0.10	0.10 ±0.01	0.10	0.09 ±0.01	0.10	0.09 ±0.01
Zr	HF-HCl	0.40	0.42 ±0.02	0.15	0.17 ±0.02	0.40	0.41 ±0.03
Cr	HF-HCl	0.40	0.43 ±0.08	0.10	0.08 ±0.03	0.20	0.20 ±0.03
Cu	HF-HCl	0.10	0.12 ±0.01	0.05	0.05 ±0.01	0.10	0.10 ±0.02
Zn	HF-HCl	0.20	0.17 ±0.02	0.10	0.09 ±0.01	0.20	0.18 ±0.02
Mg	Na_2O_2-HCl	0.20	0.22 ±0.03	0.10	0.12 ±0.01	0.20	0.21 ±0.02
Sr	Na_2O_2-HCl	0.20	0.17 ±0.02	0.05	0.04 ±0.01	0.10	0.08 ±0.01

[a]Ranges are based on the standard deviation of 10 determinations.

REFERENCES

1. M. A. Floyd, R. W. Morrow, W. B. Luzader, R. B. Farrar, and A. A. Halouma, K/TL/AT-147, Union Carbide Corporation, Nuclear Division, Oak Ridge, Tennessee, 1982.
2. B. W. Short, H. S. Spring, and R. L. Grant, GAT-T-3184, Goodyear Atomic Corporation, Portsmouth, Ohio, 1983.
3. D. E. Gordanier and B. L. Winters, KY-653, Union Carbide Corporation, Nuclear Division, Paducah, Kentucky, 1974.
4. B. Gopalan and V. A. Fassel, An ICP-AES Method for the Simultaneous Determination of Impurities in Uranium Oxide and Uranium Metal, to be submitted for publication.
5. C. A. Horton and J. C. White, Anal. Chem. 30 (1958), 1979.
6. G. F. Larson and R. E. Slagle, Y/DK-357 (preprint), Union Carbide Corporation, Nuclear Division, Oak Ridge, Tennessee, 1983.
7. J. P. Maney, V. Luciano, and A. F. Ward, Jarrell-Ash Plasma Newsletter, 2 (1979) 11.

DETERMINATION OF TRACE IMPURITIES IN URANIUM MATRICES BY INDUCTIVELY COUPLED PLASMA-ATOMIC EMISSION SPECTROMETRY

A. A. Halouma, R. B. Farrar, E. A. Hester, and R. W. Morrow, Oak Ridge Gaseous Diffusion Plant*, Union Carbide Corporation-Nuclear Division, Oak Ridge, Tennessee 37830 (U.S.A.)

EXTENDED ABSTRACT

Trace metallic impurities are one of the important factors considered in the enrichment process of uranium hexafluoride, since they affect the enrichment plant operation. Currently at the Oak Ridge Gaseous Diffusion Plant, several analytical techniques are applied to determine the impurities outlined in Vol. 36, November 14, 1971 of the U.S.A. Federal Register. The techniques include d.c. arc emission spectroscopy, atomic absorption, and spectrophotometry.

Inductively Coupled Plasma-Atomic Emission Spectrometry (ICP-AES) with its wide acceptance as a versatile technique for the determination of trace elements lends itself as a possible alternative. Our objective is to establish conditions for an ICP-AES analysis procedure that can replace several of the currently used techniques, thus achieving a faster analysis time and a reduction in analysis cost. Our proposed procedure entails the removal of the uranium from its solutions through a liquid-liquid extraction with Tri-(2-Ethyl-Hexyl)-Phosphate (TEHP)-Hexane mixture. The impurity elements are then determined using an ICP Spectrometer.

A detailed study was conducted to establish the optimum conditions for uranium extraction with TEHP, namely, (1) the effect of nitric acid concentration, (2) the effect of TEHP to Hexane ratio, (3) the effect of the fluoride ion concentration, (4) the extractability of the trace elements, and (5) the effect of residual uranium on the analyte spectral lines. The study revealed that one to five grams of uranium would be removed from a 50 ml solution containing 10% nitric acid when extracted twice with 50 ml of equal molecular volume of TEHP and Hexane (3:1 volume to volume). Under these conditions, recoveries of better than 90% were achieved for the elements Al, B, Ba, Be, Bi, Ca, Cd, Co, Cr, Cu, Dy, Eu, Fe,

---

*The Oak Ridge Gaseous Diffusion Plant is operated for the U.S. Department of Energy by Union Carbide Corporation under Contract W-7405-eng-26.

Gd, Li, Mg, Mn, Mo, Na, Ni, Pb, Ru, Sb, Si, Sm, Sn, Sr, Ti, V, W, and Zn in the aqueous phase. The recovery of Nb, Ta, and Zr required the presence of 100 μg/ml of fluoride ion in the original uranium solution. Excessive fluoride ion concentration yielded a considerable loss in the extraction efficiency of uranium. For this reason, hydrolyzed uranium hexafluoride samples were brought down to near dryness with 5 ml of concentrated nitric acid prior to extraction. The analytical wavelengths of the analyte lines were evaluated and the lines chosen were free from uranium spectral interference within the concentration range of 1 to 100 μg uranium present per ml of the solution analyzed.

The concentration of the trace elements was determined using a programmable scanning monochromator interfaced to an ICP as described below:

Monochromator   McPherson Model 2051, 1 m focal length. 1800 mm^1 holographic grating, entrance and exit slit widths were 30 μm.

Plasma   Plasma-Therm type HEP 2500D, 27.12 MHz

Nebulizer   Cross-flow.

Data Acquisition System   McPherson Model 786-A scan controller; Analog Technology Corporation Model 151 A-D convertor and a Digital Equipment Corporation PDP 11/34 minicomputer with 128K memory with dual drive RL01 hard disk.

The ICP instrument was calibrated using multi-element water standards.

Three different uranium matrices [NBL reference uranium metal, NBL reference uranium oxide (U_3O_8), and spiked hydrolyzed uranium hexafluoride (UO_2F_2)] were used to evaluate the procedure for daily routine use. The ICP results agreed with the NBL reference and spiked values with an overall relative deviation better than 10%.

MEASUREMENT OF RHODIUM AND PALLADIUM IN SELECTED HANFORD DEFENSE WASTES

M. H. West, C. S. Homi, C. H. Hunter
Rockwell Hanford Operations, P.O. Box 800, Richland, Washington 99352

ABSTRACT

Numerous radioactive, defense waste types have been generated by different fuel reprocessing plants at Hanford, Washington. These wastes have been systematically combined and concentrated using vacuum evaporators, thus significantly increasing the complexity of the wastes. High concentrations of hydroxide, aluminum species, and organic complexants (EDTA, HEDTA, citrate, etc.) currently exist in the double-shell slurry waste type. Approximately 1000 kilograms of rhodium and 790 kilograms of palladium exist within the available wastes. In order to test the applicability of the highly alkaline double-shell slurry as a potential feed stock for rhodium and palladium recovery processes, analytical methodology has been developed for precious metal analyses.

INTRODUCTION

Between 1965 and 1970, a process was developed for recovery of Rh and Pd from aged reprocessing waste at Hanford[1,2]. The Rh and Pd in these studies were recovered from aged neutralized PUREX acid wastes that had been treated for isotopic cesium removal. Since these early studies, the wastes have undergone additional processing that has increased the potential matrix interferences.

A recent study by Schulz and others[3] indicate there are two potential feed stocks for Rh and Pd at Hanford: (1) PUREX high-level waste, and (2) highly alkaline double-shell slurry. There is some chemical evidence to suggest that most of the Rh and Pd species remain soluble during evaporation of dilute supernatant and concentrate in the double-shell slurry[4]. A preliminary assessment of Rh and Pd inventories in this waste indicates that if all the Rh and Pd (1000 kg Rh, 790 kg Pd) were concentrated in the double-shell slurry (12 million gal), the overall average Rh and Pd would be approximately 0.022 g/L Rh and 0.017 g/L Pd[3].

Most classical methods of analysis for noble metals (Ru, Os, Pt, Pd, Rh, Ir) require group separation before determination of the individual metals. Within the noble metals group, further separation is necessary to isolate the metals since few specific methods are available for any noble metal[5]. Campbell and Sandoz[6,7] developed atomic emission and absorption methods based on the extraction of Rh and Pd anionic complexes with Aliquat-336 (tricaprylmonomethyl-ammonium chloride), a quaternary amine, into carbon tetrachloride. The extractions were carried out over the pH range 4-10. The measurement technology developed for neutralized PUREX wastes, which contain low levels of Al and no organic complexants, is not applicable to today's wastes.

The objective of this study was to investigate applicable analytical methods for measuring the concentration of Rh and Pd in Hanford defense wastes. Element specific instrumental techniques such as inductively coupled plasma (ICP) and atomic absorption spectrophotometry (AAS) offer advantages over less specific classical methods because measurements can, in most cases, be made without analyte separation. Both of these methods exhibit sufficient sensitivity for the requirements of this work.

EXPERIMENTAL

A 29-channel spectrometer (Applied Research Laboratories Model 137) and scanning monochromator (1-meter Czerny-Turner) were used to measure emission from the ICP source. A specially designed Babington nebulizer capable of handling high salt concentrations was employed to produce the sample aerosol.

Flame AAS measurements were made with a Varian AA-6 spectrophotometer. Both the ICP and AAS instruments are equipped to handle radioactive samples.

The mixing, collection, and transfer of representative aliquots from both double-shell slurry samples proved to be difficult. Each sample was a mixture of solids and highly viscous liquor. Aliquots were transferred from the heated sample with a large diameter glass tube and weighed. Due to the physical nature of the double-shell slurry, collection of a representative sample from the evaporator may also be questionable. The aliquots were dissolved in nitric acid with simultaneous stirring and heating. Final weight/weight dilutions of approximately 10 were used for the analyses. Transfer of aliquots from 102-AZ and 118-TX proved to be simpler. In both cases, these samples were fluid solutions containing suspended solids. Two milliliters of 102-AZ were dissolved in nitric acid and brought to a 25 mL total volume. An insoluble purple residue was noted and filtered through a 0.45 μ Millipore filter. A ten-fold dilution with dilute nitric acid was made on 118-TX. White suspended solids were observed and filtered.

RESULTS AND DISCUSSION

ICP

Analyses of the major elements in double-shell slurry number 1 were made in order to investigate matrix effects. Single matrix elements exhibited different behavior than the actual sample (10X dilution). For example, a 1% Na solution did not reproduce the behavior of the sample and a serious clogging of the nebulizer was noted. Although they did not clog the nebulizer, Al and Cr produced similar results. As a result of these observations, a synthetic major element matrix was prepared that matched the behavior of the sample and established background signals. Formulation of the synthetic matrix, simulating a 10X dilution of double-shell slurry, is shown in Table I. The salts were dissolved and brought to volume in 10% HNO_3.

TABLE I
Synthetic Double-Shell Slurry

Element	Grams/Litre	Salt	Grams/Litre
Na	48.0	$NaNO_3$	177.4
P	0.65	$NH_4H_2PO_4$	2.4
K	0.47	KNO_3	1.22
Al	5.83	$Al(NO_3)_3 \cdot 9H_2O$	81.1
Cr	0.55	$Cr(NO_3)_3 \cdot 9H_2O$	4.23

Final solution 10% (V/V) HNO_3.

The synthetic matrix was used to establish background signals for both ICP and AAS measurements. Calibration standard series for Rh and Pd were prepared in the synthetic solution.

The spectrometer Pd line at 340.46 nm proved to be satisfactory if no Zr was present (a Zr line at 340.48 nm interferes). This line was adopted for the analysis of double-shell slurry samples. In the presence of Zr, the 360.96 nm line on the monochromator must be used. Rhodium was analyzed on the monochromator at 343.49 nm.

Calibration curves were prepared in the synthetic matrix over the range 0-6 µg/mL for both Rh and Pd. Standard addition plots were prepared over the range 0-4 µg/mL for both elements. Both the calibration curves and standard additions plots (Fig. 1 and Fig. 2) were quite linear over the concentration range studied, with the Rh calibration curve having a greater

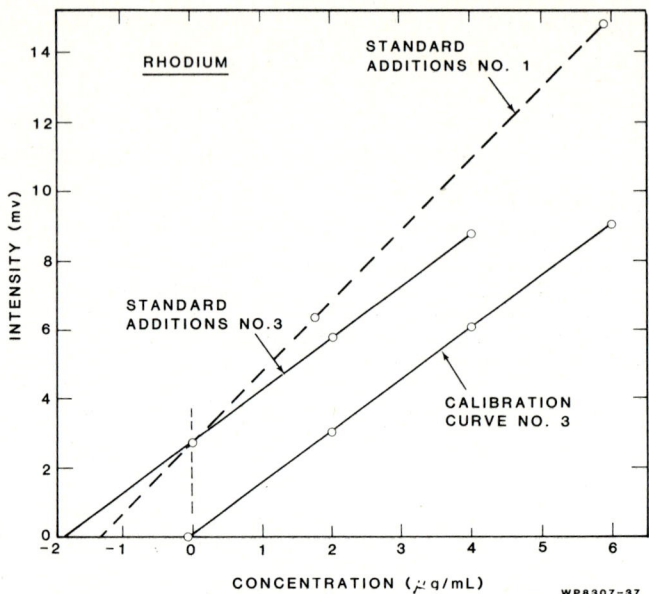

FIGURE 1. Standard Additions Plots and Calibration Curve for the Determination of Rh in Double-Shell Slurry Samples by Inductively Coupled Plasma.

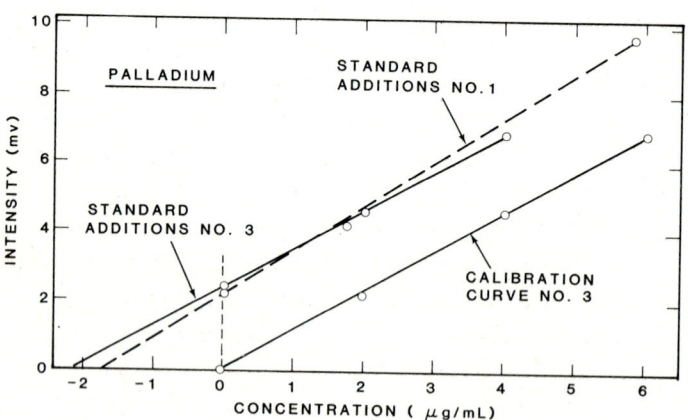

FIGURE 2. Standard Additions Plots and Calibration Curve for the Determination of Pd in Double-Shell Slurry Samples by Inductively Coupled Plasma.

slope and better detection limit relative to Pd. The results for double-shell slurry number 1 and number 3 are given in Table II.

TABLE II
Double-Shell Slurry Results

	DSS #1		DSS #3	
	Pd (µg/g)	Rh (µg/g)	Pd (µg/g)	Rh (µg/g)
ICP S.A.[††]	15.7	12.6	18.6	16.0
ICP C.C.[††]	12.6	10.5	18.3	15.7
AAS S.A.	12.7	9.8	11.6[†]	10.8[†]
AAS C.C.	7.4	8.3	8.4[†]	11.0[†]

S.A. = Standard additions
C.C. = Calibration curves
[†]Determined by flame microsampling atomic absorption spectroscopy
[††]Pd determined by 340.46 nm and Rh determined by 343.49 nm by ICP

For double-shell slurry number 3, excellent agreement was found between the values determined by ICP for both analytes using standard additions and a calibration curve. In other words, no significant matrix effects appear to exist when utilizing the synthetic matrix for purposes of calibration. On the other hand, when calculated using the calibration curve, the corresponding analyte values for double-shell slurry number 1 exhibited consistently lower levels relative to the method of additions. Since numbers 1 and 3 were sampled from the concentrator on different days, the samples are not identical and this is supported to some extent by the total slurry analysis. It was noted Pd standard additions to double-shell slurry number 1 precipitated after several days by some as yet unknown mechanism. A light precipitate was observed in these solutions. Thus, it is important to analyze the spikes immediately before sample degradation can occur.

The 102-AZ sample contains Zr so that the alternate 360.96 nm Pd line was selected on the monochromator. However, another interfering species, initially thought to be thorium, led to a dramatic increase in intensity at 360.96 nm. Further investigation demonstrated the presence of thorium but at a much lower level than originally suspected. The Rh level is approximately 7.6 µg/g.

A sample taken from 118-TX shallow supernatant proved to have a much simpler matrix than 102-AZ as no zirconium or thorium was found. In addition, the Rh and Pd levels are less than 5 µg/g for this particular sample. A synthetic matrix, containing the major components, was prepared for 118-TX in a fashion similar to double-shell slurry.

AAS

Results obtained for double-shell slurry number 1 by atomic absorption spectrophotometry are given in Table II. The concentrations of Rh and Pd reported for this technique are consistently lower than those obtained from ICP spectrometry. For both elements, absorbances decreased steadily as a function of time for a "control" sample. This was due to physical difficulties such as nebulizer clogging and more generally to salts depositing in the burner head slot during solution aspiration. This results in a continually changing absorption cell path length.

Calibration curves for the flame atomic absorption work were prepared by spiking Rh and Pd into the synthetic double-shell slurry matrix. It is necessary to utilize the synthetic matrix as a means of measuring the background (nonspecific) absorption at both the Pd (244.79 nm) and Rh (343.49 nm) analytical lines. Background correction systems currently available with the Varian AA-6 unit used in this study do not work well above approximately 300 nm.

Flame AAS microsampling is an inexpensive alternative to continuous aspiration of high dissolved solids solutions. This has been described by Voth at Varian[8] and by others[9] in the recent literature.

A rudimentary test of the microsampling method was made by micropipetting 100 µL sample aliquots into 5 mL polystyrene disposable cups and aspirating the solution by manually placing the plastic capillary tubing into the cup. A recorder was used to monitor the transient signals. A set of standards containing 0, 2, 4, and 6 µg/mL Rh and Pd in synthetic double-shell slurry was analyzed with each standard being aspirated at least four times. This sequence was then repeated and the two sets of data compared with one another. No noticeable decrease in either the Rh or Pd signal was noted at its respective analytical wavelength. Reproducibility also appeared to be quite good.

Flame microsampling AAS was applied to the determination of rhodium and palladium in double-shell slurry number 3, Table II. While the rhodium concentrations obtained from the calibration curve and standard additions plot agreed very well (Fig. 3), both values were lower than the respective ICP results. On the other hand, the palladium concentration determined from the method of additions was higher than the result predicted from the calibration curve (Fig. 4), and both values were again significantly lower than the ICP results. Several days had elapsed between analysis of double-shell slurry number 3 by ICP and flame microsampling AAS. Loss of palladium by precipitation has been observed from solutions prepared for precious metals analysis (vide supra).

Using flame microsampling atomic absorption spectroscopy, the Rh and Pd concentrations were determined for 102-AZ, Table III. The level for Rh agrees well with that found using the ICP technique while the Pd concentration is below the detection limit.

TABLE III
102-AZ Results

	Pd (μg/g)	Rh (μg/g)
ICP[†]	*	7.6
AAS S.A.	<5	8.9
AAS C.C.	<5	7.9

S.A. = Standard addition
C.C. = Calibration curve
*Interference prevented determination.
[†]Rh determined at 343.49 nm by ICP.

FUTURE WORK

Palladium lines at 342.12 nm and 351.69 nm should be investigated with the monochromator (ICP) for interferences in the 102-AZ sample. Both of these lines are nearly as sensitive[10] as the two lines used in this investigation. Also, the concentration of Th and presence of other actinides and lanthanides should be established. Elements from both of these series produce complex (ICP) spectra that can interfere with the determination of many elements. Other measurement techniques such as X-ray emission and neutron activation would be beneficial in thoroughly characterizing the interferences in 102-AZ and other waste samples.

In the same light, the presence and concentration of Ru and Tc should be checked by ICP in all samples investigated. The presence of Tc was confirmed in PUREX waste some time ago[1,2]. The economic value of Ru may alone justify further study.

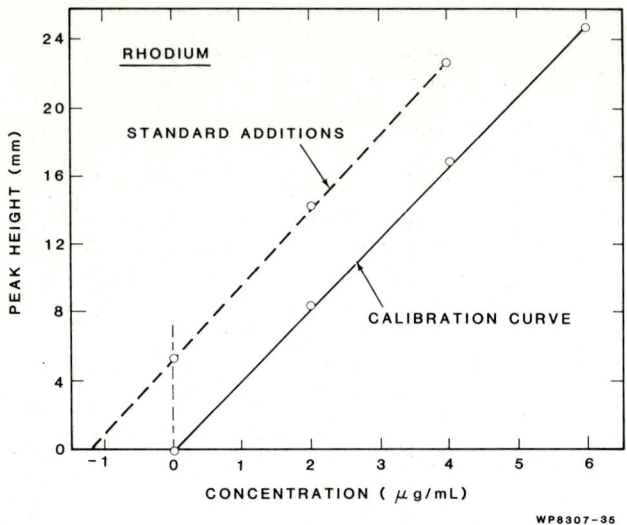

FIGURE 3. Determination of Rhodium in Double-Shell Slurry #3 by Flame Microsampling Atomic Absorption Spectrophotometry.

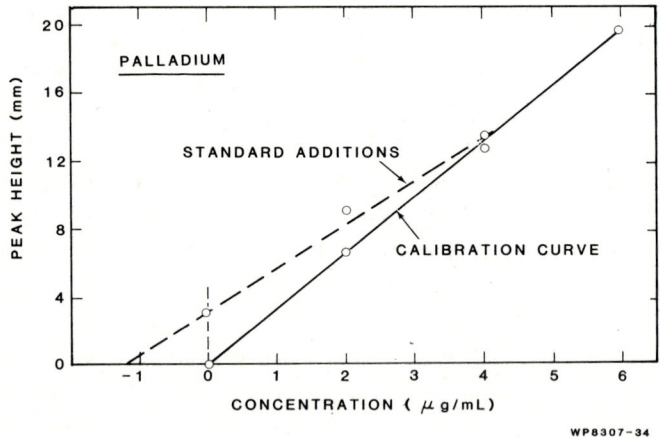

FIGURE 4. Determination of Palladium in Double-Shell Slurry #3 by Flame Microsampling Atomic Absorption Spectrophotometry.

REFERENCES

1. Panesko, J. V. (1968), "Process for Recovery of Rhodium, Palladium, and Technetium from Aged Reprocessing Wastes at Hanford," ARH-733, Atlantic Richfield Company, Richland, Washington.
2. Panesko, J. V. (1968), "Process for Purification of Fission Product Rhodium, Palladium, and Technetium Recovered from Aged PUREX Supernatant Wastes," ARH-311, Atlantic Richfield Company, Richland, Washington.
3. Schulz, W. W. et al. (1981), "Inventories of and Technology for Recovery of Am, Pm, Rh, and Pd Values at Hanford: A Preliminary Assessment," RHO-LD-170, Rockwell Hanford Operations, Richland, Washington.
4. Hoffman, W. A., Jr. (1968), "Rhodium Species in Radioactive Waste Solutions," ARH-732, Atlantic Richfield Company, Richland, Washington.
5. Interscience Publishers (1963), "The Platinum Metals," Treatise on Analytical Chemistry, Part II, Analytical Chemistry of the Elements, Volume 8, pp. 379-511.
6. Campbell, M. H. (1968), "A Rapid Determination of Rhodium and Palladium Using Liquid-Liquid Extraction with Tricapryl Monomethyl Ammonium Chloride and Flame Photometry," Analytical Chemistry, 40, p. 6.
7. Sandoz, D. P. (1966), "The Determination of Rhodium and Palladium by Atomic Absorption Spectrophotometry," ISO-593, Isochem Inc., Richland, Washington.
8. Voth, L. M. (1981), "Analysis of High Dissolved Solids Solutions by Flame Microsampling," Varian Instruments at Work, No. AA-13.
9. Urbain, H. and Martin, N. (1981), "Microsampling Technique in Flame Atomic Absorption. Application in Preparation of Platinum-Ruthenium Alumina Catalysts," Atomic Spectroscopy, 2 (4), p. 127.
10. Winge, R. K., V. J. Peterson, and V. A. Fassel (1979), "Inductively Coupled Plasma-Atomic Emission Spectroscopy: Prominent Lines," Applied Spectroscopy, 33 (3), pp. 206-219.

W.S. Lyon (Editor), *Analytical Spectroscopy*
Elsevier Science Publishers B.V., Amsterdam — Printed in The Netherlands

CONSIDERATIONS FOR THE INSTALLATION OF AN INDUCTIVELY COUPLED PLASMA FOR THE ANALYSIS OF RADIOACTIVE SAMPLES*

C.T. APEL and D.L. GALLIMORE
Group CHM-1, MS-G740, Los Alamos National Laboratory, Los Alamos, New Mexico 87545

ABSTRACT

The sensitivity, dynamic range, and sample through-put rate attributes (ref.1,2) of the inductively coupled plasma (ICP) call for its consideration as the instrument of choice for the multielement analysis of radioactive samples.
Based on our experience in handling radioactive materials, considerations will be presented concerning safety of the operator, modularity of the ICP-atomic emission spectrometer systems, reduction of the complex actinide spectra, atomization systems, drain and recovery systems, aerosol containment, heat dissipation, radiolysis effects of sample on dry-box environment, and liquid and solid sampling.

INTRODUCTION

A review by Faires (ref.3) describes many of the attributes of the ICP which make it a valuable tool to be included in the analytical chemist's kit. Its demonstrated reliability makes it economically feasible to consider the ICP for the multielement analysis of radioactive samples in spite of the problems of containment and operator safety. ICP-atomic emission spectrometry (ICP-AES) has found favor in nuclear applications at European installations such as Karlsruhe and Grenoble (ref.4). Homi (ref.5) developed one of the first ICP-AES systems for the multielement analysis of radioactive solutions in the United States. Floyd et al. (ref.6) demonstrated the necessity of separating uranium from the analyte to reduce the interference effects from its complex spectra on the determination of trace elements with ICP-AES. Fassel and Edelson (ref.7) have discussed aerosol containment and heat dissipation.

OUR EXPERIENCE AND PLANS

Planning concerning glove box design and work must be done carefully. Once the systems are inaugurated and contaminated with radioactive material, analysts may become locked in on methods (because of the expense and safety hazards involved with major modifications) for decades or longer. It is important for the

---

*Work performed under the auspices of the United States Department of Energy.

planner to consider all possible parameters and options and for the final design to be as flexible as possible.

Operator safety must be considered. The maximum permissible body burden for ^{239}Pu is 0.65 µg (ref.8). For samples with β-γ radiation, shielding must be provided. Our glove box plans call for the enclosure to be constructed from a stainless steel - lead - stainless steel sandwich. Operator windows will be of leaded glass. Optical shielding will be incorporated to provide protection from the intense uv radiation of the ICP torch. Protection from rf radiation and electric shock will also be provided. Equipment installed in the box should be easily accessed, easily repaired, and have no sharp edges.

Modularity of instrumentation is dictated by the high cost of ICP-AES equipment. All electronic and optical equipment possible should be kept outside the glove box environment to permit easy and economical repair and to provide protection from radiolysis. Only the torch, nebulizer, spray chamber, and drain system should be located in the glove box.

Reduction of the complex actinide spectra (Fig. 1) must be accomplished by separation of the actinide from the analyte before its introduction to the ICP. This also eases the performance demands on the aerosol containment system.

A block diagram of our conception of an ICP light source glove box is shown (Fig. 2).

Atomization systems should include the capability for routine solution sampling, limited volume sampling for sequential scanning, slurry analysis, and high dissolved solids sampling. A nebulizer - spray chamber - torch system alternative for the analysis of HF stabilized solutions will be included.

Waste radioactive solutions will be recovered from the drain trap by the use of a peristaltic pump to remove excess solution from the drain trap and transfer it to a residue bottle.

Aerosol containment will be accomplished through the use of easily removable filter cartridges. Frequent filter changing will prevent the build up of penetrating radiation in the area of the operator's upper body and head, and will assure free exhaust flow to ease heat dissipation from the ICP torch effluent gas.

Heat dissipation will be accomplished with a heat exchanger situated between the torch and the filter. It should be easily demounted for cleaning and contain an easily replaceable liner.

High α-emitting samples such as ^{238}Pu create radiolysis problems by collisions of α-particles with glove box construction materials and environmental gas molecules creating highly excited ions which accelerate corrosion of equipment situated inside the glove box. Chlorinated and fluorinated plastics should be avoided. Even air and water vapor in the box are known to cause problems in the presence of ^{238}Pu.

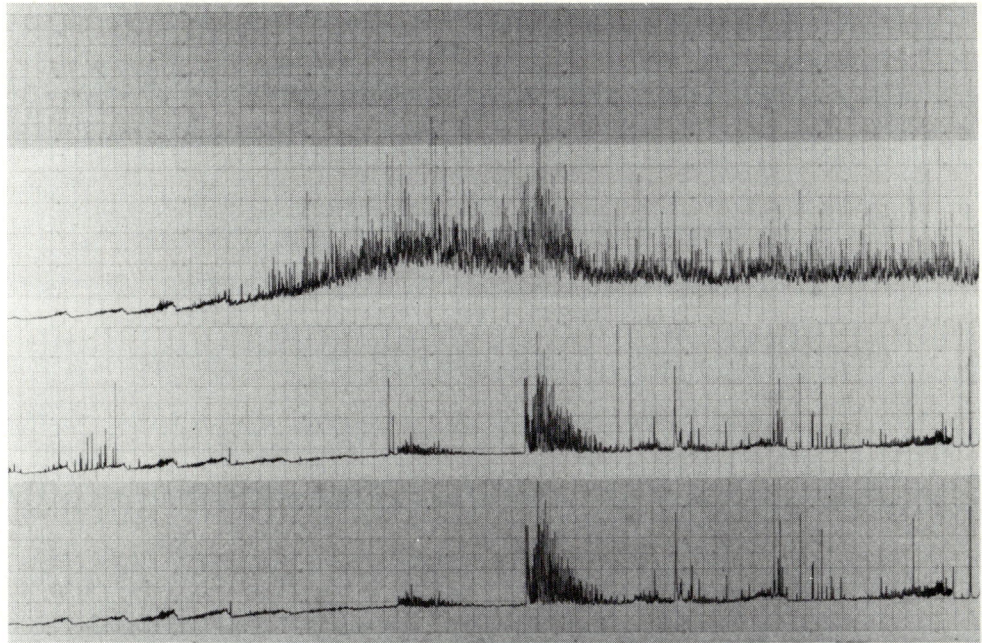

Fig. 1. ICP spectra of: bottom - 0.1 M HNO_3, Middle - Cu 10 µg/mℓ, Top - U 100 µg/mℓ. Range 200 (left) to 400 nm (right).

Solid sampling must be considered, as well as liquid sampling, because the majority of samples encountered by emission spectrochemists are solid samples. Apparatus will be included for solid sampling such as laser ablation, slurry handling, and an auxilliary arc capable of high current dc-arc or controlled waveform spark source (CWSS) excitation.

Versatility will be achieved (Fig. 3) by including in the glove box system a second polychromator with exit slits optimized for dc-arc and spark lines and capable of time-resolved recording of the spectra from the auxilliary arc, a balance box, a wet chemistry box, a storage box for contaminated instrumentation and equipment, and a glove box for the development of automated methods such as high pressure liquid chromatography and flow injection analysis.

CONCLUSION

Careful planning and proper materials selection can lead to a very effective system for the ICP-AES analyses of radioactive solutions and solids. Adequate

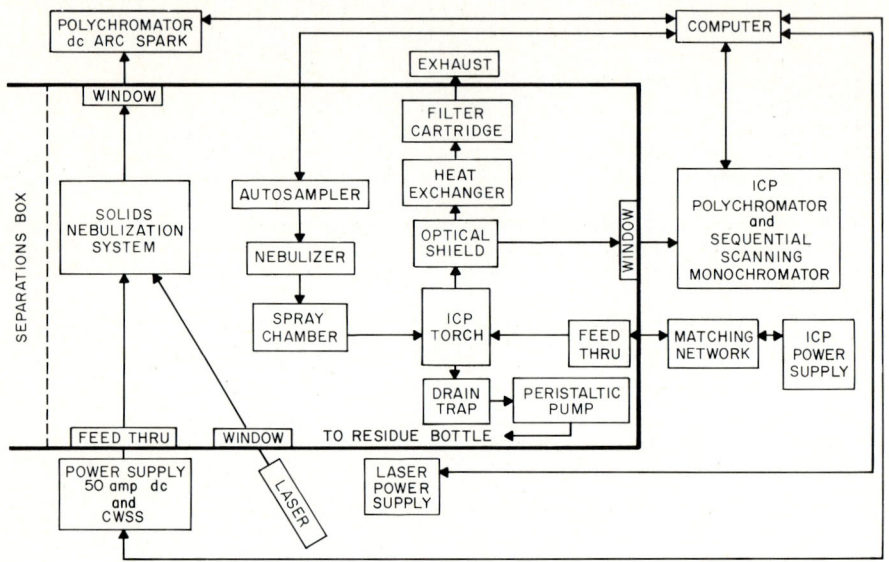

Fig. 2. Light source glove box.

protection can be provided to both the operators and the environment. Finally, careful layout of the glove box train and spectrometer complex can permit the concurrent operations of routine sample analysis and advanced methods development - operations which are usually mutually exclusive.

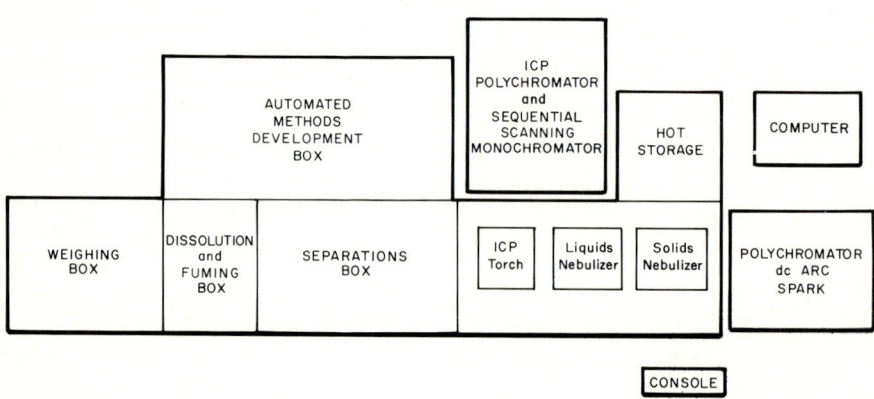

Fig. 3. Equipment and glovebox layout (top view).

REFERENCES

1. S. Greenfield, I.L. Jones, C.T. Berry, Analyst, 89, (1964) 713.
2. R. Wendt, V.A. Fassel, Analytical Chemistry 37, (1965) 920.
3. L.M. Faires, Inductively Coupled Plasma: Principles and Horizons, American Laboratory, 14, November (1982) 16-22.
4. C.E. Pietri, Private Communication, New Brunswick National Laboratory, Argonne, IL.
5. C.S. Homi, R.M. Manabe, A Facility for the Analysis of Radioactive Samples by Inductively Coupled Plasma Spectroscopy, RHO-SA-81, Rockwell International, Richland, WA.
6. M.A. Floyd, R.W. Morrow, R.B. Farrar, Inductively Coupled Plasma - Atomic Emission Spectroscopy: The Determination of Trace Impurities in Uranium Hexafluoride, K/TL/AT-64J, Union Carbide Corp., Nuclear Div., Oak Ridge, TN.
7. M. Edelson, V.A. Fassel, Private Communication, Ames Laboratory, Ames, IA.
8. J.W. Healy, Los Alamos Handbook of Radiation Monitoring, LA Report No. 4400, Los Alamos National Laboratory, Los Alamos, NM (1970).

W.S. Lyon (Editor), *Analytical Spectroscopy*
Elsevier Science Publishers B.V., Amsterdam — Printed in The Netherlands

EVOLUTION OF CONTAINMENT FACILITIES FOR SPECTROSCOPIC ANALYSIS AT ROCKWELL HANFORD OPERATIONS

J. M. HILLER
Rockwell Hanford Operations, P. O. Box 800, Richland, WA 99352 (USA)

ABSTRACT

   The analysis of radioactive material requires much thought concerning getting the job done while still maintaining a safe working environment. At Rockwell Hanford Operations, we have gone through several stages of evolution in instrumentation for spectroscopic elemental analysis, reflecting different philosophies with respect to shielding and contamination control. Atomic absorption and inductively coupled plasma emission spectroscopic systems have been used for analyzing samples in support of a fission product recovery plant, nuclear waste processing and characterization programs, and U and Pu separation plants. Design thoughts, criticisms, and lessons learned in 20 years of containment for spectroscopic analysis are presented.

INTRODUCTION

   Atomic absorption (AA) and inductively coupled plasma (ICP) emission analyses have been performed on radioactive samples at the Hanford reservation for over twenty years. In that time, several generations of containment and scrubbing systems have evolved.
   The goals of each installation design included (1) protection of the environment from gases and liquids, (2) protection of operating personnel against exposure (whole body, extremity, and internal deposition), (3) no loss of efficiency for routine or non-radioactive samples, (4) high sample throughput through minimal down time, and (5) analytical performance similar to unmodified instruments.
   Samples analyzed have come from waste management and chemical processing. The radiation levels in samples have ranged from near zero to the kCi/L level.
   The objective of this paper is to relate some experiences in the construction and operation of shielded AA and ICP units at the Hanford site. No attempt is made to exhaustively review the subject of shielded instrumentation.

INSTRUMENTATION
Generation 1: Purex AA (ca. 1965)
   The first AA modified for hot operation was a Jarrell-Ash Model 82-526 unit (1). The unit was used to support the Plutonium Uranium Extraction or Purex process. The instrument was modified, as shown in Figure 1, by separating the burner, spray chamber, and nebulizer from the remainder of the system and mounting them in a hood. Light from the hollow cathode lamp was directed through a slot in the hood, off a mirror, through the flame, off a mirror, through another slot in the hood, to a monochromator and readout device. The modifications decreased the sensitivity slightly.
   A scrubber system (Figure 2) was used to capture radioactive particulates emanating from the burner flame. A chimney resting directly on top of the burner head shielded the flame from stray air currents and directed the off-gas to the scrubber inlet. Quartz windows in the sides of the chimney permitted the passage of the light beam. Air was drawn into the chimney to dilute and reduce the off-gas temperature.
   The scrubbing solution was recirculated to reduce the amount of low-level waste. The off-gas was drawn into the scrubber by the suction created by the

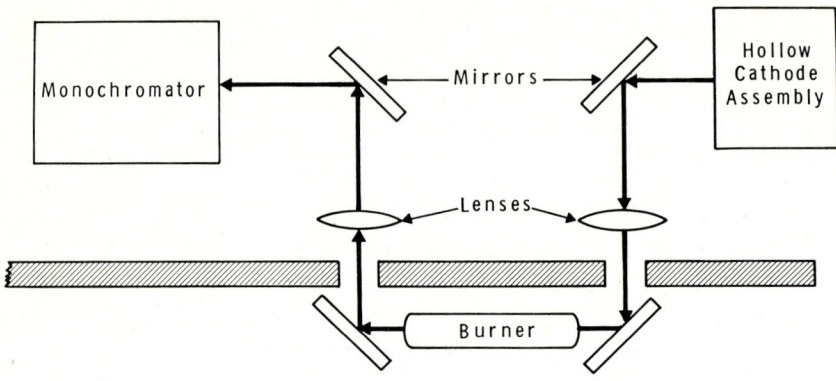

Figure 1. Optical arrangement of Purex hot AA.

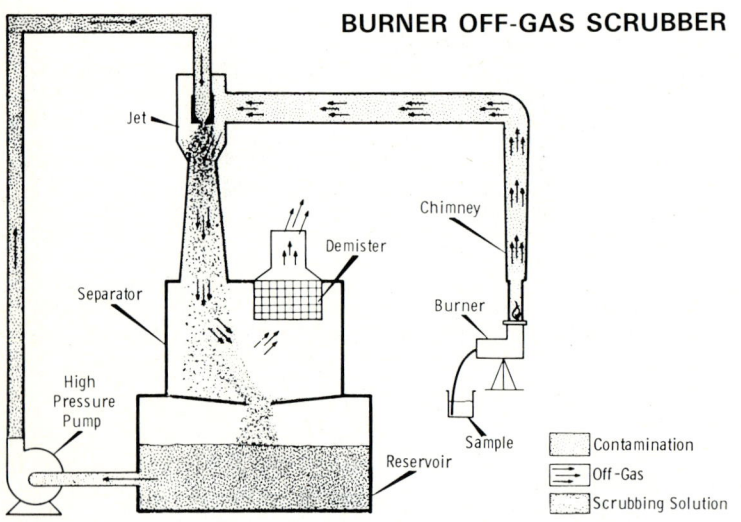

Figure 2. Purex hot AA scrubber system.

jet and, after clean-up, vented to the hood. Aqueous waste was sent to a simple drain cup. All scrubber components were fabricated from stainless steel, thereby permitting the use of caustic, acid, or neutral scrubbing solutions.

The scrubber system, however, was not extremely efficient. As indicated by Table 1, most of the alpha escaped the scrubber; beta and gamma radiation was more successfully captured.

A major collection area for alpha was on the burner head and chimney, presumably as plutonium oxide. The lesson learned from this unit was to frequently clean and/or replace the chimney. A second lesson was to keep the drain line and nebulization system clean and shielded.

This unit is no longer in service.

TABLE 1. Efficiency of Scrubber/Containment Systems

Instrument	Radiation Type	Radiation Disposition			Scrubber Efficiency
		Drain	Scrubber	Elsewhere	
Purex AA	α	95%	0.8%	4.2%	16.0%
	β, γ	95	4.5	0.5	90
Redox AA	14.6 μCi β	91	8.8	0.2	97.8
	146 μCi β	91	8.96	0.02	99.8
	2900 μCi β	91	8.6	0.4	95.5
	107 μCi α	91	8.98	0.004	99.9
Redox ICP	α,β,γ	≈90	≈10	≈0	≈100

Generation 2: Redox AA (ca. 1972)

The Redox (Reduction/Oxidation plutonium/uranium separation process, no longer active) AA (2) represents a fundamental change in design of hot instrumentation. With this unit, the AA is basically a module attached to the scrubber system. A Varian Techtron AA-6 with a Jarrell-Ash Model 82-092 monochromator was fitted to an open face hood (providing shielding) and a 6 X 6 X 10 foot stainless steel box (which serves as the off-gas scrubber system housing). The AA-6's modular form was the main reason for its selection. The optical rail passes through an alcove of the open-face hood (Figure 3). The burner assembly is seated in the alcove. In this configuration, the rest of the system remains outside the hood, unexposed to direct contamination.

OPEN-FACE HOOD AND AAS (TOP VIEW)

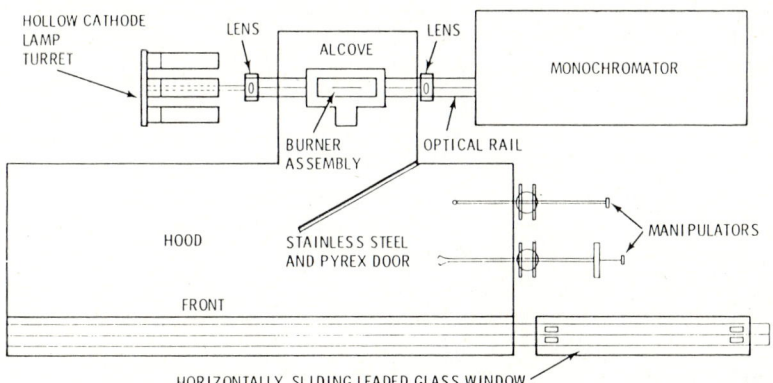

Figure 3. Redox hot AA open-face hood.

The off-gases are collected in a chimney above the burner and drawn through a short section of pipe into the Krebs scrubber located in the scrubber system housing (Figure 4). Immediately afterwards, the gases are passed through a 12-inch depth of ceramic saddles, then a condenser, and finally through double, high-efficiency particulate air (HEPA) filters. The Krebs scrubber utilizes distilled water moving through a closed cycle.

Radioactive liquid wastes, from the scrubber reservoir and the mixing chamber, are drained into a hot waste line located on the floor of the scrubber system housing. These wastes are collected and periodically jetted to the plant waste tanks where the radioactive solutions are processed.

Experience has shown that much of the radioactivity not picked-up by the scrubber ends on the chimney. Thus, the chimney was designed for easy removal and accepting simple replaceable and disposable liners. Contaminated chimneys

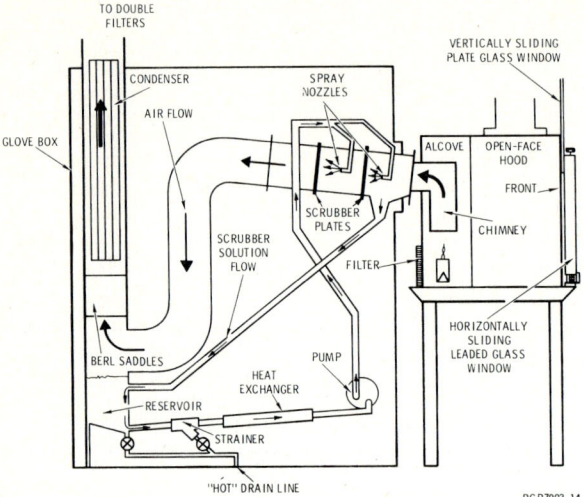

Figure 4. Redox hot AA scrubber system.

have read as high as 5 rad/hr and have been decontaminated to under 0.2 rad/hr.

The operator shielding for the instrument is somewhat elaborate. The burner sits in an alcove physically separated from the operator by a polycarbonate and stainless steel door, a one-foot deep hood, and an optional, movable, 2-in. leaded glass shield. To further reduce exposure, routine burner adjustments and sample introduction can be performed using a series of manipulators built into the containment system. The system is efficient in protecting the operators from exposure (Table 2). Much of the extremity dose comes during chimney changes and spray chamber cleaning.

This unit is in service at present.

TABLE 2. Exposure limits and experience with hot instruments.

Instrument	Extremity			Whole Body		
	DOE	RHO	Actual	DOE	RHO	Actual
Redox AA	75	15	1.6	5	3	0.72
Redox ICP	75	15	0.6	5	3	0.14

DOE: United States Department of Energy limits; RHO: Rockwell Hanford Operations limits; Exposure data reported as Rems.

Generation 3: Redox ICP (ca. 1979)

The Redox ICP (3) is a modified ARL Model 137 (ICP-Q) direct reader. As shown in Figure 5, the spectrometer was modified by attaching a custom torch box and open-face hood.

Off-gases from the ICP torch are directed to the bottom of a vertical scrubber. These gases come immediately in contact with water from the lower spray nozzle and pass through the section filled with pall rings. Water from the upper spray nozzle flows countercurrent to the off-gases in the pall ring section. Excess water returns to the reservoir where it can be recirculated to the spray nozzles or released to the contaminated drain. The system is routinely (once per shift) flushed by adding water to the reservoir and releasing the solution after circulation through the scrubber to the contaminated drain.

The exhaust gases are directed through a demister pad, through double HEPA filters, and finally to house exhaust.

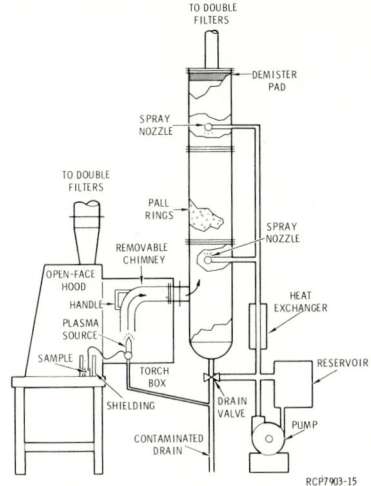

Figure 5. Redox hot ICP scrubber system.

The hood is constructed of stainless steel and glass. The torch assembly (including the matching network) rests behind the door at the back panel of the hood. Access to the side of the torch is possible through a removeable panel.

The scrubber system represents a case of over-engineering. The scrubber is largely fabricated from half-inch and greater stainless steel. The heavy steel was chosen for shielding and to dissipate heat from the plasma. In operation, it has been found that little radiation enters the scrubber--most radiation goes to waste (from the spray chamber) or plates on the chimney. The chimney was designed to be easily removed and decontaminated.

With the over-engineering has come problems. The "seal pot" (a plumbing contrivance which acts as the junction for waste and flush lines and as an optical level indicator) leaks. The seal pot is exposed to high radiation (one line comes from the nebulizer drain) as well as corrosive materials. The seal pot is also too high off the floor to permit proper drainage of the ICP torch. Work is underway to rectify the problems.

A second minor problem comes from human-compatability. The hood was constructed too deep to permit routine torch maintenance. To adjust the torch or nebulizer, one has to literally climb into the hood (naturally, with correct protective gear).

The shielding of the Redox ICP operator comes from the steel and glass panel separating the torch from the hood, moveable 2-inch leaded windows, and distance. Exposure data for the ICP operators indicate that the existing shielding is satisfactory (Table 2).

This unit is in service at present and is being upgraded.

Generation 4: Purex ICP (ca. 1984)

The Purex ICP is a Bausch & Lomb Model 3580 spectrometer which will be modified slightly by building a free-standing open-face hood around the standard torch box. The torch box will be unmodified except for moving some controls and mounting them on the back of the hood. The hood will be wider (3-arm) than the Redox hood (to permit sample preparation) and not as deep (to permit routine torch maintenance).

The experience with the other scrubber systems have shown that they were over-engineered with an emphasis on the off-gases. The Purex system (Figure 6) represents a much simpler approach. Off-gases from the torch and some hood air are directed upwards through a removable and cleanable chimney, through

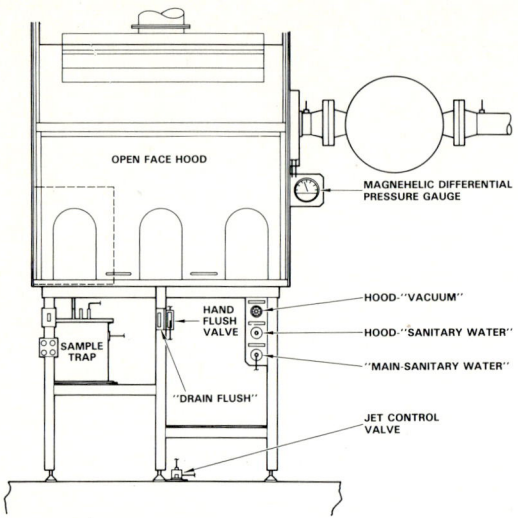

Figure 6. Purex hot ICP hood system.

removable sections of duct pipe, and finally through double HEPA filters to house ventilation. The system is fabricated to accept commercial scrubbers if experience indicates the need.

Aqueous waste from the torch spray chamber will be gravity fed to a shielded vessel mounted under the hood. The vessel will be periodically emptied and flushed into plant waste. The number of connections has been kept to a minimum to reduce the probability of leakage.

The shielding of the operator is similar to the Redox ICP. The torch will rest in its standard compartment behind a metal and glass panel. The operator will be further shielded by moveable leaded windows and the glass in the hood face.

SUMMARY and CONCLUSION

Certain lessons have been learned in constructing hot AA and ICP instrumentation. A scrubber/shielding system should be simple--excessive engineering means more areas where the system can fail. As much as possible, the system should be engineered to facilitate routine maintenance--a modular approach with access for repair and routine operation is central.

We have found that open-face hoods are satisfactory for keeping operator exposure at low levels. We have not found glove-box arrangements necessary.

REFERENCES

1. Kupfer, M.J; Haberman, H. D.; and Hayes, M. P., "Modification of the Jarrell-Ash Atomic Absorption Spectrometer for Analyses of Trace Metals in Radioactive Solutions," ISO-SA-44, Prepared for the United States Atomic Energy Commission under contract AT(45-1)2130, Atlantic Richfield Hanford Company, Richland, WA, September 26, 1967.

2. Harnley, J. M., "A Facility for the Analysis of Intensely Radioactive Samples by Atomic Absorption," ARH-SA-146, Prepared for the United States Atomic Energy Commission under contract AT(45-1)2130, Atlantic Richfield Hanford Company, Richland, WA, September 27, 1973.

3. Homi, C. S. and Manabe, R. M., "A Facility for the Analysis of Radioactive

Samples by Inductively Coupled Plasma Spectroscopy." Prepared for the United States Department of Energy under contract DE-AC06-77RL01030, September 28, 1979.

W.S. Lyon (Editor), *Analytical Spectroscopy*
Elsevier Science Publishers B.V., Amsterdam — Printed in The Netherlands

A NEW FACILITY FOR HIGH RESOLUTION INDUCTIVELY COUPLED PLASMA-ATOMIC MULTIELEMENT EMISSION SPECTROSCOPY OF RADIOACTIVE MATERIALS

M. C. Edelson and V. A. Fassel, Ames Laboratory*, Iowa State University, Ames, IA 50011

ABSTRACT

An inductively coupled plasma excitation system has been designed to operate within a total α-containment structure. The design, testing and operation of the system is discussed and some optical emission spectra of Pu-242 and U-isotopes are presented. The entire experimental process, from sample weighing to waste collection, was performed within a single glovebox. During the testing and redesign phase of this study, unanticipated failure modes were encountered, thus demonstrating the need for extensive testing of such devices before they are used with hazardous materials.

INTRODUCTION

Inductively coupled plasmas are now extensively used by analytical chemists for the vaporization, atomization, excitation, and ionization of samples. Instrumentation for observation of the ground state, excited, and ionized atoms via their atomic emission, fluorescence, and mass spectra are now provided by numerous vendors. In the design of these instruments very little attention has been devoted to the eventual fate of the fine particulates produced by the ICP in routine operation. To safely utilize ICP methods with radioactive samples it is necessary to treat the instrument exhaust gas stream, prior to its venting to the atmosphere, and remove essentially all of the entrained radionucleii. In this paper we describe the design, testing and operation of an ICP that is operated within a modified commercial stainless-steel glovebox. The ICP system is being used to investigate the feasibility of determining the elemental and isotopic concentration of U and Pu in spent fuel dissolver solutions via observation of their characteristic atomic emission spectra.

The design of the ICP enclosure and its installation was guided by the following criteria:

---

*Operated for the U.S. Department of Energy by Iowa State University under contract No. W-7405-Eng-82. This work was supported by the Office of Safeguards and Security.

A. Total containment of α-emitting radionucleii whether in solution or particulate form.
B. Minimization of the amount of apparatus that is contaminated in normal operation,
C. Provision for the in situ servicing of contaminated apparatus,
D. Avoidance of specialized in-house construction,
E. Compatibility with commercial instrumentation,
F. Minimization of the impact on the community at-large on failure of safety systems or accidental contamination.

DESCRIPTION OF FACILITY

The Alpha Containment Facility (ACF) was constructed within an existing building and utilized an L-shaped containment structure to house the glovebox-enclosed, ICP-torch box. An addition was built onto this structure to house the experimental electronics and spectrometers needed for the analysis of the ICP spectra.

The overall plan of the ACF is shown in Figure 1. Light generated by the enclosed ICP is passed through a quartz viewing port into the electronics area. The ICP operator is normally located outside of the L-shaped room during the operation of the facility. The air flow pattern for this facility is shown in Figure 2.

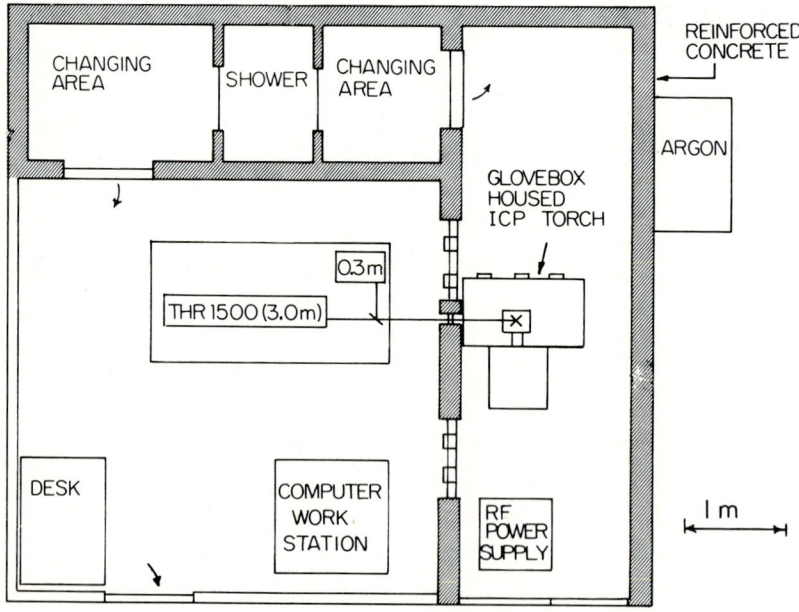

Fig. 1. Plan of the ACF. This facility is located ~3 km from the main Ames Laboratory campus site.

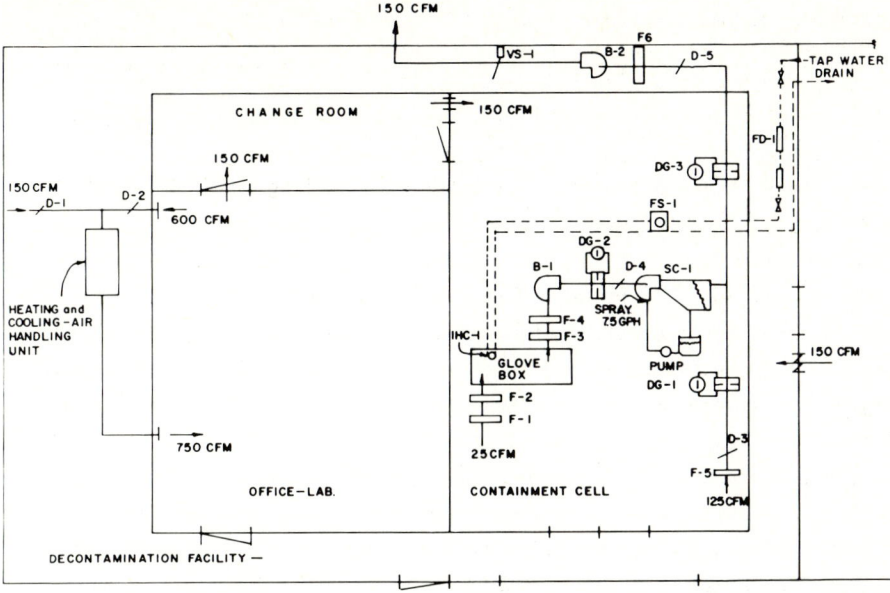

Fig. 2. Air handling plan for the ACF. B=blower, D=manual damper, DG=draft gauge with orifice, F-1=prefilter, F-2,3=HEPA filter (metal cased), F-4=20x20x15 cm HEPA filters, F-5=60x60x10 cm HEPA filter, FD-1=water demineralizer cartridge, FS-1=water flow switch (0.8 lpm), lHC-1=ICP load coil, SC-1=scrubber, VS-1=vane type air flow switch.

The ICP used for this work (Type HFS 3,000D, Plasma-Therm, Inc., Kresson, NJ) consists of three major components; a large RF power generator, a smaller automatic impedance matching network, and an enclosure that houses the RF load coil and ICP torch. Only the torch enclosure is placed inside the containment housing, which is a modified stainless-steel radiochemical enclosure obtained commercially (Keewaunee Scientific Equipment, Lockhart, TX). Although the torch-box to impedance-matching network separation was ~25 cm larger in this configuration than is normally employed, no electrical modification of the matching network was necessary. The RF power and ground return lines passed from the matching network into the torch box through a specially designed flange plate that was affixed to the rear wall of the glovebox. This flange also provided water, argon and electrical feedthroughs for the glovebox. The impedance-matching network box was rigidly connected to the torch box by a heavy-walled, stainless-steel tube that also served as a conduit for the RF power and ground return lines. By placing the matching network box on a movable platform and cutting an over-sized hole in the glovebox flange, it was possible to provide $\pm$ 5 cm motion of the torch in the glovebox in both the X and Z dimensions. This motion was useful for optical alignment of the ICP with the analysis spectrometers. Containment was maintained by the use of two flexible Gortiflex bellows (A&A Mfg. Co., New Berlin, WI) that surrounded the RF feedthrough tube, both inside and outside of the glovebox. The torch box exhaust was connected to a fire-retardant, Class I HEPA filter (Keewaunee, #2C-2276) with another Gortiflex bellows followed by a section of stainless-steel pipe. This pipe contained an in-line probe (T-Sentry-Q, Hampshire Controls, Portsmouth, NH) that continuously monitored the exhaust gas temperature and automatically turned off the RF power generator when the temperature exceeded a set limit (~60C). An ordinary glass/mercury thermometer was suspended in the glovebox to monitor the ambient temperature. No excessive heat build-up was noted during extended periods of operation.

The glovebox was equipped with a 25 CFM blower that was capable of exchanging the glovebox atmosphere approximately once per minute; fresh air was drawn first through an inexpensive fibrous filter and was then filtered by a HEPA filter prior to entering the glovebox. The specifications of the HEPA filters used in this installation were verified by the U. S. Department of Energy (DOE) Filter Certification Laboratory at Rocky Flats, CO.

AUXILIARY CONTAINMENT AND SAFETY FEATURES

The glovebox exhaust HEPA filter removed >99.9% of the particulates that impinged upon its face, but this degree of exhaust clean-up could be insufficient to enable the facility to meet current DOE regulations for maximum permissible Pu air concentration in uncontrolled areas ($6 \times 10^{-14} \mu Ci/ml$ for all isotopes except ^{241}Pu). The glovebox exhaust gases should therefore receive further treatment to remove additional particulates before being released to the atmosphere. This clean-up was accomplished by the use of a filter train consisting of an additional HEPA filter, a small gas scrubber (90-S7284 F/S 1/16, Labconco, Kansas City, MO), and a final HEPA filter.

A series of interlocks were provided to turn off the RF power in the event of: a) an interruption of the cooling water to the RF load coil; b) an abnormally high temperature in the exhaust gas stream; and c) the opening of the torch box door during operation. The last interlock could be manually defeated. A remote "OFF" button was also located in the electronics room of the ACF. Although not implemented to date, additional interlocks will be put in place to disable operation should the outer, argon-gas flow-rate fall below a set limit or should the plasma emission unexpectedly cease.

The amount of combustible materials stored in the glovebox was kept to a minimum. Only 30-50 mg of Pu (in solution) was kept in the glovebox at any time, the remainder being stored in a fire retardant safe located close to the glovebox. The room was provided with a smoke sensor located above the glovebox that was monitored by a central alarm system. The glovebox will be outfitted shortly with a halon-charged fire extinguisher that can be either thermally or manually activated. It must be emphasized that the ICP torch produces a plasma, not a flame. The plasma is sustained in argon, an inert, non-combustible gas.

TESTING

The glovebox housed ICP was moved to the ACF late in November, 1982; the first plutonium solution was introduced to the ICP on June 24, 1983. The intervening period was devoted to the testing, redesign and refabrication of the apparatus.

The first instrument failure, caused by a machinist's oversight, pointed out a serious design flaw, namely, a Teflon insulator that had been damaged could not be replaced without breaking containment. This necessitated a redesign of the apparatus and further testing.

When the present design was achieved the facility was tested for over 40 hours at plasma power levels ~10-15% higher than are normally used for ICP work (~1 KW) before it was deemed acceptable.

In-Place HEPA Filter Evaluation

In-place HEPA filter testing, performed by the Ames Laboratory Health Physics group, was accomplished in an innovative fashion. Approximately 1 kg of thorium oxide was used as a "thoron" generator and, after a suitable decay interval, the decay product ^{212}Pb was monitored before and after the glovebox exhaust HEPA filter. This test verified that the exhaust filter, after >50 hours operation at elevated temperatures, was maintaining >99.9% efficiency. After the above testing was successfully completed, limited amounts of ^{242}Pu, a relatively low-activity Pu isotope, were introduced into the ICP. Air samplers were employed to monitor the room activity and the activity in the air stream directly above the glovebox exhaust HEPA filter during a two-hour ICP run. The chief source of activity in the 99.93% pure ^{242}Pu sample resulted from a 0.035% ^{241}Pu impurity that could be readily detected by gamma ray spectroscopy. No ^{241} activity was detected in samples taken at the two sampling sites indicating that the HEPA filter was working efficiently and also that additional mechanisms, such as "plating out" on the exhaust tubing, may be operating to further cleanse the exhaust stream of particulates.

OPERATION NOTES

The glovebox described above is the only location at the Ames Laboratory currently cleared to handle Pu, consequently some chemical activities, such as sample dissolution and weighing of samples must be performed within the glovebox. Thus, the glovebox volume must be efficiently used to perform the above operations and to store the apparatus needed to support these operations. The nebulizer chosen for this work has a simple concentric design (TR-50-C0.5, JE

Meinhard Assoc., Santa Ana, CA) that features a low uptake rate (~0.3 ml/min) to minimize sample solution volume and radioactive particulate generation. A peristaltic pump (Minipuls 2, Gilson Med. Electronics, Middletown, WI) is directly connected to the nebulizer, but is connected to the sample reservoir and rinse solution through a remotely switchable 3-way valve (38-082 Stream switching valve, Rainin Inst. Co., Woburn, MA). In this manner the ICP can be turned on and the optics aligned while water is nebulized into the plasma. After the Pu solution is put in place, the operator can leave the room and the Pu remotely connected to the nebulizer for introduction to the ICP. Argon gas flows are controlled by a three-channel mass flow controller (MKS Instruments Inc., Burlington, MA).

The light emitted by the ICP is collected by a lens contained within the glovebox and focused onto the slit of a 3 meter spectrometer (THR-1500, Jobin-Yvon (ISA), Longjumeau, France). Examples of the spectra produced by this instrument are shown in Figure 3.

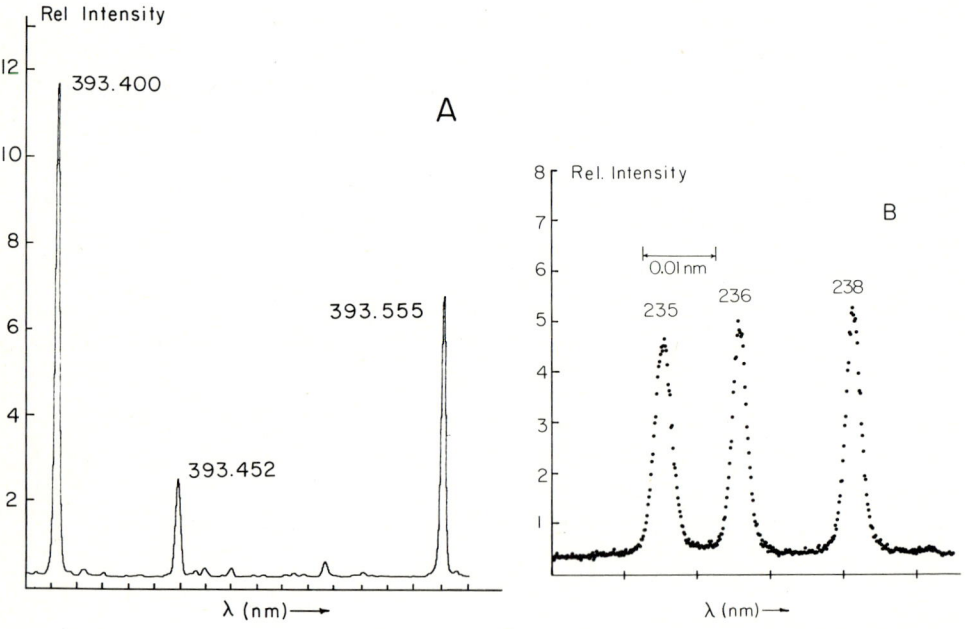

Fig. 3. A) Spectrum of ~1000 ppm Pu-242 (99.93% isotopic purity) near 393.4 nm.
B) Spectrum of ~650 ppm U-isotope mixture, U-238 peak at 424.437 nm, U-235/U-236 splitting = ~0.01 nm.

ACKNOWLEDGEMENTS

The authors wish to thank the following for their very important contributions to the work described above; D. Devine and L. West (Plasma-Therm, Inc.) and E. Mainka (Kernsforschungszentrum, Karlsruhe) for discussions of ICP-glovebox interfacing; N. Cent, R. Struss, and M. Core (Ames Laboratory) for supervision of the facility design and construction; G. Wells and T. Johnson (Ames Laboratory) for the construction of the glovebox-housed ICP instrument; G. Holland and R. Prior (Ames Laboratory) for the electrical interlock system implementation; M. Voss, R. Staggs, L. Mathison, G. P. Jones, P. Millis and K. Malaby (Ames Laboratory Health Physics Group) for the constant interest; C. Pietri (New Brunswick Laboratory) for many discussions; and J. Goleb (Office of Safeguards and Security, DOE) for his patience.

W.S. Lyon (Editor), *Analytical Spectroscopy*
Elsevier Science Publishers B.V., Amsterdam — Printed in The Netherlands

PLANS FOR INDUCTIVELY COUPLED PLASMA ATOMIC EMISSION SPECTROSCOPY (ICP-AES) ANALYSIS OF IMPURITIES IN PLUTONIUM MATERIALS AT ROCKY FLATS

C. E. MICHEL AND G. E. BROWN
Analytical Lab, Rockwell International, P.O. Box 464, Golden, CO 80401 (USA)

ABSTRACT

Due to the complex emission spectra of plutonium, ICP analysis of impurities in plutonium materials requires an effective and efficient separation of the impurities of interest from the plutonium matrix. The use of dihexyl N-N diethylcarbamoylmethylenephosphonate (DHDECMP) in the extraction and separation of plutonium and americium from impurities of interest has been evaluated and shown to meet these requirements.

Topics discussed in this paper are: 1) A proposed method for the determination of impurities in plutonium metal; 2) Recovery of plutonium from DHDECMP and recycle use of DHDECMP; 3) Instrument and equipment plans.

INTRODUCTION

In recent years, the inductively coupled argon plasma spectrometer (ICP) has been developed as a technique for trace elemental analysis. Because of its advantages; including high sensitivity, ease of sample preparation, and small sample size, ICP is a potential technique for use with nuclear materials.

At Rocky Flats, an Instruments SA/JY-48 simultaneous ICP has been installed for the primary purpose of analyzing impurities in plutonium materials. A radioactive containment hood has been designed. Once the hood is installed, nuclear materials will be analyzed.

PLUTONIUM EXTRACTION BY DHDECMP

After evaluation of plutonium separation possibilities, liquid-liquid extraction was chosen, using dihexyl N-N diethylcarbamoylmethylenephosphonate (DHDECMP) as the extractant. There has been considerable interest in bifunctional organo-phosphorous compounds for separation and purification of actinide elements, primarily americium. Of these compounds, DHDECMP has proven most useful. McIsaac et al. (1) at Idaho Falls, Schulz et al. (2) at Hanford, and Navratil et al. (3,4) at Rocky Flats have been leaders in the development and use of DHDECMP.

DHDECMP complexes +3, +4, and +6 valence actinides (1), while Group I, Group II, the transition elements, and other elements of interest (i.e. Al, Si) are either weakly complexed or not complexed. Table I lists the elements of interest in plutonium metal. Presently, these elements are being analyzed by either Atomic Absorption or Emission Spectroscopy. It is estimated that these elements can be determined more quickly and, in most cases, more accurately by ICP than with our present methods.

TABLE 1
Elements of Primary Interest in Pu Metal

Al*, B, Be, Ca, Cr, Cu, Fe*, Ga*, K, Mg
    Mn, Mo, Ni*, Pb, Si*, Sn, Ti
*Elements of special importance

The procedure developed for plutonium metal dissolution and impurity separation is listed in Table 2. After the separation of the plutonium and americium from the impurities of interest, a small amount of hydrofluoric acid is added to the aqueous phase to dissolve silicon and other refractory elements.

TABLE 2
Dissolution and Separation of Plutonium with 30% DHDECMP in Xylene

1) Transfer metal sample to a 50 ml polyethylene centrifuge tube with screw cap.
2) Dissolve in minimum amount of HCl (Pu^0 to Pu^{+3}).
(4 ml of 2N HCl for 0.5 gm Pu metal)
3) After dissolution, add 10 ml of 8N HNO_3. Allow to stand 1 hour (Pu^{+3} to Pu^{+4}).
4) Add 10 ml of 30% DHDECMP in xylene.
5) Mix 30 seconds on vortex mixer.
6) After phase separation, siphon off top organic phase.
7) Repeat steps 4 through 6 using 5 ml of 30% DHDECMP in step 4.
8) Add 0.1 ml of 10N HF to aqueous phase. Mix and allow to stand for 1 hour.
9) Analyze aqueous phase for elements of interest.

The extraction of plutonium and americium using 30% DHDECMP in xylene shows excellent transfer from the aqueous phase to the organic phase. Two 30 second contacts remove >99.95% of the plutonium and 90 to 95% of the americium from a 0.5 gram plutonium sample in approximately 6N nitric acid while leaving most of the impurities in the aqueous phase. The aqueous phase can then be analyzed directly by ICP. Evaluation of the proposed procedure has been successfully demonstrated in determining impurities in depleted and normal uranium. Uranium +6 is also efficiently complexed and extracted by DHDECMP.

PLUTONIUM RECOVERY - DHDECMP RECYCLE

It is desirable to recover the plutonium and to recycle the DHDECMP because of the high cost of DHDECMP ($750 per gallon for 84% purity) and because the plutonium can be converted to a relatively pure form for easy processing and recovery.

The procedure developed for recovery of plutonium and recycle use of DHDECMP is shown in Figure 1. An equal volume of 0.75M oxalic acid - 2N nitric acid, heated to 50°C, is slowly added to the Pu-organic phase, also at 50°C, while rapidly mixing the organic phase using a magnetic stirrer. Precipitation at 50°C produces a more dense crystalline precipitate than precipitation at room temperature (5), greatly improving the filtration. A small portion of the organic phase will accompany the plutonium oxalate precipitate and appears, by phase separation, in the filtrate after the filtration and rinsing steps are completed. This smaller portion is combined with the major portion of the organic phase prior to treatment with sodium carbonate. Greater than 99.5% of the plutonium and 98% of the americium is stripped from the organic phase after one precipitation when the plutonium and americium concentrations are 33 g/L and 5×10^{-3} g/L, respectively. The sodium carbonate wash removes most of the remaining plutonium, americium, and radiolysis products while the nitric acid wash neutralizes and removes any remaining sodium carbonate and acid equilibrates the DHDECMP making it ready for recycle use.

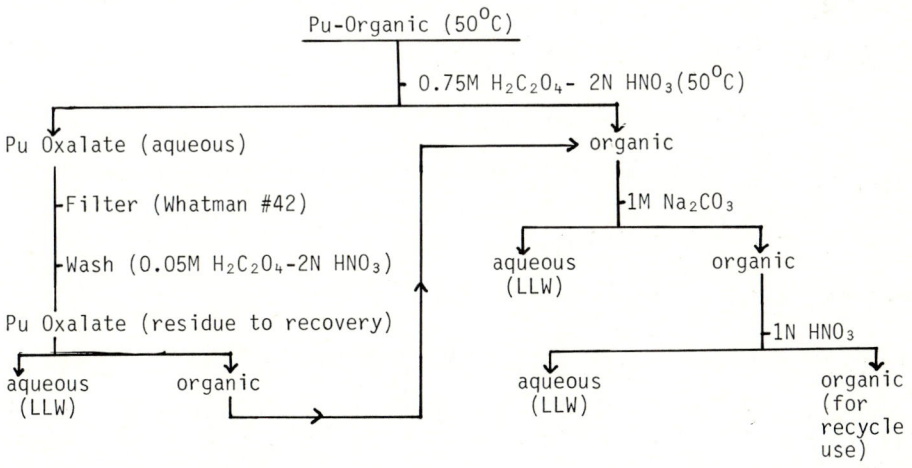

LLW = Low Level Waste
Organic = 30% (V/V) DHDECMP in xylene

Fig. 1. Recovery of Pu from 30% DHDECMP in Xylene

It is important to strip the plutonium and americium from the DHDECMP as soon as possible, preferably within 24 hours, since radiolysis does occur. The decomposition products inhibit stripping from the organic phase. The longer the plutonium and americium stay in contact with the DHDECMP, the less efficient stripping will be.

Recycle use of DHDECMP has proven successful with plutonium. Two recycle uses have shown extraction efficiencies to be as good as using fresh material. Roughly 90% of the DHDECMP can be recovered in each recycle use. The remaining 10% is lost due to solubility and phase separations.

THE INSTRUMENT

The Instruments SA/JY-48 inductively coupled plasma atomic emission spectrometer analyzes 39 elements simultaneously, including plutonium. The wavelength used for plutonium is 2996.40Å. The instrument also has Group I and external scanning monochromators, which can be set by the operator to any wavelength for additional analysis capabilities. The external scanning monochromator covers the wavelength range 2500-6500Å, while the Group I monochromator covers the range from 3000 to 11000Å.

The JY-48 has a Plasma-Therm torch assembly and radio-frequency generator. Currently, an acid resistant cross-flow nebulizer and demountable torch is being used. Samples may be introduced into the plasma either manually or using an autosampler and peristaltic pump.

RADIOACTIVE CONTAINMENT HOOD

Because of the efficient separation procedure developed, the solutions analyzed will be quite low in plutonium (<15 µg/ml) and americium (<1 µg/ml). Safety approval has been obtained to operate in an open faced containment hood that meets industry standards for design and safety. Plans are to feed the extracted sample solutions from an autosampler directly into the torch. The autosampler will be placed in an adjacent glove box, with the sample tubing passing through into the radioactive containment hood. It is hoped that this will minimize contamination in the containment hood. However, if necessary, the hood design will permit attachment of glove ports and gloves.

The radioactive containment hood will surround the torch assembly and matching network. The Group I monochromator will be inside the hood while the scanning monochromator will be outside the rear of the hood. It will be optically coupled to the plasma torch using a light pipe and quartz window. A second quartz window on the side of the box will allow light to pass from the torch into the spectrometer.

The radioactive containment hood has been designed, but not fabricated. We have, however, installed and evaluated a cardboard "mock-up" of the hood.

Cooling water for the torch will be provided by a Lepel recirculating water cooling system containing critically safe tanks. Argon will be piped into the torch assembly through a penetration in the hood wall. Exhaust air will enter through the front of the hood, then will be exhausted through a HEPA (High Efficiency Particulate Air) filter into the building contaminated air system.

CONCLUSIONS

DHDECMP has proven to be an effective complexing agent for separating plutonium and americium from many impurities. Perhaps its biggest drawbacks are cost and availability in high purity form. Recycle usage of the extractant is a practical method of overcoming the cost disadvantage. Separation of uranium from many impurities using DHDECMP and subsequent ICP analysis of the impurities has been successfully demonstrated.

FUTURE WORK

Future plans, after the radioactive containment hood is installed, include calibration of the ICP with radioactive standards, checks for plutonium interference on the channels of interest, precision studies on the DHDECMP extraction and subsequent ICP analysis, and final development of the procedure for impurity analysis of plutonium metal.

REFERENCES

1. L. D. McIsaac, J. D. Baker, J. F. Krupa, R. E. LaPointe, D. H. Meikrantz, and N. C. Schroder, Study of Bidentate Compounds for Separation of Actinides from Commercial LWR Reprocessing Waste, ICP-1180, May 1979.
2. W. W. Schulz and L. D. McIsaac, Bidentate Organophosphorus Extractants: Purification, Properties, and Applications to Removal of Actinides from Acidic Waste Solutions, ARH-SA-263, May 1977.
3. J. D. Navratil and W. W. Schulz (Eds.), Actinide Separations, ACS Symposium Series 117 - 1980.
4. J. D. Navratil - Rocky Flats Plant, Rockwell International, Personal Communication - Summer 1982.
5. L. L. Martella - Rocky Flats Plant, Rockwell International, Personal Communication - Summer 1982.

NUCLEAR

UTILIZATION OF THE INTENSE PULSED NEUTRON SOURCE (IPNS) AT ARGONNE NATIONAL LABORATORY FOR NEUTRON ACTIVATION ANALYSIS*

R. R. HEINRICH,[1] L. R. GREENWOOD,[1] R. J. POPEK,[1] and A. W. SCHULKE, JR.[2]

[1]Chemical Technology Division, Argonne National Laboratory, 9700 South Cass Avenue, Argonne, Illinois, 60439 (U.S.A.)

[2]IPNS Division, Argonne National Laboratory, 9700 South Cass Avenue, Argonne, Illinois, 60439 (U.S.A.)

ABSTRACT

The Intense Pulsed Neutron Source (IPNS) neutron scattering facility (NSF) has been investigated for its applicability to neutron activation analysis. A polyethylene insert has been added to the vertical hole VT3 which enhances the thermal neutron flux by a factor of two. The neutron spectral distribution at this position has been measured by the multiple-foil technique which utilized 28 activation reactions and the STAYSL computer code. The validity of this spectral measurement was tested by two irradiations of National Bureau of Standards SRM-1571 (orchard leaves), SRM-1575 (pine needles), and SRM-1645 (river sediment). The average thermal neutron flux for these irradiations normalized to a 10 μamp proton beam is 4.0×10^{11} n/cm^2-s. Concentrations of nine trace elements in each of these SRMs have been determined by gamma-ray spectrometry. Agreement of measured values to certified values is demonstrated to be within experiment error.

INTRODUCTION

The motivation of this study was initially generated several years ago at the closing and decommissioning of Argonne's CP-5 reactor. Although this facility was primarily utilized for basic neutron research, it did provide the analytical chemist with the very sensitive and powerful capability of neutron activation analysis. Because of economic considerations, the Intense Pulsed Neutron Source (IPNS) design could not consider sample activation as a high priority item. However, several neutron beam holes which extended into the IPNS target reflector region were provided and these, although limited in sample volume and neutron flux, could conceivably be used for sample activation.

The challenge in using this high energy accelerator-based neutron source focuses upon knowing the neutron spectral distribution within the sample volume since neutrons of all energies are generated within the target. Thus, before meaningful activation analysis could be done, careful neutron dosimetry of the neutron spectrum had to be accomplished. This paper describes those efforts

---

*This work was performed under the auspices of the U.S. Department of Energy.

and indeed demonstrates that meaningful neutron activation analysis can be done at the IPNS facility.

INTENSE PULSED NEUTRON SOURCE (IPNS)

The IPNS is based upon a rapid cycling synchrotron (RCS) which accelerates protons to 500 MeV (ref. 1). The RCS produces 100-ns pulses of protons at repetition rates up to 30 Hz. The protons from the accelerator are transported to one of two target assemblies, the Radiation Effects Facility (REF) and the Neutron Scattering Facility (NSF), which have been optimized for radiation effects and neutron scattering, respectively. The proton beam time is divided between the two facilities such that the NSF receives the proton beam about 75% of the operating time. For this reason, two vertical irradiation thimbles were installed in the NSF. The target in each assembly is made of 238-U. Neutrons with all energies up to the incident proton energy are produced by the processes of spallation and fission, at a rate of $\sim$20 neutrons/proton.

The REF (ref. 2) also consists of a 238-U target, two vertical irradiation thimbles, and a horizontal irradiation thimble, all surrounded by a Pb neutron reflector. The neutron reflection and the high-energy (n, 2n) reactions in Pb increases the neutron flux and also reduces the neutron gradient in a direction perpendicular to the target axis. The two vertical irradiation thimbles are located on either side of the target at positions of maximum flux. Each has a liquid He cryostat which can operate at temperatures between 4 and 1000 K. The horizontal irradiation thimble is located on an axis parallel to and directly below the target and operates at ambient temperatures. The neutron spectrum in the REF is considerably harder than in the NSF in order to maximize damage in materials which are under study.

In the NSF, the fast neutrons produced by fission and spallation reactions in the uranium target are thermalized by moderators (with a high hydrogen density) located above and below the target. Two types of moderators have been used. The first type consisted of four ambient-temperature polyethylene moderators with an inner graphite reflector. The second type consisted of four 100 K circulating liquid methane moderators with an inner 100 K beryllium reflector. Currently the polyethylene-type moderator is being used. Twelve horizontal beam tubes and one vertical neutron beam tube view selected faces of the moderators to provide neutron beams to instruments. Each horizontal beam tube contains a 36-inch thick steel beam gate. Two vertical holes (VT3 and VT4) and two horizontal holes (F-6 and H-2) may be used for sample irradiations. A polyethylene insert has been added to the vertical hole VT3 to enhance the neutron flux and has resulted in an increase of the thermal component by a factor of two.

NEUTRON SPECTRUM DETERMINATION

The multiple-foil activation technique was used to determine the neutron energy distribution in the VT-3 position with and without the polyethylene liner. Twenty-eight different neutron-activation reactions were measured at the maximum flux location. The foil activities were measured with HPGe and Ge(Li) detectors over decay periods of several half-lives for each reaction. Peak integrations and Compton-background subtractions were done by means of standard computer codes. Corrections for cover foils, neutron and gamma self-shielding, and decay during and after the irradiation were made for the activation products. The STAYSL computer code (ref. 3) was used to adjust the neutron spectrum (100 energy groups) by fitting the foil activities. This method calculates the activity for each activation reaction using ENDF/B-V (ref. 4) nuclear cross-sections and an initial input neutron spectrum calculated with the HETC (ref. 5) computer code. The spectral adjustment procedure follows a least-squares approach where uncertainties and covariances are assigned to the measured activities, cross sections, and starting flux spectrum. The output flux spectrum and covariance matrix are determined by a matrix inversion procedure which minimizes the chi-square parameter. The neutron spectrum for the VT3 polyethylene lined hole is shown in Fig. 1. Collapsed neutron spectra of both the poly and bare VT3 is presented in Table 1. Both are normalized to a 10-μamp proton beam.

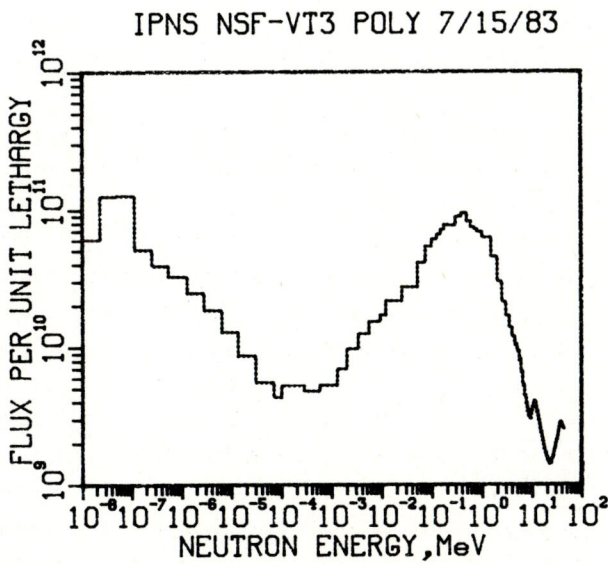

Fig. 1. Neutron spectrum IPNS NSF VT3 (poly) vertical hole.

TABLE 1

IPNS-NSF-VT3 neutron fluxes normalized to a 10-µamp proton beam current

Energy, MeV	Flux, x 10^{11} n/cm^2-s	
	Bare	Poly
Total	8.67	10.17
>0.1	3.00	3.28
>1.0	0.893	0.880
Thermal (<0.5 eV)	2.00	4.42
0.5 eV - 0.1	3.65	2.50
0.1 - 1.0	2.10	2.40
1.0 - 5.0	0.805	0.755
5.0 - 10.0	0.0418	0.0521
10.0 - 20.0	0.0243	0.0268
20.0 - 30.0	0.00811	0.00874
30.0 - 44.0	0.0119	0.0131
>44.0[a]	0.0262	0.0287

[a] Calculated HETC (ref. 5).

EXPERIMENTAL

Standard Reference Materials (SRM)

Three National Bureau of Standards Standard Reference Materials (SRM) were used for the determination of various trace element concentrations. These were SRM-1571 (orchard leaves), SRM-1575 (pine needles), and SRM-1645 (river sediment). Sample sizes ranged from 240 mg for the smallest sample to 480 mg for the largest. SRM-1571 and SRM-1575 were oven dried at 85°C for about two hours and then stored in a desiccator until weighed and encapsulated. SRM-1645 was oven dried at 50°C for about two hours and also desiccator stored. Samples were prepared for the irradiations by encapsulating in 1-cm (dia.) by 2.5-cm polyethylene vials that had been previously cleaned in distilled water and ethanol. After the vials were clean, handling during the weighing and packaging processes was done using clean polyethylene gloves.

Flux monitors of nickel and cobalt were placed in horizontal positions at the bottom, middle, and top of the irradiation package, which consisted of each of the SRMs and an empty (blank) vial. Vertical flux monitors of iron, nickel, and cobalt were strategically placed in the interior of this package so that correlations could be made with the horizontal monitors and also measure the vertical flux gradients. All flux monitors were encapsulated in polyethylene tubing to avoid sample cross-contamination.

Irradiation Conditions

Two irradiations were made in the polyethylene-lined vertical tube VT3 of the NSF. The first was a relatively short irradiation of 16.9 hours which enabled the determination of short-lived activation products. The second irradiation spanned 12.0 days and enabled the determination of long-lived activation products. The time-averaged neutron flux of the short irradiation was calculated to be 7.47×10^{11} n/cm^2-s and for the long irradiation, 6.41×10^{11} n/cm^2-s. The measured thermal neutron fluxes for the two irradiations were 3.2×10^{11} and 2.8×10^{11} n/cm^2-s, respectively. Average proton beam currents were measured to be 8.64 μamps for the short and 6.54 μamps for the long irradiation.

Gamma-Ray Spectrometry

The counting of the SRM samples from the short irradiation started about four hours after the end of the irradiation and utilized three absolutely calibrated germanium detectors (30-45 cm^3). Each sample was counted on all detectors to reduce any calibration biases. Counting of each sample for the particular nuclide of interest was repeated at least three times within three half-life periods from the end of the irradiation. The principal gamma-ray of each nuclide was counted for sufficient length of time to provide counting statistics ranging from 0.1-5%, and the precision for replicate counts was required to be within the 5% range. Counting geometries were adjusted such that errors due to dead-time and random summing were negligible. Peak integration and identification was done by conventional computer techniques (ref. 6).

RESULTS AND DISCUSSION

The comparison of measured SRM trace element concentrations to NBS values is presented in Table 2. The excellent agreement demonstrates primarily the accuracy of the neutron dosimetry and the methodology used in determining the IPNS neutron spectrum. Uncertainties on the measured values are estimated to be ±7% which includes estimated errors in the neutron cross-sections, spectrum unfolding, detector calibration, counting statistics, and neutron flux gradient corrections. Of the nine trace elements listed, only the As, Rb, and Sb appear to be biased high. This could be attributed to the reaction cross sections used since uncertainties for these reactions are listed at about twice the magnitude as uncertainties for the reactions for the other elements.

Conclusions are that meaningful neutron activation analysis can be performed at the IPNS in the VT3 (poly) position, but this is contingent upon the fact that adequate neutron dosimetry be provided also.

TABLE 2

Comparison of measured to certified trace element concentrations

	Element Concentration, ppm unless indicated					
	SRM-1571 (Orchard Leaves)		SRM-1575 (Pine Needles)		SRM-1645 (River Sediment)	
Element	Value	This Work	Value	This Work	Value	This Work
Na	82 ± 6	82 ± 6	--	38 ± 3	5500[a]	4892 ± 342
Sc	(.06)[b]	0.06 ± 0.004	0.03[a]	0.038 ± 0.003	2[a]	1.9 ± 0.1
Cr	2.3[a]	2.5 ± 0.2	2.6 ± 0.2	2.3 ± 0.2	2.96 ± 0.28%	2.99 ± 0.21%
Mn	91 ± 4	89 ± 6	675 ± 15	654 ± 46	785 ± 97	795 ± 56
Fe	300 ± 20	290 ± 20	200 ± 10	188 ± 13	11.3 ± 1.2%	10.4 ± 0.7%
Co	0.2[a]	0.27 ± 0.02	0.1[a]	0.14 ± 0.01	8[a]	8.1 ± 0.6
As	14 ± 2	18 ± 1	0.21 ± 0.04	--	66[a]	109 ± 8
Rb	12 ± 1	17 ± 1	11.7 ± 0.1	16.5 ± 1.2	--	58.5 ± 4.1
Sb	--	5.7 ± 0.4	0.2[a]	0.4 ± 0.03	51[a]	60 ± 4

[a] Uncertified; information value only.
[b] Ref. 7.

REFERENCES

1. C. W. Potts, IEEE Trans. Nucl. Sci., NS-28 (1981) 2104.
2. R. C. Birtcher, T. H. Blewitt, M. A. Kirk, T. L. Scott, B. S. Brown, and L. R. Greenwood, J. of Nucl. Mater., 108 (1982) 3-9.
3. F. G. Perey, Least Squares Dosimetry Unfolding: The Program STAYSL, ORNL-TM-6062 (1977); modified by L. R. Greenwood (1979).
4. Evaluated Nuclear Data File, Version V, National Neutron Cross-Section Center, Brookhaven National Laboratory (1979).
5. K. C. Chandler, T. W. Armstrong, Operating Instructions for the High-Energy Nucleon-Meson Transport Code (HETC), ORNL-4744, Oak Ridge National Laboratory (1972).
6. R. Gunnink and J. B. Niday, Computerized Quantitative Analysis of Gamma-Ray Spectrometry, UCRL-51061, Vol. I-V, Lawrence Livermore Laboratory, March, 1972.
7. E. T. Kucera and R. R. Heinrich, J. Radioanalytical Chem., 32 (1976) 137.

HEAVY-ION INDUCED X-RAY SATELLITE EMISSION AS A CHEMICAL PROBE

T.M. ROSSEEL, J.M. DALE, L.D. HULETT, H.F. KRAUSE, S. RAMAN, C.R. VANE and
J.P. YOUNG
Oak Ridge National Laboratory, Oak Ridge, TN 37830

ABSTRACT

Advances in energy technology often require corresponding advances in materials fabrication and characterization. Among the new techniques being developed for the improved characterization of materials is an x-ray fluorescence method which uses heavy ions for excitation. High resolution measurements of heavy-ion excited x-ray spectra have revealed a series of prominent satellite lines in addition to the normal emission lines. It has been shown that these satellites display intensity variations indicative of the chemical state or environment of the emitting target atom and the projectile velocity. In order to evaluate heavy-ion induced x-ray satellite emission (HIXSE) as a chemical probe, we have examined a series of sulfur compounds and titanium, vanadium and molybdenum alloys and compounds. Results will be presented which demonstrate the chemical sensitivity of this technique, the range of elements which can be analyzed and the potential for applications to real chemical and materials problems.

INTRODUCTION

New materials with improved physical properties such as strength and ductility are being developed for use in the field of energy technology. Future advances in this field may depend upon the characterization of not only the elemental composition of these materials but also the chemical nature or bonding of their constituents. Among the new techniques being developed for the chemical characterization of materials is an x-ray fluorescence method which uses fast, heavy-ions for excitation. In this paper, the analytical potential of heavy-ion induced x-ray satellite emission (HIXSE) as a probe of the chemical environment of an element in a solid will be examined in terms of its chemical sensitivity, the range of elements which can be analyzed and by the application of the technique to chemical and materials problems. Future trends for HIXSE will also be discussed.

---
*Research sponsored by the Office of Energy Research, U. S. Department of Energy, under Contract W-7405-eng-26 with the Union Carbide Corporation.

Chemical Sensitivity

The relaxation of an atom with a core level vacancy can occur through the emission of an x-ray. Without any knowledge of how the vacancy was produced, the identity of the emitting atom can be determined by the characteristic energy of the radiation. For example, an emission line at 4.93 keV corresponds to the K_β (1s → 3p) transition of titanium. The intensity of the line, however, depends in part upon the efficiency or cross section for producing a K-vacancy in the titanium target. While different excitation sources such as electrons, photons, and charged particles have different x-ray production efficiencies, each has unique analytical advantages. Fast (10^9 cm/sec) heavy-ions, for example, have a high cross section for producing multiple inner and outer shell vacancies. This results in a dramatic increase in the relative intensity of the satellite lines associated with the normal diagram line. Fig. 1 shows the high resolution titanium K_β satellite spectrum produced by the bombardment of titanium metal with 36 MeV Cl ions extracted from the Oak Ridge (model EN) tandem Van de Graaff. The peaks in this spectrum correspond to the Ti K_β x-rays emitted from configurations having n L-shell vacancies. As shown in Fig. 2, the analytical advantage of multiple vacancy production is that the intensity distribution of the satellite lines is sensitive to the chemical environment of the emitting atom. The Ti $K_\beta L^n$ satellite spectra of Ti metal and TiO_2 have been corrected for mass absorption effects, projectile energy loss, and concentration effects.

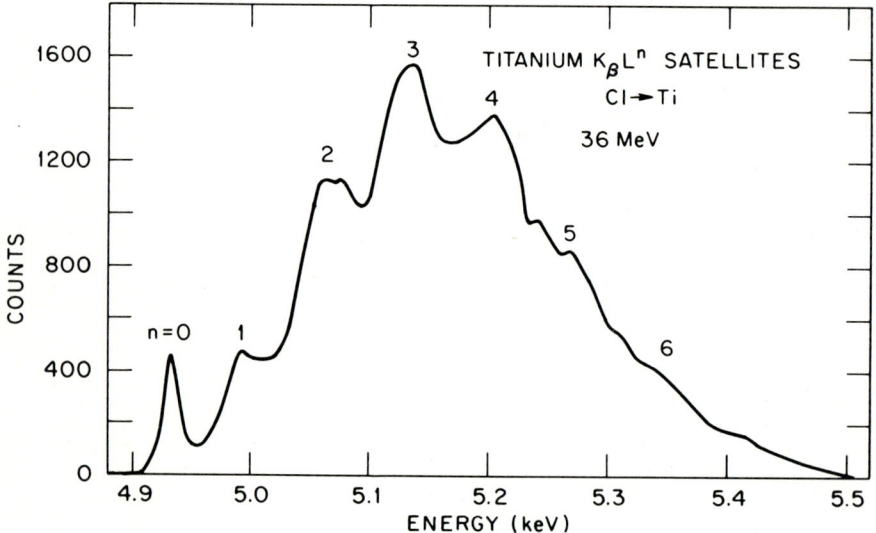

Fig. 1. The corrected titanium $K_\beta L^n$ satellite spectrum produced by the collision of 36 MeV Cl ions with titanium metal.

The observed spectral variations between Ti metal and TiO_2 are the consequence of differences in the refilling rates of the L vacancies prior to the emission of the $K\beta$ x-ray. Because the outer shell electrons are also removed during the collision process, the refilling of the L shell is believed to proceed through direct inter-atomic Auger transitions or by inter-atomic electron transfer to the valence level of the ion followed by rapid intra-atomic Auger cascading (ref.1,2). A change in the chemical environment or valence electron distribution will alter the electron transfer rates and thus change the observed satellite x-ray yield.

Initial attempts to correlate the $K\alpha L^n$ satellite intensity distribution of third row elements with the chemical environment used a parameter called the average L-vacancy fraction P_L (ref.2). It is defined as the <u>apparent</u> number of L-vacancies as determined by the weighted sum of the relative satellite intensities divided by the total number of possible vacancies. This parameter was shown to be linearly related to the bulk valence electron density (the number of valence electrons/$Å^3$) of the target material. We have recently shown in a study of the chemical effects on the $K\alpha L^n$ satellites of a series of sulfur compounds that individual satellite lines are more strongly influenced by the chemical environment than the aggregate satellite intensity distribution (ref.3).

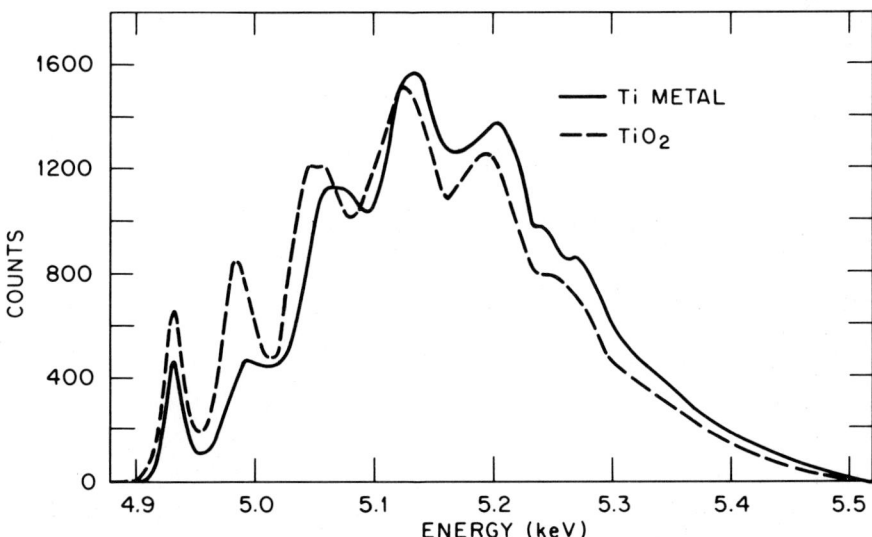

Fig. 2. The corrected titanium $K\beta L^n$ satellite spectra produced by the collision of 36 MeV Cl ions with titanium metal (solid line) and TiO_2 (dashed line).

Range of Elements

The utility of HIXSE as a chemical probe depends in part upon the range or variety of elements for which this technique is sensitive. Although it has been conclusively shown that the K_α transition is sensitive to the chemical environment of the second and third row elements, few studies have explored other regions of the periodic table (ref.4). Because the sensitivity of the K_α transition decreases as the number of shells between the L-shell and valence shell increases, the expansion of this technique to the transition metals, for example, is contingent upon the evaluation of the chemical shifts associated with other transitions. We have begun the necessary evaluation by exploring the $L_{\alpha,\beta}$ transition of molybdenum (ref.5) and the K_β transition of titanium and vanadium.

The corrected molybdenum L x-ray spectra produced by the collision of 1 MeV/amu Cl ions with various molybdenum bearing targets displayed variations in the intensity distribution of the L_α and L_β satellites ($L^n M^m$). Amplification of those variations was obtained by calculating difference spectra using the molybdenum metal spectrum as the subtrahend. It was found that a specific region of the difference spectra displayed a systematic correlation with the bulk valence electron density as shown in Fig. 3. It can be seen that

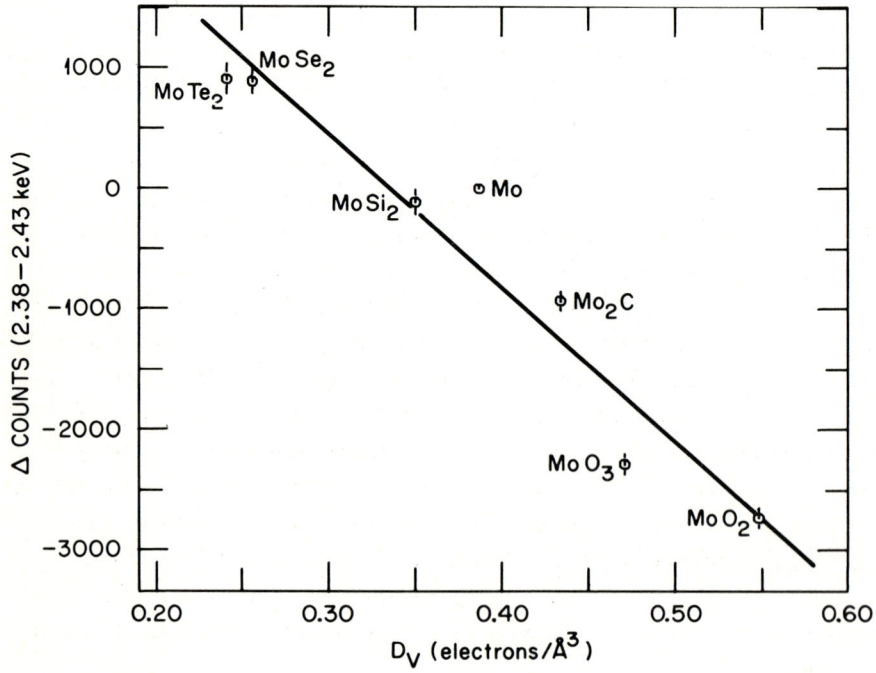

Fig. 3. The integrated difference counts in the 2.38-2.43 keV region vs the valence electron density of molybdenum compounds.

as the bulk valence electron density increases, the x-ray yield, relative to molybdenum metal, decreases. This indicates that the refilling mechanism, in agreement with the K_α results for second and third row elements (ref.2), occurs through inter-atomic electron transfer. Clearly the L x-rays can be used to probe the chemical state or environment of intermediate Z elements.

The heavy-ion excited K_β spectra of a series of titanium and vanadium targets like the titanium spectra shown in Fig. 2, exhibited variations in their satellite intensity distribution for different chemical states. Systematic correlations await improved counting statistics.

Applications

The application of HIXSE to the characterization of the chemical environment of the constituents of alloys is another component of its development as an analytical probe. We have explored this avenue by examining a series of binary molybdenum alloys and a complex group of vanadium alloys (ref.6). In addition to observing variations in the L x-ray satellite spectra of molybdenum compounds, variations were also detected in the spectra of a series of molybdenum-titanium bcc alloys. We found, however, that a strong correlation with the bulk valence electron density exists only for a different region of the satellite difference spectra than that observed for compounds. The slope of the correlation was also reversed. These observations may be due to differences in the distribution of valence electrons between bonding partners. The valence electrons involved in metallic bonding, for instance, are more delocalized than those involved in ionic or covalent bonding. In addition, these systematic variations may be delineating metallic valence states. Just as a valence state can be used to describe inorganic compound formation, metal valence states may be useful for describing and predicting bonding and crystal formation of alloys. Future studies will explore this possibility.

A more severe test of the ability of HIXSE to probe the valence environment of alloys was the examination of the K_β spectra of a group of long range ordered (LRO) alloys with the same composition ($V(Co_{.43}Ni_{.57})_3$) but different crystal structures. These materials are of interest because in the cubic form they possess unusual strength and ductility at high temperatures. Using the difference spectra approach described previously, we have shown that HIXSE can differentiate between an LRO alloy in a disordered cubic (fcc) form and an ordered hexagonal ($BaPb_3$) form. Although it won't replace diffraction methods, the ability to detect changes in the valence electron distribution due to crystallographic alterations demonstrates the potential of HIXSE to probe metallic bonding in alloys.

Future Trends

The future of heavy-ion induced x-ray satellite emission (HIXSE) as a chemical probe will depend upon the expansion of the work described in this report. We will continue to use HIXSE to examine metallic bonding and metal valence states. Specifically, we will attempt to probe the valence states of a series of lanthanide metals by examining their L x-ray satellite spectra. The L x-ray spectra of zirconium alloys and compounds will also be evaluated. We expect these results to concur with the molybdenum results. In order to extend the range of elements which can be probed, we will appraise the chemical sensitivity of the M-transition for high Z elements (>68). New applications will also be evaluated. It was recently shown by Deconninck and Van den Broek (ref.7) that the heavy ion excited fluorine x-ray satellite spectra could be used to determine the formation of FeF_3 produced by F^- implantation of iron. This suggests that implanted ions could serve as a probe of the chemical environment of a host lattice. By judiciously choosing the probe ion, the sensitivity of HIXSE to the host could be maximized.

Finally, preliminary studies (ref. 8) have indicated that the chemical sensitivity of the $K\alpha$ satellite spectra could be enhanced by increasing the resolution of the x-ray spectrometer. The satellite lines shown in Fig. 1 are actually composed of a large number of multiplet lines corresponding to the angular momentum coupling of the unpaired electrons. These individual transitions should be even more sensitive to changes in the refilling mechanism induced by the chemical environment than the peaks representing the overall vacancy configuration. A high resolution spectrometer is currently under development.

References

1 R.L. Watson, T. Chiao and F.E. Jenson, Phys. Rev. Lett. 35 (1975) 254-257.
2 R.L. Watson, A.K. Leeper, B.I. Sonobe, T. Chiao and F.E. Jenson, Phys. Rev. A15 (1977) 914-925.
3 C.R. Vane, L.D. Hulett Jr., S. Kahane, F.D. McDaniel, W.T. Milner, G.D. O'Kelley, S. Raman, T.M. Rosseel, G.G. Slaughter, S.L. Varghese and J. P. Young, Proceed. Third Int. Conf. on Particle Induced X-ray Emission (PIXE), Nucl. Instrum. Methods (in press).
4 S. Raman and C.R. Vane, Proceed. Third Int. Conf. on Particle-Induced X-ray Emission (PIXE), Nucl. Instrum. Methods (in press).
5 T.M. Rosseel, J.M. Dale, H.W. Dunn, L.D. Hulett Jr., S. Kahane, H.F. Krause, S. Raman, G.G. Slaughter, C.R. Vane and J.P. Young, Proceed. Third Int. Conf. on Particle-Induced X-ray Emission (PIXE), Nucl. Instrum. Methods (in press).
6 T.M. Rosseel, J.P. Young, J.M. Dale, L.D. Hulett Jr., H.F. Krause, S. Raman C.R. Vane, A. Das Gupta, and C.T. Liu, J. Phys. F (submitted).
7 G. Deconninck and S. Van den Broek Conf. Applications of Accelerators in Research and Industry, IEEE Trans. Nucl. Sci., NS-28 (1980) 1404-1406.
8 S. Raman, E. Kallne, J. Kallne, T. Nakajima, C.W. Nestor Jr., P.H. Stelson, C.R. Vane and T.A. Walkiewicz, Proceed. Third Int. Conf. on Particle-Induced X-ray Emission (PIXE), Nucl. Instrum. Methods (in press).

ANALYTICAL MEASUREMENTS OF ACTINIDE MIGRATION IN A LABORATORY-SIMULATED
BASALT HLW REPOSITORY*

D. L. BOWERS, T. J. GERDING, S. M. FRIED, G. F. VANDEGRIFT, AND M. G. SEITZ
Chemical Technology Division, Argonne National Laboratory, 9700 South Cass
Avenue, Argonne, Illinois, 60439, U.S.A.

ABSTRACT

Described are the analytical methods and results used to determine the migration of actinides in a flowing-groundwater laboratory-simulation of a basalt repository for high-level nuclear waste. The radiochemical methods developed and employed to measure actinide activities in groundwater samples and component surfaces include the use of gamma, X-ray, and alpha spectroscopy, with associated preparation and calibration. These methods include double isotope spiking with Np-235 and Pu-236 for determination of Np-237 and Pu-239 in groundwater, gamma-spectroscopic measurements of rock surfaces for spatially profiling neptunium activity, and a leaching procedure for quantitating the actinide activities on those rock surfaces.

INTRODUCTION

Dispersal of actinides by flowing groundwater is the major concern related to the isolation of high-level nuclear waste (HLW) in a basalt repository. The analytical methods described in this paper are ones developed for an experimental study at Argonne National Laboratory (ANL) on the effect on radionuclide migration of interactions between flowing, simulated groundwater and basalt repository components in laboratory analog experiments. The term "analog" is used because the repository components are placed in the water stream analogously to the configuration that would occur from a hydraulic breach of the repository. Radionuclide migration from a breached nuclear-waste repository by groundwater flow depends on the leaching of radionuclides from solid waste and on the chemical reactions that occur as a radionuclide moves away from the repository. Therefore, migration involves the interactions of leached species and the groundwater components with (1) the waste form and canister, (2) the engineered barrier, and (3) the geologic materials surrounding the repository. Some of these interactions would occur in the radiation and thermal fields centered on the solidified waste. Rather than

---

*Work supported by the Nuclear Regulatory Commission under Contract FIN A-2230.

trying to predict what the important interactions are and then studying them individually, we consider a combination of potential interactions, using these analog experiments.

A schematic of the apparatus for an analog experiment is shown in Fig. 1. In an experiment, groundwater is pumped through the first vessel, which contains basalt chips, bentonite, and the glass waste form; then through the second vessel, which contains more bentonite and basalt chips; and then through a narrow basalt fissure in the third vessel.

Six analog experiments, with a duration of approximately 120 days each, have been run to test effects on radionuclide migration of materials of apparatus construction, gamma radiation, and hydrothermal aging of repository components. Samples of groundwater exiting the experiment at the outlets of the rock core vessel and vessels one and two were taken periodically during each experiment and at the end of each experiment. After the apparatus was disassembled, the rock core and bentonite samples from various places along the flow path were analyzed for radionuclide content.

The following sections describe three analyses that were employed in this study. These analyses are (1) $[^{237}Np]$-$[^{239}Pu]$ measurements in solution,

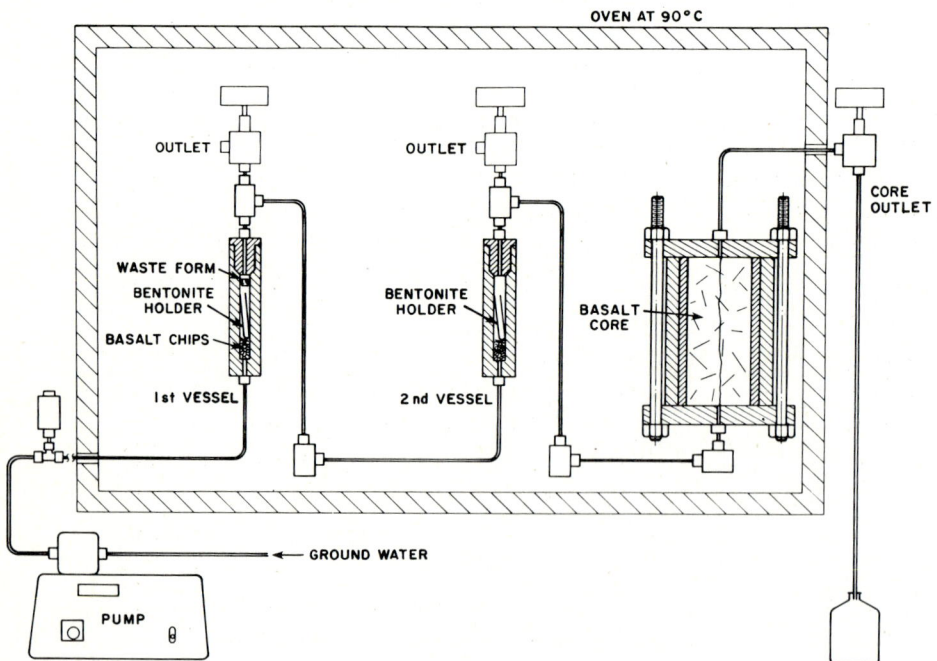

Fig. 1. Apparatus for analog experiments

(2) gamma-spectroscopic measurements of rock surfaces for spatially profiling neptunium-237 activity, and (3) a leaching procedure for quantitating the actinide activities on those rock surfaces. Comprehensive discussions of these experiments and the results of this program can be found elsewhere (refs.1-3).

METHODS AND RESULTS

Determination of [^{237}Np] and [^{239}Pu] in Groundwater

To determine the concentrations of ^{237}Np and ^{239}Pu in the groundwater, it is necessary to separate these elements from other dissolved solids in solution because, upon evaporation, these solids (∼730 mg/L) interfere with determining the alpha disintegration rate using a silicon barrier alpha pulse height detector. The separation procedure is a traditional solvent extraction technique (ref.4) using methyl isobutyl ketone (hexone) as the organic solvent. Because, as with many chemical separations, 100% recovery cannot be assumed, it is necessary to spike the solution with a different isotope of the same element to monitor the recovery of a radionuclide. The spikes chosen for these analyses were ^{235}Np and ^{236}Pu. The ^{235}Np has a relatively long half-life (396 days) and can be easily detected by its 13.6 keV X-ray. The ^{236}Pu ($t_{1/2}$ = 2.8 year) has a high energy alpha decay of 5.7 MeV which is easily resolved from the ^{239}Pu (5.1 MeV) and ^{237}Np (4.8 MeV).

The choice of ^{235}Np over the more conventional ^{239}Np as a yield monitor was made because of its availability and relative ease of detection (ref.5). Neptunium-239 has to be milked from its parent ^{243}Am and its short half-life ($t_{1/2}$ = 2 days) necessitates the procedure to be completed quickly. (The characteristic gamma-ray decay of the ^{239}Np spike is counted to measure the chemical yield of neptunium.)

Although ^{235}Np spiking eliminates most of the problems associated with that of ^{239}Np, there is still one precaution that must be addressed, its use. A sufficient amount of the ^{235}Np must be used for good counting statistics and to overshadow the contribution to the ^{235}Np signal by X-rays emitted from the other actinides that may be present, *i.e.*, the 13.3 keV X-ray emitting ^{237}Np with a branching ratio of 59%, ^{236}Pu with a 13.6 keV X-ray and branching ratio of 13%, and ^{234}Pu with a 13.6 keV X-ray and branching ratio of 4.4%. Consequently, care must be taken to limit the amount of these interfering isotopes in the analysis. Fortunately, with the high sensitivity and low backgrounds of alpha pulse counting, it is possible to measure low activities of these isotopes with high accuracy.

The procedure is as follows: An aliquot (varied from 0.1 to 50 mL, depending on the expected activities of ^{237}Np and ^{239}Pu) was spiked with a known amount of ^{235}Np and ^{236}Pu. Three drops of concentrated HF, three drops of concentrated HClO$_4$ and 2-3 mL of $8M$ HNO$_3$ were added. The HF-HClO$_4$ were added to ensure isotopic exchange of the actinides. The sample was evaporated to dryness in a Teflon beaker; $8M$ HNO$_3$ was used to redissolve the sample and it was again taken to dryness. The solids were again dissolved in 0.5-0.7 mL $8M$ HNO$_3$ and added to ~6 mL $2.8M$ Al(NO$_3$)$_3$ salting solution containing 0.1 mL $0.25M$ KMnO$_4$ (the KMnO$_4$ was added to ensure that neptunium and plutonium were both in the (VI) oxidation state necessary for extraction). The resultant solution was mixed on a Vortex mixer with a milliliter of hexone for a minimum of three minutes and then centrifuged. The hexone phase was separated from the aqueous phase and dried on a planchet, taking care that none of the aqueous salting solution was on the counting plate. The planchet was then measured for ^{235}Np content (using a high-purity, high-efficiency germanium detector with a beryllium window) and placed on an silicon barrier alpha-pulse detector for determination of ^{237}Np, ^{239}Pu, and ^{236}Pu.

A variation of this procedure was used for some of the samples analyzed in the analog program. In this variation, the sample and spikes were first reduced with HBr, taken to dryness, and oxidized with the HF-HNO$_3$-HClO$_4$ before extraction with hexone. Also, the extraction step used NH$_4$NO$_3$/KBrO$_3$ instead of the Al(NO$_3$)$_3$-KMnO$_4$. The data produced from this method were in general agreement with the standard method. Table 1 compares a few results of these two methods; there is no trend evident between the measured actinide activities and (1) the method employed to measure them or (2) the chemical yield.

The data in Table 1 are representative of the analyses performed, where the range of activities varied from background (0.001 d/m/mL) to 90 d/m/mL. The yields varied between 20 and 85%, with a mean value of ~50%.

Basalt Core Profile of ^{237}Np

The final component in the analog flow system is a basalt rock core which has been split axially to provide a fissure through which fluid flows. The simulated groundwater flowing into the rock core contains low concentrations of radioactive materials leached from the glass waste form; most of this activity was deposited on the rock surface. The distribution of radioactive materials retained on the rock surface was measured by counting the gamma activity using a Ge(Li) detector through a 8-mm by 58-mm opening in a 4-mm

TABLE 1
Examples of ^{237}Np-^{239}Pu groundwater analyses.

Example	Method	^{237}Np d/m/mL	% Yield	^{239}Pu d/m/mL	% Yield
A	1	0.076	38	0.05	35
	2	0.087	45	>0.001	47
B	1	1.5	42	6.8	48
	1	1.7	32	7.1	38
	1	2.6	47	9.3	52
	2	0.8	82	9.3	75
C	1	77	44	62	62
	1	89	39	64	52
	1	81	42	64	49

thick lead plate. The lead plate shielded the detector against radiation from other than the exposed part of the rock's surface. The 86.5 KeV gamma of ^{237}Np was the only signal detectable from the rock surface; ^{152}Eu, ^{133}Ba, and ^{241}Am were looked for but their activities were below detectability. The highly irregular surface geometry of the rock faces precludes quantitative data being obtained by this method; however, the qualitative data do allow comparison of the ^{237}Np distribution on the rock core face from various analog experiments.

Figure 2 shows the profile of ^{237}Np distribution obtained from gamma measurements on one rock core. The relative values of the measured activities may be in error by as much as 20% due to the irregular surfaces of the rock fissures. Even with this high uncertainty, it is clear that the adsorption of the ^{237}Np is essentially complete in the first one third of the rock core.

Core Leaching and Mass Balance of ^{237}Np

Although gamma counting of the rock core surface can give a relative profile of ^{237}Np activity, it cannot be used as a quantitative measurement. To quantitate the amount of neptunium and plutonium sorbed on the rock core, it was necessary to leach these elements from the rock core. The leach solution that was utilized was one developed for removing plutonium from sediments using sodium citrate/sodium dithionate solution (ref.6), followed by leaching with strong acid solution. The results of this analysis were coupled with results of ^{237}Np groundwater analyses to check the mass balance of the first analog experiment.

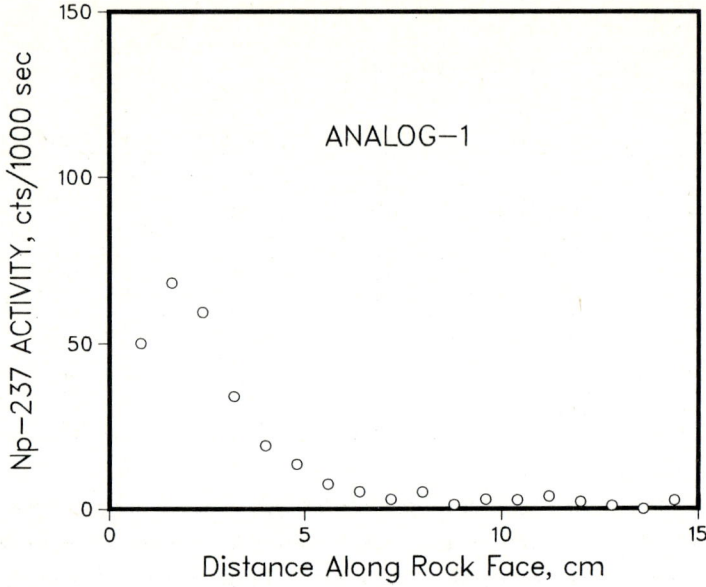

Fig. 2. Distribution of neptunium along the face of the basalt fissure from an analog experiment.

A solution consisting of 0.3M sodium citrate-0.1M sodium dithionate was used to leach the fissured face of the half core. The split core was placed fissured-face down in a trough-like Pyrex vessel, leach solution added to ~0.5 cm above the face, and heated at ~70 C for two hours. The leachate was transferred to a 200-mL volumetric flask and an aliquot was analyzed for ^{237}Np and ^{239}Pu using the same procedure as described previously.

To determine complete leaching of these isotopes, a portion of the core was counted on a Ge(Li) detector for the characteristic 86 keV gamma decay of ^{237}Np and compared to the activity level of the unleached core. A second leach was performed using 1M HCl-0.5M HNO$_3$ under the same conditions as the first leach. The core was recounted on the Ge(Li) detector and the results indicated that 82.2% of the ^{237}Np was removed. The leach data for this experiment are presented in Table 2. The measured content of ^{239}Pu of the second leach was at background levels, which implies that all the plutonium on the rock core was removed in the first leach. The calculated amounts of the ^{237}Np on the rock core, determined independently from the two leaching steps, compared well. The calculations were made by comparing ^{237}Np in the respective leachate solutions to the fractional loss of ^{237}Np activity measured on the rock core face.

TABLE 2

Leaching data for ^{237}Np and ^{239}Pu from a basalt core used in an analog experiment.

Leach number	Volume of leachate, mL	[^{237}Np] in leachate, dpm/mL	[^{239}Pu] in leachate, dpm/mL	% ^{237}Np activity remaining on core after leaching	Total actinide deposited on core, dpm	
					^{237}Np	^{239}Pu
1	200	7.16	0.70	57.1	6670	680
2	100	13.34	≤0.007[a]	17.8	6730	

[a] No ^{239}Pu activity was detectable above background levels.

The concentration of ^{237}Np entering the rock core was ∼5.7 dpm/mL and was ∼0.03 dpm/mL at the rock core's exit. The total volume of groundwater passing through the rock core was 1675 mL. These data would predict 9500 dpm of ^{237}Np was retained in the core vessel (on either the rock surface or on the bentonite at the face of the core inlet). Previous results of gamma spectroscopic analyses of bentonite samples in the experiment showed there was ∼1600 dpm ^{237}Np in bentonite deposited at the core inlet. When this amount is subtracted from the total ^{237}Np activity that entered the core holder, the predicted ^{237}Np activity on the rock fissure is 7900 dpm. This compares well to the measured value of 6700 dpm.

No mass balance of ^{239}Pu was attempted. The comparison of ^{239}Pu sorption is difficult because of its association with bentonite. Since bentonite was found throughout the apparatus and especially near the inlet core face, to find a relevant mass balance for ^{239}Pu would entail a destructive analysis of all bentonite samples to measure its alpha activity.

CONCLUSIONS

We have developed an effective and efficient method for ^{237}Np-^{239}Pu content in groundwater or other waters with high dissolved solids, and demonstrated that it is possible to spatially profile and quantitate sorbed radioisotope activity on rock surfaces. In laboratory analog experiments, these analyses have shown the correlation between radionuclide retardation with changes in groundwater composition and hydrothermal alteration of basalt repository components. These methods are applicable to the study of other potential repository sites.

REFERENCES

1. G. F. Vandegrift, D. L. Bowers, T. J. Gerding, S. M. Fried, C. K. Wilbur, and M. G. Seitz, The Interaction of Groundwater and Fresh Basalt Fissure Surfaces and Its Effect on the Migration of Actinides, ACS Symposium Series, Geochemical Behavior of Disposed Radioactive Waste, in press.
2. M. J. Steindler, *et al.*, Fuel Cycle Section Quarterly Progress Reports, July-September 1981, ANL-81-82, May 1982; October-December 1981, ANL-82-18, May 1982; January-March 1982, ANL-82-34, December 1982; April-June 1982, ANL-82-58, March 1983; July-September 1982, ANL-82-78, March 1983; October-December 1982, ANL-83-19, August 1983; January-March 1983, ANL-83-68, in press; April-June 1983, ANL-83-78, in press.
3. M. G. Seitz, G. F. Vandegrift, D. L. Bowers, and T. J. Gerding, Effect of "Aged" Waste Package and Basalt on Radioelement Release, presented August 30-September 1, 1983 at NRC Contractors Meeting in Reston, VA; to be published as part of a NUREG document.
4. G. A. Burney and R. M. Harbour, Radiochemistry of Neptunium, NAS-NS-3060, 1974, p. 161.
5. I. Ahmad, J. Hines, and J. E. Gindler, Electron Capture Decay of ^{237}Pu, ^{235}Np, and ^{236}Np, Physical Review C, 27 (1983) 2239.
6. D. N. Edgington, J. J. Alberts, M. A. Wahlgren, J. O. Karttunen, and C. A. Reeve, Plutonium and Americium in Lake Michigan Sediments, Transuranium Nuclides in the Environment, IAES-SM-199/47, 1976.

ANALYTICAL RESEARCH AT THE INSTITUTE OF RADIOCHEMISTRY, NUCLEAR RESEARCH CENTER KARLSRUHE, FRG

E. Gantner, E. Mainka, H. Ruf and H. J. Ache

Kernforschungszentrum Karlsruhe, Institut für Radiochemie, Postfach 3640, 7500 Karlsruhe, FRG

ABSTRACT

A review will be presented of the various analytical techniques applied to the solution of chemical problems within the ongoing research in energy technology at the Institute of Radiochemistry, Nuclear Research Center Karlsruhe, FRG. Three typical applications, neutron monitoring of solutions of fissile materials polarographic and voltammetric methods for the determination of traces of Pt, Tc and Re, and ICP-spectroscopy of uranium solutions will be discussed.

Introduction

Analytical Research at the Institute of Radiochemistry (IRCH), KFK, can be divided into three categories:

Special Analytical Service

This includes mainly the analysis of nuclear fuel and specifically the quality control of non-irradiated nuclear fuel, in which case the determination of the concentration of uranium and plutonium, as well as the isotopic composition of uranium and plutonium are the major objectives. A second equally important task performed in this category is the analysis of spent nuclear fuel for process control and nuclear safe guard.

Development of Modern Analytical Techniques

Research in this area is done to accomplish the following goals:

1. Reduction of required amount of (highly radioactive) sample material to simplify handling procedures and avoid expensive facilities (hot cells etc.) by developing simultan (multichannel) emission spectrometers and mass-spectrometric techniques using new designs of ion sources.

2. Reduction of analysis-time and labor by the development of automated analytical instrumentation.

3. Improvement of sensitivity and range of application by adapting laser-raman-spectroscopy or laser-fluorimetry to the solution of analytical problems in nuclear technology.

4. In-line monitoring of radioactive materials in high-radiation areas (neutron monitors).

5. Surface Analytical Techniques.

R + D projects which require primarily the use of analytical techniques.

These techniques and procedures will be discussed by using appropriate examples from the Institute of Radiochemistry research program.

NEUTRON MONITORS

1. Hafnium monitor

Hafnium plates are used in pulsed extraction columns in the Purex process in reprocessing plants firstly as sieve plates to accomplish the required thorough mixing of the aqueous and organic phases and secondly as heterogeneous neutron absorbers to prevent the solutions from becoming inadvertently critical. In such a high radiation area the number of devices to monitor the presence and intactness of these hafnium plates is rather limited and one of the methods proposed was to perform neutron-transmission measurements. As shown in fig. 1 the technique consists basically of a suitable neutron emitter, in this case a Cf-252 source (10^8 n/sec.), moderated and collimated, a neutron counter (BF_3) 2.5 cm i.d., length 20 cm with a Cd-window (10 mm thick) and an appropiate scanning device which allows the measurement of the neutron count rate along the axis of the column, whose inner diameter was 30 cm.

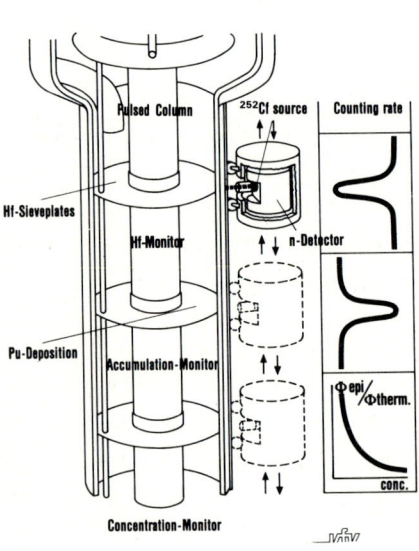

Fig. 1 Neutron Monitor

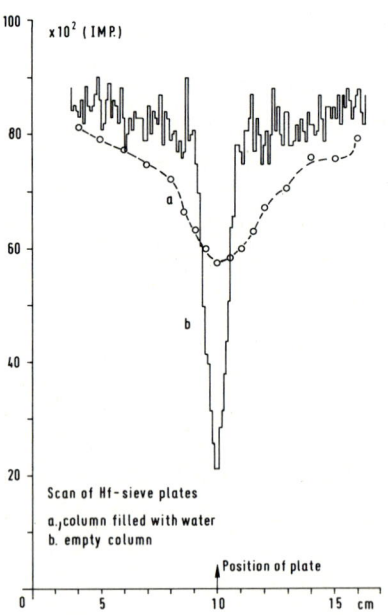

Fig. 2 Scan of Hf-sieve plates
a) column filled with water
b) empty column

In order to test the method a series of experiments was performed with an empty stainless steel column and a column filled with water, both containing Hf-sieve plates (2 mm thick) fig. 2, where the neutron counts are plotted as a function of column length shows the results obtained in these experiments, in which the Cf-source and detector were arranged on opposite sides of the column. In the presence of water the resolution is relatively small caused by the neutron scattering properties of the water, leading to a rather diffuse picture which allows an identification of an individual plate only if the distance to the nearest plate exceeds 10 cm. The max. reduction of the neutron countrate is about 30 %, compared with that what is observed in the absence of a Hf-plate. The same kind of measurement done in an empty column (fig. 3) shows a drastically improved resolution which allows the identification of individual Hf-sieve plates with a minimum distance of 1.5 cm between adjoining plates. In this case the neutron countrate is reduced to about 30 % of its original value, i. e. as measured in absence of a Hf-plate. Additional measurements show that the thickness of the plate in the range from to 3 mm has no effect on the neutron countrate.

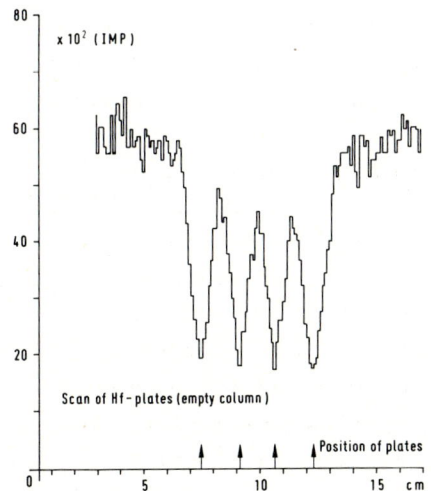

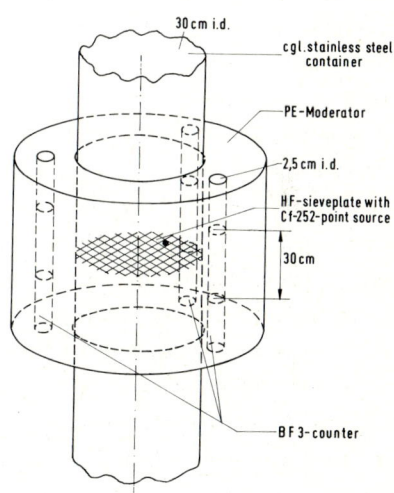

Fig. 3 Scan of Hf-plates (empty column)

Fig. 4 Experimental Arrangement of Accumulation Monitor

Summarizing it can be said that this simple technique allows a determination whether a Hf-sieve plate is still in its position but does not provide any information not about the degree of corrosion, which might have recured.

2. Accumulation monitor

Pu accumulations in form of Pu-phosphates etc. on the sieve plates of a pulsed extraction column have to be monitored in order to avoid any criticality problems, e. g. in a column (30 cm i. d.) containing Hf-sieve plates and a solution whose Pu concentration is about 100 g/l the required minimum detectable limit would be an accumulation of 45 g Pu per plate. The technique proposed to accomplish this goal involves the passive neutron measurement of spontaneous fission neutrons and neutrons originating from (α, n) nuclear reactions. The experimental arrangement is shown in fig. 4. The cylindrical stainless steel column (30 cm i. d.) containing a Hf-sieve plate is surrounded by a polyethylene ring (7 cm thick) for neutron moderation which in turn is equipped with three BF_3 counters which are placed in a 120° angle arrangement either in a vertical or horizontal position relative to the stainless steel cyclinder. A relatively weak Cf-252 point source (ca. 10^6 n/sec.) was used to simulate the Pu accumulation on the plate. It was placed in about 50 positions on the plate and in each position the count rate was registered. The assumption was made that the nuclear fuel (PWR) had a burn up of 33 000 MWd/to, that the neutron emission of the Pu in the fuel is identical with that in the accumulations on the plate (767 n/sec g Pu) and that the fraction of neutrons originating from Cm is negligible compared with the total number emitted by the accumulated Pu. The measurements showed that in the case of an empty column and a random distribution of the Pu a countrate of 120 imp/min g Pu was obtained. To achieve the same countrate of 120 imp/min 1.2 - 1.3 g Pu were required if the Pu was positioned as a point source in the center of the plate and only 0.5 - 0.6 g Pu if placed on the periphery of the plate. From these results a lower limit of detection of 1 g Pu could be deduced if the time of measurement is limited to 1 min. In a column filled with water the corresponding numbers are: random distribution of Pu = 130 imp/min. g Pu; point cource, in the center 130 imp/min. = 2.1 - 2.5 g Pu and on the periphery 0.3 - 0.35 g Pu. Thus the limit of detection is approximately the same as in an empty column. The experimental results can be summarized as follows: Pu-accumulations can be estimated within the factor of two, in the presence of larger amounts of Pu the estimate will be more accurate. The detection limit is about equal or less than 1 g Pu/plate at counting times of about 1 min. No clear answer can be obtained as to the shape of the Pu distribution on the plate. A possible uncertainty is introduced if the Pu accumulation is contaminated with an unknown amount of other neutron emitters, especially Cm.

3. Concentration monitor for fissile material in solutions in pulsed extraction columns

The principle of this technique is based on calculations by G. Schulze and H. Würz (Nuclear Technology 35, 663 - 670 (1977), in which these authors showed that the following correlation exists between the epithermal neutron flux multiplication factor M_{epi} and the concentration of fissile material C_{fiss} even in the presence of neutron poisons (fig. 5).

$M_{epi} = f(C_{fiss}/\Sigma_{abs}^{th})$

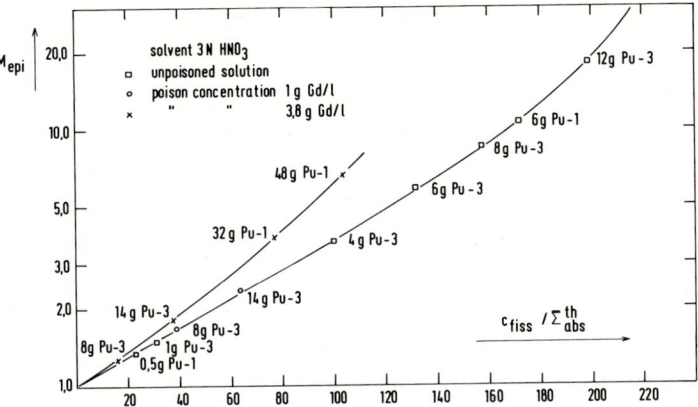

Fig. 5 Calculated epithermal neutron flux multiplication factor M_{epi} for 3 N nitric acid solutions containing different fissile material and different Gd poison concentrations

where Σ_{abs}^{th} is the macrospic thermal absorption cross section of the solution, which in turn is a function of the ratio of the thermal and epithermal neutron flux ϕ_{th}/ϕ_{epi} as shown in fig. 6

$\phi_{th}/\phi_{epi} = f(\Sigma_{abs}^{th})$

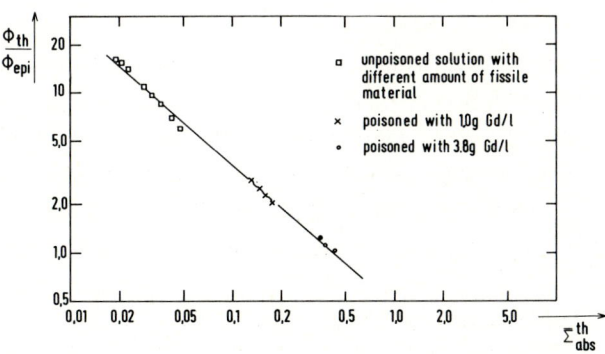

Fig. 6 Calculated dependence of ϕ_{th}/ϕ_{epi} on the macroscopic thermal absorption cross section Σ_{abs}^{th} of solutions containing fissile material and Gd as homogeneous poison. The values are given for cylinders of diameter 30 cm with source and detector at the surface of the cylinder

This function can be determined by calibration using solutions of fissile material containing various amounts of neutron absorbers (poisons) such as Gd. By measuring M_{epi} and $\emptyset_{th}/\emptyset_{epi}$ in a column C_{fiss} can be determined. These theoretical predictions have to be verified by experiments involving calibration measurements with enriched uranium solutions (20 % U-235) containing various amounts of Gd as neutron poison. The results can be converted in an appropiate manner to Pu solutions. The actual measurements which includes the handling of nitric acid solutions of enriched uranium (8 kg enriched U) will be carried out in a α-Box, which is partially shielded with paraffin. A Cf-252 neutron source (ca. 10^8 n/sec.) will be used for the active measurements. A cylindrical stainless steel column (30 cm i. d.) with Hf-sieve plates contains the uranium solutions. One unshielded BF_3 counter will be used for the measurement of $\emptyset_{therm}$, a Cd-shielded BF_3 counter for the measurement of $\emptyset_{epi}$ and a third BF_3 placed in a polyethylene shield determines M_{epi}.

A series of experiments with U solutions with concentrations ranging from $C_U = 2$ g U/l to 200 g U/l in the presence of up to 2 g Gd/l as homogeneous and Hf-sieve plates as heterogeneous neutron absorbers will be carried out.

DETERMINATION OF TRACE AMOUNTS OF PT, TC AND RE BY ELECTROCHEMICAL TECHNIQUES

1. Determination of Pt in presence of U

In modern reprocessing plants where U (VI) is reduced electrolytically in nitric acid solutions by utilizing Pt-electrodes, it is important to determine how much Pt is dissolved during the course of this process. The most sensitive methods for the detection of Pt in aqueous solutions are based on the catalytic effect of certain Pt-complexes such as the Pt(II) ethylenediamine compound to achieve a hydrogen evolution at the polarized mercury drop in an alkaline medium. In the presence of greater amounts of uranium this technique cannot be applied because in the required alkaline solution uranium forms a precipitation which interferes with the electrochemical process. Therefore a technique was developed by which the traces of Pt were precipitated in an acidic solution via reduction with H_2SO_3 in the presence of small amounts of tellurium (0.5 mg) in form of tellurite as carrier. The precipitation is isolated from the bulk solution and after dissolution in an alkaline medium the platinum is polarographically determined via the Pt (II) ethylenediamine complex. This method proved to be highly selective, only selenium will infere, this however was not present in the solutions under study. Tellurium gives no polarographic signal under these experimental conditions. The experimental results are shown in figure 7, where the DP-polarographic signal of 27 ng Pt is shown at about -1.6 V(vs Ag/AgCl) after its seperation from 441 mg U via reductive precipitation with 0.5 mg Te carrier added. For comparison the signals from a solution with no Pt and 400 ng and 800 ng Pt present are shown (all in 5 ml electrolyte solution). In solutions whose uranium contents (U IV or U VI) exceeds 1 mg/ml the chemical seperation of the Pt can be avoided if Pt is present as $(PtCl_4)^{2-}$ in 0.1 NHCl. In these cases Pt can be detected selectively and with a high sensitivity using voltammetric techniques with a rotating glass carbon electrode. The halfwave potential is

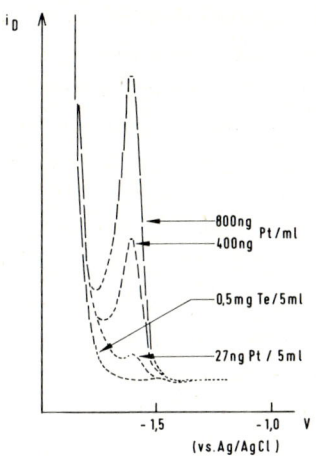 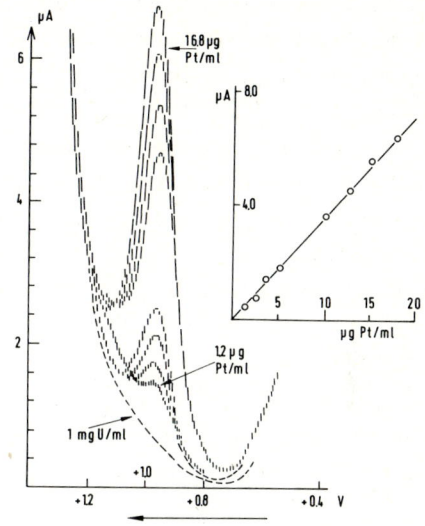

Fig. 7 DP-Polarograms of Pt(II)-EDA complexes after seperation from 441 mg uranium by co-precipitation with tellurium

Fig. 8 DP-Voltammograms v. Pt-II in presence of uranium at the GCE in 0,1 N HCL

under these conditions + 1.0 V (Ag/AgCl). Fig. 8 shows a series of voltammograms obtained in solutions which contained 1 mg U/ml. Further investigations showed that Pt IV can be measured polarographically using a dropping mercury electrode in KOH-triethanolamin solutions in presence of U, i. e. without any prior chemical seperation if the U concentration which has to be in the VI oxidation state is limited to ten to twenty times the amount of Pt. The halfwave potential (ca. - 0.6 V) is in 0.3 M triethanolamin-KOH solution close to that of U, however, the Pt-concentration dependent DP signal is directed opposite to that of U, i. e. there will be no overlapping but if there is any effect at all it will lead to a slight reduction of the amplitude of the signals. Fig. 9 shows the Pt signals for solutions containing various amounts of Pt and 25 µg U. The reverse signal is probably the result of the desorption of negatively charged ions from the mercury drop.

2. Determination of trace amounts of Tc

In previous studies the determination of trace amounts of Tc was attempted by using inverse polarography with the socalled Kemula electrode in which case the limit of detection was about 0.3 µg Tc if dilute NaOH was the electrolyte . Our investigations showed that this limit could not be improved by applying mercury thin film or glass carbon electrodes or gold and silver as carrier to this technique. On the other hand our studies revealed that a significant increase in the sensitivity could be achieved by utilizing inversvoltammetry in connection with a glass carbon electrode, avoiding any mercury presence. After the enrichment step which included electrolysis at a seperation potential of - 0.8 V (vs Ag/AgCl), one can register in the acidic solution following the dissolution in the DP-mode a well detectable signal which obviously consists of a main current peak

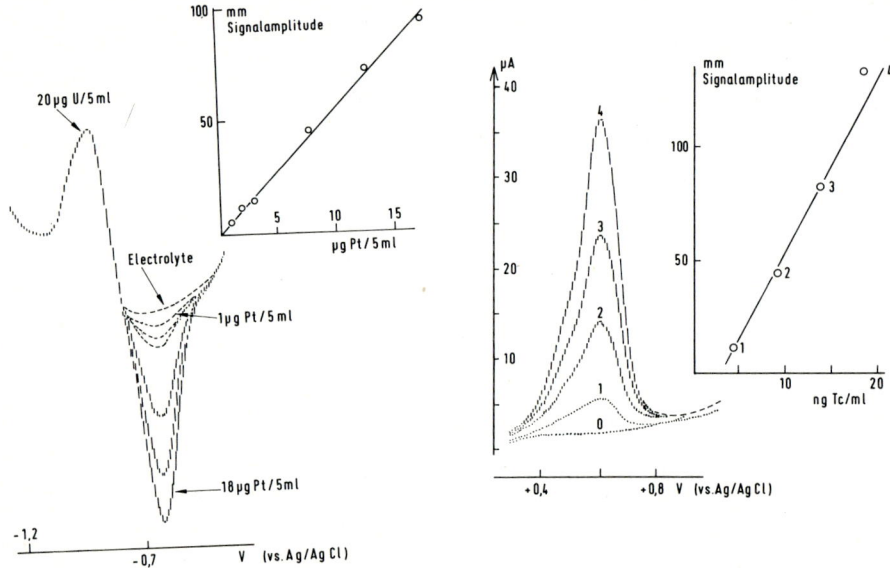

Fig. 9 Pt in presence of U at the DME in Triethanolamine-KOH

Fig. 10 DP-Inversvoltammograms of TcO_4^- at the GCE
Electrolyte: 0.06 N $HClO_4$

at + 0.56 V and a smaller adjoining peak at + 0.50 V (Fig. 10). The electrolyte was 0.06 M perchloric acid solution. This current peak at 0.56 V proved to be linearly proportional to the Tc concentration as low as 3×10^{-8} M TcO_4^- with excellent reproducibility. No signal is obtained from Mn, Rh and other metals under these experimental conditions. We assume that this current at 0.56 V is due to the oxidation of the Tc(O) metal (previously incorporated into the electrode) to pertechnetate since the redox potential for this process is + 0.47 V (vs Ag/AgCl) in acidic solution.

3. Determination of trace amounts of Re

Although some polarographic and voltammetric method have been reported for the determination of Re, no inversvoltammetric method including prior enrichment by electrolysis has been discussed in the literature. Our studies using this latter technique showed, that Re present in form of perrhenate in 0.01 N sulfuric acid can be concentrated on an amalgameted gold electrode via electro reduction at a potential of - 0.8 V (Ag/AgCl). Subsequent anodic polarisation of the electrode in the DP-mode in the same solution yields a well developed oxidation peak at - 0.27 V, which allows the quantitative determination of extremely small concentrations of Re. The inverse voltammograms obtained after 2 min of enrichment, are shown in fig. 11. The gold electrode is amalgameted by electrolytic reduction of the mercury under stirring from a solution of 20 ml 0.05 N KNO_3, which contains 160 - 500 µg of Hg^+ ions.

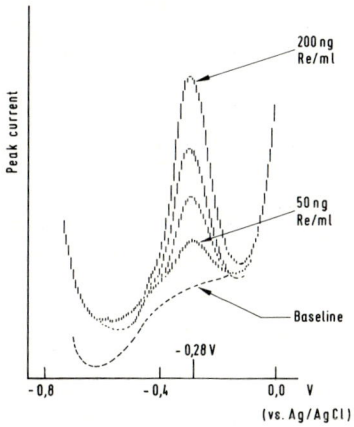

Fig. 11 DP-Inversvoltammograms of 0,01 N H_2SO_4 acidic ReO_4^--solution at the amalg. Au-electrode

ICP-EMISSIONSPECTROSCOPY OF URANIUM-PLUTONIUM SOLUTIONS

The quality control of MOX (U-Pu-Mixed Oxides) nuclear fuel elements which is routinely done at the IRCH includes about 1 000 samples per year in which 28 trace elements have to be analyzed. In the past 23 trace elements had been seperated from the U-Pu-matrix via fractionated distillation and analyzed by emissionspectroscopy. The four rare earth elements Sm, Gd, Eu and Dy as well as Th, however, could not be seperated by this method and an additional step was required based on anion exchange. With the introduction of ICP-emission spectroscopy a direct determination of all 28 trace elements in the U-Pu matrix was attempted. The experiments, however, showed that the resolution of the available sequential spectrometer (6 Å/mm) was not sufficient to determine 1 ppm of these elements as required in the specifications in the presence of the large number of spectral lines of the matrix elements (U and Pu). Thus a 3.5 m grating spectrograph which could achieve a resolution of 1 Å/mm was attached to the α -box in which the ICP excitation source was placed, but even with these improved resolution no satisfactory sensitivity could be obtained e. g. for the determination of Gd. To illustrate the difficulties involved it should be printed out that at a nuclear fuel concentration of 500 µg/ml in the range from 240 to 450 nm approximately 2400 spectral lines could be registered. It seemed therefore more advantageous to resort to a prior chemical seperation of the elements under study from the matrix, which involved in contrast to the previous procedure now only one seperation step for all 28 trace elements.

It is based on the extraction of the 8 M HNO_3 solution with a mixture of tributylphosphate (TBP) and Kerosin (volume ratio of 1 : 4). The procedure is very simple and effective. 500 mg of the mixed oxide are dissolved in 10 ml of 8 M HNO_3 and extracted four times with the organic mixture (160 ml total), which leads to an almost quantitative transfer of U and Pu into the organic phase.

U and Pu can be re-extracted into the aqueous phase and returned to the nuclear fuel element production process. Replacing the two previously used seperation techniques by one and taking advantage of the very high sensitivity of the ICP-method compared with e. g. excitation by spark source, the required amount of material for the analysis is reduced by the factor of 20, which leads to a drastic minimization of waste if one considers that 1 000 samples are processed per year. The analyzer is again the 3.5 m spectrograph in connection with photographic films and automated computerized data analysis of these films.

COMPARISON OF BURNUP METHODS FOR (Th,U)O_2 FUEL

L.W. Green, R.M. Cassidy, W.J. Edwards and C.H. Knight
Chalk River Nuclear Laboratories, Atomic Energy of Canada Limited Research Company, Chalk River, Ontario, Canada, K0J 1J0

ABSTRACT

Analysis methods based on thermal ionisation mass spectrometry (TIMS), high performance liquid chromatography (HPLC) and instrumental neutron activation (INA) were investigated. Fission products ^{139}La, ^{140}Ce, ^{142}Ce, $^{145+146}$Nd and the sum of the major lanthanides (masses 139 to 152) were evaluated as fission monitors. A TIMS-$^{145+146}$Nd method was selected as a reference method, and a HPLC-^{139}La method showed good agreement and similar precision. Results obtained by INA-^{142}Ce, HPLC-(139-152) and TIMS-^{148}Nd methods showed significant differences from those obtained by the above two, for various reasons. The weight and initial composition of the fuel was shown to give a sufficiently accurate estimate of the initial heavy element content; this combined with the HPLC-^{139}La method yielded a fast, low-cost burnup method.

INTRODUCTION

Thorium fuel cycles have a number of advantages over uranium fuel cycles (ref. 1), such as more efficient use of fertile material. For the CANDU^{TM} reactor, some thorium fuel cycles are compatible with existing reactors designed for natural uranium fuel, and the technology to fabricate thorium-based CANDU^{TM}-type fuels is being developed. A major component of testing these experimental fuels is their burnup determination, which is the topic of this paper.

Fission in Th based fuels is dominated by fission of the fissile material (^{235}U or ^{239}Pu) mixed with the starting material and of ^{233}U formed during irradiation. The resultant U isotopic composition and fission product yields are substantially different than those of more conventional fuels, and conventional burnup techniques such as the ^{235}U-depletion or ^{148}Nd methods (ref. 2) can not be applied directly. Consequently, in our laboratory, various analysis methods and fission monitors were investigated. Thermal ionisation mass spectrometry (TIMS), high-performance liquid chromatography (HPLC) and instrumental neutron activation (INA) were selected for the investigations; the latter two because of their potential for rapid, low-cost, automated analyses and the first because it is the recognized technique against which others may be compared.

The fission products ^{139}La, ^{140}Ce, ^{142}Ce, $^{145+146}$Nd, ^{148}Nd and the sum of the major lanthanides (masses 139 to 152) were identified as potential fission monitors. They all satisfied the criteria for fission monitors (ref. 3,4) and, additionally, ^{139}La, ^{140}Ce and the sum of 139-152 have nearly equal yields from ^{233}U and ^{235}U. For TIMS, ^{148}Nd and $^{145+146}$Nd were selected because of their common use with uranium fuels (ref. 5) and because of difficulties with mass spectrometry of the others, described later. For HPLC, ^{139}La and the sum of the major fission-product lanthanides were chosen because these do not require isotopic analysis. Cerium-142 was selected as a fission monitor for INA because of the high specific activity and ease of detection of the activation product, ^{143}Ce.

EXPERIMENTAL SECTION

Fuel Sampling and Dissolution. The fuel was a 36 element CANDU^{TM}-type bundle of (Th,^{235}U)O$_2$ fuel that had been irradiated in the NRU reactor for 2 years and allowed to decay for 2 years. In a hot cell, fuel sections were removed from the fuel sheath by mechanical vibration, accurately weighed, dissolved in 0.05 M HF/13 M HNO$_3$ and diluted with water to a Th concentration of 1 mg/g (ref. 6).

Mass Spectrometric Analysis. Classical anion-exchange chromatography with HCl-acetone and HNO$_3$-methanol eluents was used to provide Th, U and Nd separated fractions (ref. 6). Th and U were isotopically analyzed in nanogram quantities with a centre filament technique (ref. 7) whereas the conventional side filament technique was used for Nd (ref. 6). The mass spectrometer was a Nuclide single stage 90° magnetic sector instrument with an electron multiplier detector.

For isotope dilution analysis the samples were mixed with the appropriate spike (^{230}Th, ^{235}U or ^{150}Nd), heated to near boiling, and evaporated to dryness several times to ensure chemical and isotopic equilibrium.

HPLC Analysis. Several column systems were considered and a dynamic cation-exchange system was selected because it previously showed superior resolution and greater choice of separation conditions (ref. 8). With ^{139}La as the fission monitor, 100 µL of the diluted fuel solution was injected onto a 5 µm Supelcosil LC-18 column (ref. 9). The mobile phase contained 0.18 M α-hydroxyisobutyric acid (HIBA) and 0.005 M sodium 1-octanesulfonate at pH 4.6. The lanthanides were detected by visible absorption at 635 nm after postcolumn reaction with Arsenazo III.

When the sum of the major lanthanides was used as the fission monitor, $\sim$ 1 g of the diluted fuel solution was quantitatively spiked with a lanthanide chemical yield monitor, such as Gd (III), and the actinides were removed by anion-exchange in HCl-acetone (ref. 9). The fission product fraction was then analysed as described above for ^{139}La, except that gradient elution from 0.03 M to 0.3 M HIBA over 15 min was used.

INA Analysis. Cerium was separated from $\sim$ 1 g of the diluted fuel solution in the same manner as the major lanthanides, described above, except that for HPLC lithium 1-octanesulfonate was used instead of the sodium salt, to avoid spectral interferences. The cerium fraction was evaporated to dryness in a thin, high purity, quartz irradiation tube which was then flame-sealed. An aliquot of a natural Ce standard solution was evaporated and sealed in a similar quartz tube. The Ce fraction and standard were simultaneously irradiated for 1 h in a thermal neutron flux of $\sim 3 \times 10^{14}$ neutrons $cm^{-2}s^{-1}$, then counted with a Ge(Li) spectrometer about 24 hours after irradiation. Fission product ^{144}Ce was used as a chemical yield monitor and ^{140}Ce was used as a check on natural Ce contamination.

RESULTS AND DISCUSSION

In this work burnup is expressed as atom percent fission: the number of fissions per 100 initial heavy atoms (ref. 3). Burnup computations required effective fission yields for each monitor; these were estimated from the absolute fission yields for ^{233}U and ^{235}U (ref. 10) and from the fraction of fissions in each of these fissile nuclides. The yields of masses 139 through to 152 were used for the sum of the major lanthanides. Estimates of the fission fractions were obtained from an irradiation simulation by the reactor physics computer code LATREP (ref. 11) and are listed along with the effective yields in Table 1.

Table 1.
Fission fractions and effective yields for a $(Th,U)_2$ fuel

Fuel Ring	Fission Fraction		^{139}La	^{140}Ce	^{142}Ce	$^{145+146}Nd$	^{148}Nd	139-152
	^{233}U	^{235}U						
Outer	0.304	0.696	6.38	6.34	6.12	6.60	1.54	48.4
Middle	0.223	0.777	6.38	6.33	6.05	6.68	1.58	48.6
Inner	0.202	0.798	6.38	6.32	6.04	6.70	1.59	48.6

The simulation treated each fuel element of a ring identically (the bundle contained three concentric rings), thus the estimated fission fractions, effective yields and burnups were identical for elements from the same ring.

Figure 1 shows a mass spectrum of a Nd fraction. In addition to the peaks for Nd, the spectrum shows large peaks for Ba (138), Ce (140,142) and Pr (141). The Ba and Ce peaks were due to Ba and Ce contaminants in the rhenium filament, and prevented the use of ^{139}La and ^{140}Ce as fission monitors for mass spectrometry. The ^{141}Pr peak was due to incomplete separation of Pr from Nd. The relative standard deviations of $^{145+146}Nd$ and ^{148}Nd concentration determinations were 0.8%, determined by seven replicate analyses through the entire procedure.

Figure 2 shows a chromatogram of a fuel sample as prepared for La determination. The La peak is well resolved from the large Th and U peaks, whereas peaks for other lanthanides are merged with peaks for the latter two. Well resolved peaks were obtained for La, Ce, Pr, Nd and Sm when the actinides were removed

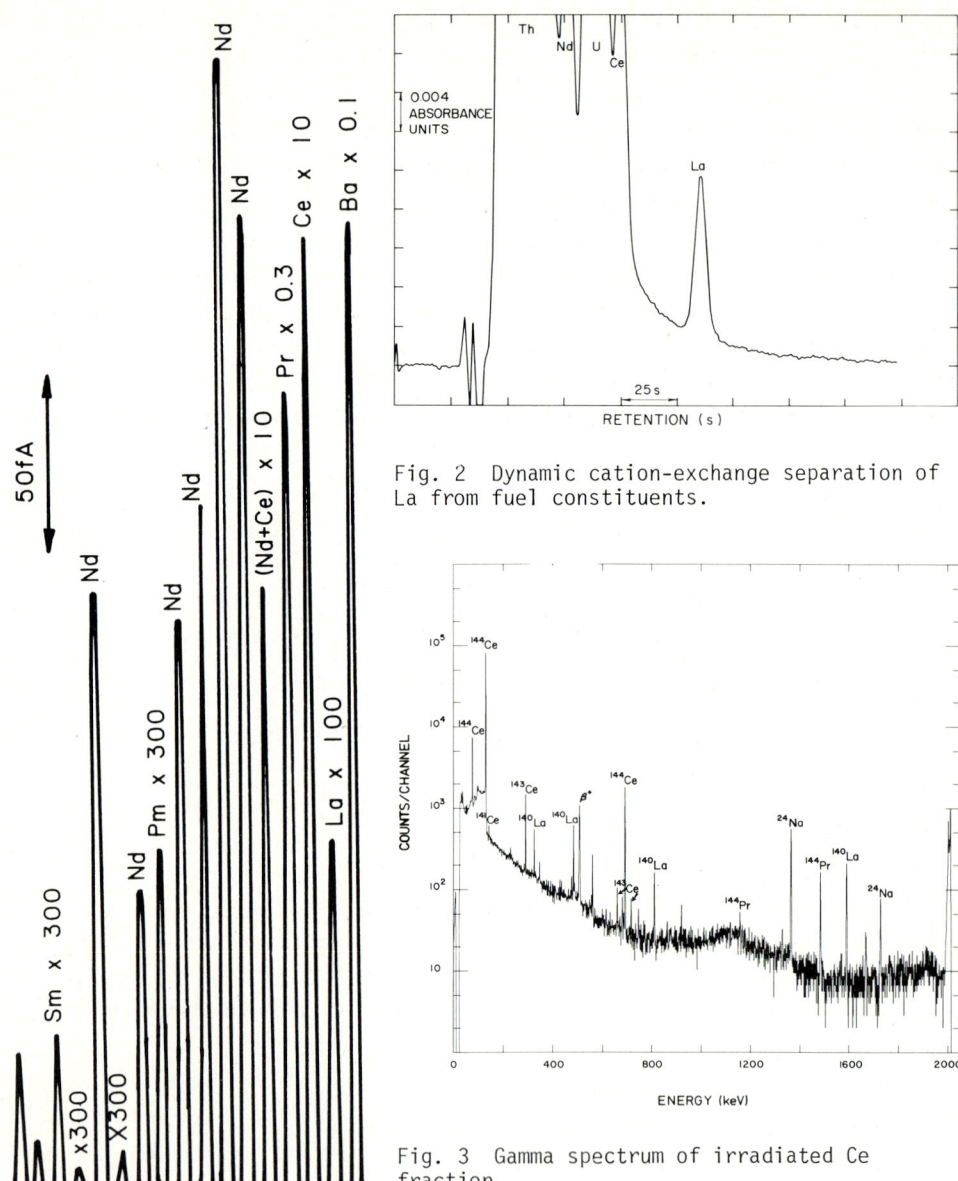

Fig. 1 Mass spectrum of spiked Nd fraction. Peaks were multiplied by the given scale factors.

Fig. 2 Dynamic cation-exchange separation of La from fuel constituents.

Fig. 3 Gamma spectrum of irradiated Ce fraction

and gradient elution was used. The La peak is 100% ^{139}La, since no other long-lived isotopes of La are produced in significant quantities in fission. The relative standard deviation of ^{139}La determinations was 1%, determined by seven replicate analyses that included recalibration.

Figure 3 shows a gamma spectrum of an irradiated Ce fraction. The 293.3 keV ^{143}Ce and the 133.5 keV ^{144}Ce peaks are well defined and have ample intensity. However, the 145.4 keV ^{141}Ce peak is weak and on the tail of the large ^{144}Ce peak; this makes ^{140}Ce a poor fission monitor for INA. The ^{140}La peaks are likely due to incomplete separation of Ce from fission-product ^{139}La, and subsequent formation of ^{140}La in the neutron activation step.

Burnup results for the various methods are listed in Table 2; for burnup computations the mass spectrometrically determined Th and U concentrations were used in all cases, except where noted. Relative standard deviations were estimated from propogation of the uncertainties in the concentrations. Possible sources of systematic error are the fission yields and, for ^{139}La, a slight depletion by neutron capture. The differences in burnup for fuel elements in different rings are due to the flux gradient across the fuel.

Table 2.

Atom percent fission in fuel bundle BDL 417 AAX

Ring	Element	Atom Percent Fission						
		TIMS $^{145+146}$Nd	TIMS $^{145+146}$Nd^W	TIMS ^{148}Nd	HPLC ^{139}La	HPLC 139-152	INA ^{142}Ce	LATREP
Outer	4	2.27	2.29	2.33	2.27	2.1	2.3	2.17
Outer	14	2.23	2.27	2.30	2.23		2.5	2.17
Middle	30	1.41	1.40	1.46	1.42	1.5	1.6	1.42
Inner	36	1.09	1.11	1.11	1.05	.96	1.3	1.15
rsd		1.5%	0.8%	1.5%	1.5%		5%	

^{W}fuel weight used in burnup computation (see text)

Based on its previous acceptance with U and (U,Pu) fuels (ref. 6), the TIMS-$^{145+146}$Nd method was chosen as a reference method against which others were compared. The agreement between this method and the LATREP estimates are within the accuracy expectations of the code. The HPLC-^{139}La results show good agreement with the TIMS-$^{145+146}$Nd results; an estimated t-test value of .75 is lower than the tabulated value of 2.776 for 90% confidence and indicates that there was not a significant difference between the two data sets (ref. 12). The TIMS-^{148}Nd results are consistently higher than those of the reference method and an estimated t value of 7.3 indicates that a systematic difference between the two methods is highly probable. The difference is likely due to a ^{147}Nd neutron capture effect (ref. 5,6).

The HPLC sum of the major lanthanide results show poor agreement with those of the reference method. Irreproducible recoveries of the lanthanides in the actinide removal process is suspected, because the yields of the individual lanthanides varied relative to each other each time through the process, and the burnup varied when different chemical yield monitors were used (Er, Tb and Gd).

The INA-^{142}Ce results, except for element 4, are high compared to the

TIMS-$^{145+146}$Nd results. Cerium 140/142 ratios did not indicate significant contamination with natural Ce. Incomplete separation of Ce from fissile isotopes and subsequent production of ^{143}Ce by fission during the activation step is a possible source of error, but the gamma spectra did not show evidence of the other associated fission products. This may be due to the small quantities required to cause the error. A t-test between this and the reference method did not clearly indicate a systematic difference; more data are required.

For the second set of $^{145+146}$Nd burnups (Table 2), the initial heavy element content was estimated from the weight and initial composition of the fuel, instead of from Th and U concentrations. A t-test value of 1.55 indicated satisfactory agreement with the reference method for the limited number of samples. The weight based approach offers a large time saving because Th and U determinations are not required.

Of the methods examined in this study, the HPLC-^{139}La method offers the shortest analysis time and the least exposure to radioactivity, without sacrificing accuracy or precision. Coupled with the weight-based burnup calculation, time and cost savings of at least ten fold can be obtained. The applicability of this method may be limited to relatively low burnup fuels (% fission < 10%), because of significant weight loss of the fuel due to fission gas release and significant loss of ^{139}La by neutron capture at high burnups. These losses can be estimated and corrected for, but higher uncertainties will result. Nevertheless, there are many fuel samples for which the method is applicable.

REFERENCES

1. S. Banerjee, E. Critoph and R.G. Hart, Can. J. Chem. Eng., 53 (1975) 291-296.
2. Annual Book of ASTM Standards, Part 45, Nuclear Standards, American Society for Testing Materials, Philadelphia, 1980.
3. J.E. Rein, in Analytical Chemistry in the Nuclear Fuel Cycle, Status of Burnup Measurement Methodology, International Atomic Energy Agency, 1972, pp. 449-472.
4. R.P. Larsen, M.T. Lang, J.J. McCown and E.R. Ebersole, Proceedings of 8th Conference on Analytical Chemistry in Nuclear Technology, Gatlinburg, Tenn., Oct. 1964.
5. W.J. Maeck, R.L. Tromp, F.A. Duce and W.A. Emel, Idaho Chemical Programs Report ICP-1156, Allied Chemical Corp., Idaho Falls, 1978.
6. L.W. Green, C.H. Knight, T.H. Longhurst and R.M. Cassidy, Mass Spectrometric Determination of Burnup of Thorium-Uranium Dioxide Fuel, submitted for publication.
7. L.W. Green, N.L. Elliot and T.H. Longhurst, Separation and Mass Spectrometry of Nanogram Quantities of Thorium and Uranium from Thorium-Uranium Dioxide Fuels, in press.
8. R.M. Cassidy and S. Elchuk, Anal. Chem., 54 (1982) 1558-1563.
9. C.H. Knight, R.M. Cassidy, B.M. Recoskie and L.W. Green, Dynamic Ion-Exchange Chromatography for Determination of Number of Fissions in $(Th,U)O_2$ Fuel, submitted for publication
10. R. Kinsey (Ed.), Brookhaven National Laboratory Report BNL-NCS-17541 (ENDF-201), Brookhaven, New York, 1979.
11. G.J. Phillips and J. Griffiths, Atomic Energy of Canada Limited Report AECL-3857, Atomic Energy of Canada Limited, Chalk River, Ontario, 1971.
12. H.A. Laitinen and W.E. Harris, Chemical Analysis, McGraw-Hill, New York, 1975.

IRRADIATION AND GAMMA-RAY SPECTROMETRIC PARAMETERS FOR ^{129}I NEUTRON ACTIVATION ANALYSIS

J. H. KAYE, R. S. STREBIN, JR., F. P. BRAUER AND W. A. MITZLAFF

Radiological Sciences, Pacific Northwest Laboratory, P. O. Box 999, Richland, Washington 99352 USA

ABSTRACT

This paper describes the influence of reactor neutron irradiation facilities on low-level ^{129}I analysis methodology. Three reactors are compared - the HFIR at ORNL, the N-Reactor at Hanford, and a 1-MW research reactor at Washington State University, Pullman, WA. Parameters compared include sensitivity for ^{129}I and natural iodine and interference effects due to high levels of ^{127}I in irradiated samples. Selection of detector systems for off-site use at HFIR is discussed. A comparison of neutron activation analysis with other sensitive detection methods for measurement of ^{129}I is given.

INTRODUCTION

This paper discusses our experience using the High Flux Isotope Reactor (HFIR) at the Oak Ridge National Laboratory (ORNL) for neutron activation analysis of ^{129}I, and compares this reactor with the N-Reactor, 35 miles north of Richland, WA, and the 1-MW research reactor at Washington State University (WSU), Pullman, WA. Parameters to be compared will include 1) the effect of neutron flux on ^{129}I and ^{127}I measurement sensitivity and on interference due to production of ^{129}I from natural iodine, 2) irradiation containers 3) irradiation and cooling time limitations and 4) selection of gamma-ray spectrometer systems. A comparison of neutron activation analysis with other methods for measurement of ^{129}I is also given.

EXPERIMENTAL

1. HFIR Facility at ORNL

The High Flux Isotope Reactor at ORNL is a 100-MW, beryllium-reflected, light-water cooled and moderated reactor, used for the production of transuranium isotopes and for other irradiation experiments. It operates 24 hours a day for 21-23 days, then is shut down for about 18 hours for refueling due to burnup of the highly enriched ^{235}U fuel. Longer shutdowns are occasionally required for control plate changeout, maintenance and inspection. The reactor employs a unique flux trap design which provides a thermal neutron flux of up to 2.9 x 10^{15}n-cm^{-2} sec^{-1}. A compressed air operated rabbit system in the beryllium reflector allows irradiation in a thermal neutron flux of 5×10^{14} n-cm^{-2}-sec^{-1} and was used for our irradiations (ref. 1). Graphite irradiation capsules are loaded and unloaded in a hot cell; a large radiochemical hood is immediately adjacent to this hot cell.

The graphite irradiation capsule can accomodate at most two of our quartz irradiation ampoules of 4mm O.D. All irradiations were performed for us by Lamont C. Bate of ORNL. The post-irradiation chemistry and counting was done by Pacific Northwest Laboratory (PNL) personnel.

Counting was done in a room one floor above where the rabbit facility was located. A 15-cm diameter NaI(Tℓ) well crystal with provision for beta gating was used. Data was acquired with a Canberra Series 85, 8192-channel pulse height analyzer (PHA). The PHA was operated in recycle mode, with simultaneous output of regions of interest to a Texas Instruments Silent 700 teletype and total spectrum readout to a digital cassette. This data was taken back to PNL for processing. A downward pointing, intrinsic Ge detector was also set up to allow identification of interferences in purified samples.

2. Research Reactor at WSU, Pullman, WA

This research reactor is located in Pullman, WA, 150 miles by automobile from PNL. It is a 1-MW, light-water moderated, swimming pool-type reactor which utilizes a TRIGA core. In the irradiation position used by PNL a thermal neutron flux of $\sim 10^{13}$ n-cm^{-2}-sec^{-1} is available. The reactor operates for 8 hours a day, currently three days each week. Ampoules to be irradiated are loaded into a plastic container which can accomodate a large number of ampoules. Irradiations are generally done for $\sim$ 8 hours; however, some 16-hour irradiations have also been made. Samples are returned by a 4 hour automobile trip to PNL for postirradiation chemistry and counting.

3. Hanford N-Reactor

This reactor is used both for electric power generation and production of plutonium. It is graphite moderated, light-water cooled and generates 800MW of electrical power. It is located next to the Columbia River, 35 miles north of PNL. It is operated 24 hours a day, 7 days a week. However, it is shut down every summer for 2-3 months for scheduled maintenance, and has other outages for a variety of reasons.

Neutron irradiations may be performed in the core at a flux of $\sim 1 \times 10^{14}$ n-cm^{-2}-sec^{-1}, or in the reflector at a flux of 5×10^{12} n-cm^{-2}-sec^{-1}. A titanium irradiation container is used which can accomodate fourteen 6-mm O.D. quartz ampoules. The rabbit system is hydraulically operated, using water pressure, and discharges underwater. The ampoules may be de-canned from the irradiation container underwater, which minimizes radiation exposure.

DISCUSSION

Sensitivity

The achievable sensitivity for ^{129}I is directly related to the number of atoms of ^{130}I present in the sample when it is counted. Table I gives the sensitivity in units of picocuries of ^{130}I (corrected for decay to end of irradiation) per 10^8 atoms of ^{129}I. The table shows that the ^{129}I sensitivity at HFIR for a 400 minute irradiation is 2.5 times better than for a 1250 minute irradiation at N-Reactor, and the ^{129}I sensitivity for a 100 minute irradiation at the HFIR is about the same as for a 1250 minute irradiation at the N-Reactor. The ^{129}I sensitivity for the reflector at N Reactor is about the same as for the WSU reactor.

The natural iodine (^{127}I) content of a sample may be determined from measurement of ^{128}I produced by the reaction $^{127}I(n,\gamma)$ ^{128}I, or by measurement of ^{126}I produced by fast neutrons via the $^{127}I(n,2n)$ ^{126}I reaction. The threshold for the latter reaction is 11 MeV. The fourth column in Table I gives the sensitivity for ^{127}I in units of picocuries of ^{126}I per μg of ^{127}I. ^{127}I detection is not difficult even for a 10 minute irradiation at the HFIR.

TABLE I. Relative Iodine Activation Analysis Sensitivities at Three Reactors

REACTOR	IRRADIATION (MIN)	^{129}I $\left(\dfrac{\text{pCi }^{130}I}{10^8 \text{ atoms }^{129}I}\right)$	^{127}I $\left(\dfrac{\text{pCi }^{126}I}{\mu g\ ^{127}I}\right)$	^{129}I from ^{127}I $\left(\dfrac{\text{Atoms }^{129}I}{\mu g\ ^{127}I}\right)$
WSU	480	0.28	13	
N (Reflector)	1437	0.28	4.2	
N (Core)	1250	4.8	140	2.3×10^6
HFIR	10	2.1	21	
HFIR	100	4.3	53	
HFIR	400	12	180	3.6×10^7

Interference from Natural Iodine

For samples which contain microgram amounts of natural iodine, multiple capture of neutrons can lead to measurable ^{129}I production via the reactions $^{127}I(n,\gamma)$ $^{128}I(n,\gamma)$ ^{129}I (ref. 2). The last column of Table I gives the amount of ^{129}I produced in this way at the N-Reactor and HFIR. For a 400 minute irradiation at HFIR, 16 times as much ^{129}I is produced as at the N-Reactor. For such samples irradiation in a lower neutron flux is desirable, although once the ^{127}I level has been measured a correction can be made to the ^{129}I estimate. Measurement of the ^{129}I produced by irradiation of a standard containing only natural iodine is required for doing this.

Choice of Detector Systems

For use at the HFIR, it was desired to have high counting efficiency, but the degree of discrimination against other nuclides achievable by use of multiple gamma coincidence techniques was not desired. In fact, for irradiation of blanks it was desirable to not discriminate against other nuclides because we wanted to identify what they were. Therefore a 15-cm NaI(Tℓ) well detector was used, with provision for beta gating for background reduction. This detector has good resolution, 8.1% FWHM for ^{137}Cs at 662 keV, and good efficiency.

It was decided to use NaI(Tℓ) rather than Ge because Compton events in the Ge detector due to ^{126}I in the sample, produced from natural iodine as described earlier, can interfere with detection of low levels of ^{130}I. As mentioned earlier, we did set up an intrinsic Ge detector to aid in identifying interferences.

CONCLUSIONS

The HFIR has better sensitivity both for ^{129}I and ^{127}I due to the higher thermal neutron flux available. On the other hand, 24-hour irradiations are routinely available at the N-Reactor, which provide a sensitivity only a factor of 2 lower than at HFIR. Since only one or at most two samples can be run at a time at the HFIR, we did not run a comparator standard with each set of samples as is customarily done, but instead relied on the constancy of the reactor power level over a period of many days. As pointed out earlier, the higher thermal neutron flux can be detrimental for samples which contain high levels of natural iodine, although corrections can be made once the amount of natural iodine is known and is not excessive.

The HFIR has far greater sensitivity for ^{129}I and ^{127}I than the WSU reactor, and is clearly an excellent backup facility. The single radiochemical hood and single detector at this off-site location at the present time are somewhat limiting, however. Cooperation from the staff at ORNL has been outstanding.

In Table II a comparison is made between activation analysis and other methods currently being developed for measurement of ^{129}I. Some of these methods have not yet been applied to actual environmental samples, but may become competitive with activation analysis at some future date. A higher degree of pre-measurement chemical purification is currently required for other methods than for neutron activation analysis.

TABLE II. Sensitive Methods Available or Under Development for Measurement of ^{129}I.

METHOD	SAMPLE FORM	MINIMUM AMOUNT OF ^{129}I MEASURABLE (atoms ^{129}I)	$^{127}I/^{129}I$ ISOTOPIC RATIO	COMMENTS
Neutron Activation Analysis	I_2	10^7	$\leq 2 \times 10^9$ (a)	air, water, soil, vegetation samples analyzed (ref. 2-4)
Thermal Ionization Mass Spectrometry	AgI	5×10^7	$\leq 10^8$	air, water samples analyzed (ref. 5-6)
Thermal Ionization Mass Spectrometry	NaI	10^7	$\leq 5 \times 10^7$	only data on pure NaI published to date (ref. 7)
Tandem Accelerator Mass Spectrometry	AgI	3×10^6	$\leq 10^{13}$	meteorites, lunar rock samples measured (ref. 8-9)
Laser Induced Fluorescence	I_2	2×10^{10}	---	only data on pure I_2 published to date (ref. 10)

(a) assuming 1μg ^{127}I in sample, irradiated in core of N-Reactor

REFERENCES

1. T. M. Sims and J. H. Swanks, High Flux Isotope Reactor (HFIR) Experiment Facilities and Capabilities, Oak Ridge National Laboratory, 1979, 65pp.

2. F. P. Brauer, R. S. Strebin, Jr., J. E. Fager and J. H. Kaye, J. Radioanal. Chem., 72, No. 1-2 (1982) 501-514.

3. F. P. Brauer and R. S. Strebin, Jr., Standard Materials for Iodine Activation Analysis, in Nuclear Activation Techniques in the Life Sciences 1978, IAEA-SM-227/65, 1979, pp. 27-35.

4. F. P. Brauer and J. H. Kaye, IEEE Trans. on Nucl. Sci. 1974, NS-21 (1974) 446.

5. R. A. Rankin, R. A. Nielsen, F. A. Hohorst, E. E. Filby and W. A. Emel, Abstract FOD 7, 31st Annual Conference on Mass Spectrometry and Allied Topics, Boston, MA, May 9-13, 1983.

6. R. A. Rankin, F. A. Hohorst, R. A. Nielsen, E. E. Filby and W. A. Emel, Div. of Nucl. Chem. & Tech. Abstract No. 20, 186th ACS National Meeting, Washington, D. C., Aug 28-Sep 2, 1983.

7. J. J. Stoffels, Radiochem. Radioanal. Letters, 55, No. 2 (1982) 99-106.

8. D. Elmore, H. E. Gove, R. Ferraro, L. R. Kilius, H. W. Lee, K. H. Chang, R. P. Beukens, A. E. Litherland, C. J. Russo, K. H. Purser, M. T. Murrell and R. C. Finkel, Nature, 286 (1980) 138-140.

9. K. Nishiisumi, D. Elmore, M. Honda, J. R. Arnold and H. E. Gove, Abstract for 46th Meteoritical Society Meeting, Mainz, West Germany, Sept. 1983.

10. R. W. Goles, R. C. Fukuda, M. W. Cole and F. P. Brauer, Anal. Chem 53 (1981) 776-778.

ACKNOWLEDGEMENTS

The authors want to thank the staff at ORNL and WSU for their excellent cooperation, and in particular Lamont C. Bate for allowing us to share the radiochemical facility at HFIR. The authors gratefully acknowledge the support of this work by the United States Department of Energy. Work was performed under Contract DE-AC06-76RLO-1830.

MONITORING OF RADIONUCLIDES AND METALLIC IMPURITIES IN THE
KNK-II PRIMARY SODIUM

H.H. Stamm[1] and K.Ch. Stade[2]

[1]Kernforschungszentrum Karlsruhe, Institut für Radiochemie, Karlsruhe, F.R. Germany

[2]Kernkraftwerk-Betriebsgesellschaft m.b.H. Eggenstein-Leopoldshafen, F.R. Germany

ABSTRACT

The first German experimental LMFBR, KNK-II at the Karlsruhe Nuclear Research Center, is regularly monitored for impurities in its primary sodium system. Representative sampling from the liquid alkali metal is an even more critical step than from many other solvent systems.

Radionuclides in KNK primary sodium are determined by gamma spectrometry. After decay of ^{24}Na, the main activities are due to ^{22}Na, ^{54}Mn, ^{65}Zn and ^{110m}Ag. Fission products (^{137}Cs, ^{134}Cs and ^{131}I) have been detectable only after fuel element failures.

Fe, Cr, Ni, Mn, Co, Cu, Pb and Zn in sodium are determined by flameless atomic absorption spectroscopy. Detection efficiencies achieved are 2 to 10 ng metal/g Na.

INTRODUCTION

KNK-II is an experimental, sodium-cooled fast breeder reactor, located in the nuclear research center KfK near Karlsruhe/Germany. Like FFTF at Hanford, KNK-II is a loop-type LMFBR. It was operated until 1974 with a thermal core as KNK-I. After some reconstruction, it went into operation with its first fast core in 1977, and was shut down in 1982 at a maximum fuel burn-up of 100 000 MWd/t. In August 1983 the reactor achieved its full power (58 MW$_{th}$, 21 MW$_e$) again with the second fast core.

Like EBR II at Idaho Falls, KNK-II is widely used as an experimental facility. Therefore, the operation mode of the plant is governed by the experimental program rather than by energy production. Radionuclides and metallic impurities in the primary sodium coolant are determined for plant surveillance as well as for an extensive experimental radiochemistry program [1].

SODIUM SAMPLING

As with every analytical method, a reliable and representative sampling procedure is an absolute necessity for the determination of impurities in liquid sodium. That means first of all that the total sample composition has to remain unchanged until the analytical work begins. Because of the very low solubility of most of the impurities in liquid sodium at low temperatures, this is the main problem of sodium sampling.

When a sample is taken at a temperature of 320°C for instance, it has to cool down to room temperature before it can be removed from the sampling station. During that cooling period the impurities are segregating towards the outer layers of the sample and to the walls of the sampling vessel. This segregation effect was observed even for tracer amounts of such readily soluble radionuclides like ^{137}Cs and ^{110m}Ag where limited solubility gives no explanation. Several different sampling procedures have been tested during the past 30 years in order to overcome these segregation problems. The "overflow method" has proved to be the most reliable procedure[2].

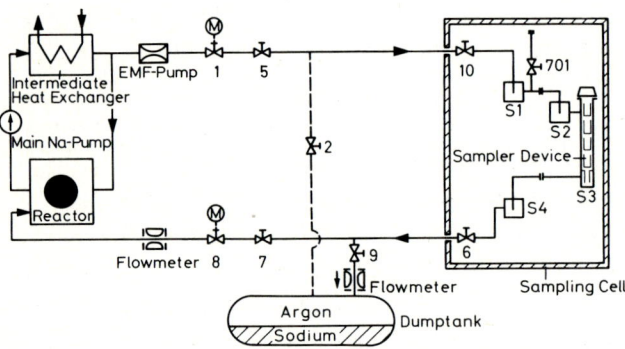

Fig. 1. KNK-II Primary Sodium Sampling Station

At KNK-II, an overflow sampling station is included parallel to one of its two primary heat transfer loops (Fig. 1). The sampling is performed in a special sampling loop, which branches off from the main primary system after one of the two intermediate heat exchangers; that restricts sampling temperatures at full power operation to 320°C. During down-times of the reactor, the sodium is circulating at 200°C, and sodium sampling is accomplishe at 200°C, too. A vertical sampling tube may contain up to seven crucibles with capacities of four to five grams of sodium each. The sodium return pipe passes an electromagnetic flowmeter before entering the reactor vessel again. To get the sodium flowing through the sampling station, the valves from 1 to 8 have to be opened, and the electromagnetic pump has to be activated. After the intended flushing time, valves 1, 2, 7 and 8 are closed, valves 9 and 10 are opened, and the excess sodium flows into the dump tank. About 30 grams of primary sodium are left in the seven crucibles inside the sampling station, they contain

1.9 Ci ^{24}Na (7.2 x 10^{10} Bq ^{24}Na) at full power operation of the plant. Since the sampling tube cannot be removed unless the activity has dropped down to a µCi level, it is necessary to wait for about 20 half-lives (12.5 days) of ^{24}Na, before the lead-shielded sampling cell may be opened. After cooling down to room temperature, freeze seals in the siphons S1 to S4 are protecting the samples and the station against moisture and oxygen. The sampling tube is now disconnected from the sampling station, and transfered to the laboratory. The crucibles are removed from the sample holder inside an inert gas glove box. The amount of sodium in each crucible is determined with an analytical balance.

PROCESSING OF THE SODIUM SAMPLES

The further processing of the sodium-filled crucibles depends on the analytical aim. No processing at all is necessary when gamma emitting radionuclides have to be determined by means of gamma spectrometry (most frequent determination). In that case each crucible is enclosed in a small, gastight plastic container and transfered to the counting room.

When the determination of metallic impurities is intended, the bulk sodium has to be removed from the crucible as a first step. The easiest way of removing metallic sodium from its non-volatile impurities is vacuum distillation. Still inside the argon glove box, each sample is placed separately into a glass distillation flask. The closed distillation flasks are evacuated to a pressure below 6×10^{-8} bar. The heating has to be done very carefully in order to avoid spattering. High frequency induction heating has proved to be superior to direct resistance heating.

RADIONUCLIDES IN PRIMARY SODIUM

Gamma-emitting radionuclides are determined directly by gamma spectrometry without any previous radiochemical separation. Ge(Li) detectors are used in connection with two multichannel analysers. The evaluation is done on-line by a PDP-11/34 minicomputer with the Canberra program SPECTRAN-F.

Results

Table 1 is a compilation of analysis results from the five past years. Depending on the operation mode of the plant, 15 to 20 sodium samplings have been accomplished per year. Most prominent radionuclides in the KNK primary sodium are ^{22}Na, ^{54}Mn, ^{65}Zn and ^{110m}Ag. ^{58}Co and ^{182}Ta are not regularly found in the samples, ^{51}Cr, ^{59}Fe and ^{60}Co only occasionally. Fission products originated from two fuel pin failures (April 1979, May/Sept. 1980), and from purposely defected experimental fuel pins (1981/1982), ^{131}I decayed according to its half-life after the failed fuel was removed from the core. Long-lived ^{137}Cs, however, kept on circulating with the sodium. It was reduced only to a certain amount by evaporating into all cover gas plenums (upper part of the reactor vessel, pumps, etc.).

TABLE 1

Radionuclides (nCi/g Na) in KNK-II Primary Sodium

Nuclides and Half-Lifes		Sampling Reactor Power (MWth)	September 1979 54	June 1980 55	August 1981 52	July 1982 29	August 1983 25
^{22}Na	2.60 a		284	378	424	600	377
^{51}Cr	27.70 d		81	84	19	-	-
^{54}Mn	312.2 d		478	811	357	359	42
^{58}Co	70.78 d		32	27	2.7	1.4	-
^{59}Fe	45.1 d		5.4	2.7	-	-	-
^{65}Zn	244 d		132	270	162	173	41
^{110m}Ag	249.9 d		78	108	105	141	46
^{124}Sb	60.3 d		35	54	38	38	3.3
^{182}Ta	114.43 d		27	22	27	-	0.4
^{131}I	8.02 d		-	3.5	-	-	-
^{134}Cs	2.06 a		-	-	2.2	-	0.6
^{137}Cs	30.17 a		5.4	8.1	43	27	21

TABLE 2

Radionuclides (nCi/g Na) in 6 Sodium Samples Sampling (4/4/82): 4 h at 320°C in Stainless Steel Cruc.

Holder Position	1	2	3	4	5	6	Average $\bar{x}$	Relative Standard Deviation S(%)
Sample Weight (g)	4.241	3.957	4.139	3.762	4.060	3.983	4.024	4.1
^{22}Na	532.4	538.4	532.6	538.5	527.5	540.5	535.0	0.9
^{54}Mn	128.0	75.4	128.1	133.2	132.0	127.3	120.7	18.3
^{65}Zn	130.3	119.0	123.6	127.3	122.9	123.0	124.4	2.7
^{110m}Ag	132.5	137.9	138.3	139.7	132.4	136.8	136.3	2.3
^{124}Sb	37.6	38.1	37.9	39.1	38.7	38.1	38.3	1.4
^{125}Sb	3.3	3.1	4.4	3.5	3.7	3.1	3.5	14.0
^{131}I	60.5	43.0	52.4	66.2	73.4	81.5	62.8	22.3
^{137}Cs	41.1	45.6	34.7	44.8	43.8	52.1	43.7	13.1
^{124}Sb/^{125}Sb	11.4	12.3	8.6	11.2	10.5	12.3	11.1	12.6

TABLE 3

Radionuclides (nCi/g Na) in 6 Sodium Samples Sampling (3/3/82): 4 h at 200°C in Stainless Steel Cruc.

Holder Position	1	2	3	4	5	6	Average $\bar{x}$	Relative Standard Deviation S(%)
Sample Weight (g)	4.316	3.970	5.176	5.316	5.221	5.085	4.847	11.6
^{22}Na	513.3	522.0	511.0	506.5	508.9	506.3	511.3	1.2
^{54}Mn	5.5	37.9	22.9	6.8	7.4	11.0	15.3	83.6
^{65}Zn	79.8	113.5	102.1	94.9	97.4	99.5	97.9	11.2
^{110m}Ag	131.5	135.6	130.9	127.9	130.7	129.6	131.0	2.0
^{124}Sb	18.6	26.4	22.5	21.6	19.5	23.9	22.1	13.0
^{125}Sb	1.6	3.0	1.6	1.8	1.9	2.1	2.0	26.5
^{137}Cs	11.9	9.0	10.2	7.3	11.1	8.2	9.6	18.3
^{124}Sb/^{125}Sb	11.6	8.8	14.1	12.0	10.3	11.4	11.4	15.8

The reproducibility of the measured activity concentrations have been tested several times. Table 2 shows as an example values obtained from six crucibles of one sampling at 320°C. The relative standard deviation S (last column) may be regarded as an index for the reproducibility of the values. Most of the activation products (nuclides above the dashed line) have relative standard deviations better (smaller) than the S-value of the sample weights. These are radionuclides generated by neutron irradiation of metallic elements readily soluble in liquid sodium at the temperature given. ^{54}Mn, however, produced by a (n, p) reaction from ^{54}Fe, is probably circulating with the coolant in form of small particles. Its relative standard deviation is higher than the S-values of the "soluble" activation products. If the solubility is determinant for the reproducibility, lower sodium sampling temperatures should result in worse (higher) relative standard deviations. Table 3 confirms this speculation: The S-values of ^{54}Mn, ^{65}Zn and ^{124}Sb are remarkably higher at 200°C than at 320°C. ^{110m}Ag is readily soluble in liquid sodium even at 200°C, its standard deviation remains low. From these results it has to be concluded that sodium sampling should be done at the highest possible Na temperature in the system. Therefore, at the German prototype LMFBR at Kalkar, SNR-300, sodium sampling will be possible at 540°C.

METALLIC IMPURITIES

Sample preparation and measurements

At KNK-II, flameless AAS is used for the determination of metallic impurities in the liquid sodium coolant. Several crucible materials (Mo, Ta, V, W) were tested first with respect to blanks and detection limits. Finally, tantalum was chosen as crucible material with the lowest blanks for most of the metallic impurities (Table 4).

TABLE 4

Blanks and Detection Limits of Metallic Impurities in Liquid Sodium (4 g Na Samples in Ta Crucibles)

Impurity	Blank (ng/g Na)	Det.Limit (ng/g Na)
Cr	< 2	2
Mn	6	1
Fe	30	7
Co	< 3	3
Ni	20	9
Cu	50	1

After distilling off the bulk sodium as described earlier, the residue in the crucible is dissolved in 2 ml of hot HCl, and the solution is dilluted to 20 ml with double distilled water. 20 µl of this dilluted solution is injected into the graphite furnace (HGA-76, automatic sample injection divice AS-1) of the atomic absorption spectrometer (Perkin-Elmer Model 420). Since 1981/82, the impurity determination is extended to Ca, Ag, Zn, Pb and Sn. Ca, Ag and Sn are determined from the vacuum distillation residue as described. Lead and zinc, however, are volatile under the conditions of Na high-vacuum distillation. Therefore, it is necessary for these determinations to dissolve the bulk sodium in an alcohol/water mixture, to neutralize with HNO_3, and to add water to a total of 500 ml. Again, aliquots of 20 µl are injected into the graphite furnace. Because of the high salt concentration in these solutions, the detection limits for lead (0.3 µg/g Na) and zinc (0.5 µg/g Na) are much higher than for impurities determined from the distillation residue (table 4). However, these values might be acceptable for routine analyses [3].

Results

TABLE 5
Metallic Impurities in KNK Primary Sodium (1978)

Date of Sampling	Sampling Temperature (°C)	Oxygen Content (µg/gNa)	Metallic Impurities (ng/gNa)					
			Fe	Cr	Mn	Ni	Co	Cu
January, 1	200	9	5900	54	113	80	<3	116
February, 2	320	6	100	2	6	40	<3	500
March, 3	320	7	300	32	7	<9	<3	<1
April, 9	200	6	1100	18	8	10	<3	54
April, 18	200	10	4300	54	40	80	<3	<1
June, 3	200	11	3000	120	61	100	<3	<1
July, 4	215	4	2200	140	49	80	<3	136
September, 3	200	n.m.*	1000	20	12	40	<3	<1
October, 28	205	13	7900	220	620	90	<3	8

* n.m. = not measured

Table 5 gives a compilation of analysis results for one year (1978) of reactor operation. Fe was always the predominant metallic impurity. Its amount in the sodium was lower by a factor of 10 in samples taken at 320°C compared to 200°C samples. An increase of the oxygen content resulted in a higher iron content. Similar dependencies were noticed for Cr, Mn, and Ni, too. Increasing Fe values were always associated with an increase of Cr, Mn and Ni. The amounts of Fe, Cr and Mn in the primary coolant were always higher than their known solubilities [4]. Therefore, it has to be assumed that Fe, Cr, Ni and Mn are mainly suspended in the liquid alkali metal in form of small particles. This speculation is supported by high relative standard deviation in parallel samples (see table 6). The source of such particles is the wall material of components and piping, a Nb-stabilized, ferritic steel (2 1/2 Cr - 1 Mo). The composition of this steel is very similar to the "relative composition" of the impurities in the primary sodium (see table 6). Co was never detectable in KNK-II sodium. The wide variations of the Cu values are not explainable so far. Cu should be present as

TABLE 6

Metallic Impurities in 6 Na Samples (ng/g Na) Sampling (7/10/82): 4 h at 320°C in Tantalum Crucibles

Holder Position	1	2	3	4	5	6	Average $\bar{x}$	Relative Standard Deviation S(%)	Composition (%) of Impurities	KNK Structure Material
Sample Weight (g)	4.501	4.886	4.523	4.891	4.465	4.727	4.666	4.2		
Iron	1200	400	600	400	1200	980	800	47.2	91.7	93.8-95.4
Chromium	28	36	18	14	40	34	28	37.1	3.2	2.0- 2.5
Manganese	18	47	16	6	19	7	19	78.3	2.2	0.4- 0.8
Nickel	<9	<9	<9	<9	<9	<9	-	-	<1.0	0.3- 0.8
Cobalt	<3	<3	<3	<3	<3	<3	-	-	<0.3	-
Copper	<1	<1	<1	<1	10	16	13	-	-	-

TABLE 7

Pb and Zn (μg/g Na) in KNK Primary Sodium (1982)

Metallic Impurity	Sampling Reactor Power [MWth]	March 6 —	March 13 —	May 29 58	July 23 30	August 11 30	October 25 —
Zinc (Zn)		3.5	3.5	7.5	7.0	4.0	2.0
Lead (Pb)		2.0	2.0	3.0	5.0	4.0	2.0

metal and readily soluble in liquid sodium [5].

Ca is determined only occasionally. Its concentration is varying from 40 to 100 ng/g Na, and it is much lower than the Ca content of the original KNK sodium delivered by the manufacturer. Ag, Pb, Sn and Zn are metallic impurities with high solubilities in liquid sodium (4). Silver is found regularly in all sodium samples with a mean value of 1.5 to 2 μg/g Na. Table 7 gives results of six samplings for Pb and Zn. The standard deviations of the mean values of all the better soluble metals are smaller (better) than the S-values of Fe, Cr, Mn and Ni. However, as already mentioned earlier, higher sodium sampling temperatures would improve the reproducibility. So far, the highest possible sodium sampling temperature at KNK-II is 320°C, because the sampling loop is branching off from the main loop <u>after</u> the intermediate heat exchanger (IHX). In order to improve the analytical results further, the branch of the sampling loop should be moved to a point <u>before</u> the IHX. The reproducibility of analysis results for metals with low solubilities (ppb) like Fe, Cr, Mn and Ni, however, cannot be improved by increasing the Na sampling temperature. They are mainly transported in form of suspended particles at all possible temperatures. In order to improve the accuracy of the Fe-, Cr-, Mn-, and Ni-values, parallel samples have to be taken. Outliers should be eliminated before mean values are calculated as analysis result. Future experiments will demonstrate if stainless steel filters in the entrance of the sampling device are helpful in this respect.

CONCLUSIONS

Sodium sampling for the determination of readily soluble radionuclides and metallic impurities should be done at the highest possible sodium temperature.

The reproducibility is sufficient to rely on one crucible per sampling for radionuclide determination. Radiochemical separations are not necessary for the determination of gamma emitting radionuclides.

Metallic impurities with high solubities in liquid sodium (Zn, Pb, Ag) are present in concentrations of several ppm. Metals of low solubilities (ppb level) like Fe, Cr, Mn and Ni are mainly circulating with the flowing sodium in form of small particles. For reliable results several crucibles of one sampling have to be used for analysis, and the results should be treated with statistical methods.

REFERENCES

1 H. H. Stamm, H. Richard, K. Ch. Stade, Proc. 2nd Int. Conf. Liquid Metal Technology in Energy Production, Richland/WA, April 20 - 24, 1980, pp. 17/58 - 66.
2 J. M. F. Rohde, M. Hissink, L. Bos, J. Nucl. Energy 24 (1970) 503 - 508
3 S. P. Awasti, H. U. Borgstedt, J. Nucl. Mat. 116 (1983) 103 - 111
4 T. D. Claar, Reactor Technol. 13 (1970) 124 - 146
5 T. Berry et al., Analyst 108 (1983) 189 - 195

SURFACE CHARACTERIZATION OF LEADSCREWS TAKEN FROM THE TMI-2
REACTOR VESSEL

K. J. Hofstetter[1], H. Lowenschuss[2], and V. F. Baston[3]
[1]GPU Nuclear Corporation, Middletown, Pennsylvania
[2]Eidg. Institut fur Reaktorforschung, Wurenlingen, Switzerland
[3]Physical Sciences Incorporated, Idaho Falls, Idaho

ABSTRACT

Analyses of leadscrews have revealed a 25-40 µm tightly adhered surface layer containing ^{137}Cs and ^{134}Cs (up to 1000 µCi/cm^2). A second more loosely bound layer (70-90 µm) is also present containing principally core debris. Dissolution tests show ∿90% of the radiocesium is associated with the tightly adhered layer. Surface analyses (SEM, microprobe and Auger spectrometry) and macroanalyses (emission spectrometry, XRF and XRD) confirm the presence of corrosion materials and core debris on the surface.

INTRODUCTION

The leadscrews from several Control Rod Drive Mechanisms (CRDM) were removed from the reactor vessel at Three Mile Island - Unit 2 (TMI-2). The leadscrews were removed to permit in-core camera inspections, to characterize the contamination on reactor internal surfaces and to assess damage to the core. Portions of these leadscrews have been subjected to extensive laboratory examinations and tests as a means of anticipating conditions that may be encountered during reactor head and fuel removal operations.

EXPERIMENTAL

Three leadscrews, one from the center CRDM, one from an outer CRDM and one from approximately mid-radius, were removed from the reactor vessel. The middle threaded portion of the center leadscrew (made of 17-4PH stainless steel) was cut into three short sections. Two of these sections were sent to laboratories for detailed metallographic analyses while one section was retained for chemical characterization studies. The chemical tests involved soaking one leadscrew section (approximately one foot long) in successive solutions arranged in increasing order of chemical aggressiveness.

The solutions used were: 1) demineralized water, 2) borated water (3500 ppm boron as boric acid adjusted to pH 7.5 with NaOH and 1% TRITON X-100 surfactant), 3) sodium carbonate and hydrogen peroxide (5.0 Wt.% Na_2CO_3 and 1% H_2O_2), 4) a standard two step decontamination process-APC ((i) 10% NaOH - 3% $KMnO_4$ and ii) oxalic acid (25 g/l) - dibasic ammonium citrate (50 g/l)), and 5) nitric and hydrofluoric acid (10 Wt.% HNO_3 - 0.1M HF).

Each solution was analyzed for alpha, beta, and gamma emitting radionuclides in both the liquid and insoluble phases. The soluble portion was also analyzed for elemental content by source excited X-ray fluorescence (XRF). The radionuclides remaining on the leadscrew after each soaking operation were determined by gamma-ray spectrometry.

TEST AND ANALYSES RESULTS

The chemical and radiochemical data obtained from the leadscrew soaking tests have been evaluated in an attempt to better understand the character of the leadscrew surface (ref.1). The analysis of demineralized water used for the first leadscrew soak showed low boron, sodium and radionuclide concentrations. The radiochemical assay confirmed the presence of a large source of soluble radionuclides (principally ^{137}Cs and ^{134}Cs) on the leadscrew surface that could not be accounted for by evaporated coolant. The demineralized water contained other radionuclides (^{144}Ce, ^{125}Sb, ^{106}Ru, and ^{60}Co). The quantity of radionuclides remaining on the leadscrew after the 24-hour demineralized water soak did not change, however. The same was true after the 24-hour soak in borated water. Gamma-ray spectroscopy on the leadscrew before and after each soak operation revealed that the distribution of radionuclides on the leadscrew remained unchanged until the nitric-hydrofluoric acid soak. The inventory of radionuclides remaining on the leadscrew section is shown in Fig. 1.

As the entire leadscrew was immersed in successive solutions, it was assumed that each solution would have removed all species removed by the preceding solutions and thus the experiments were cumulative. Using this hypothesis, accumulated concentrations in the soak solutions, decay corrected to the date of TMI-2 accident, are shown in Fig. 2.

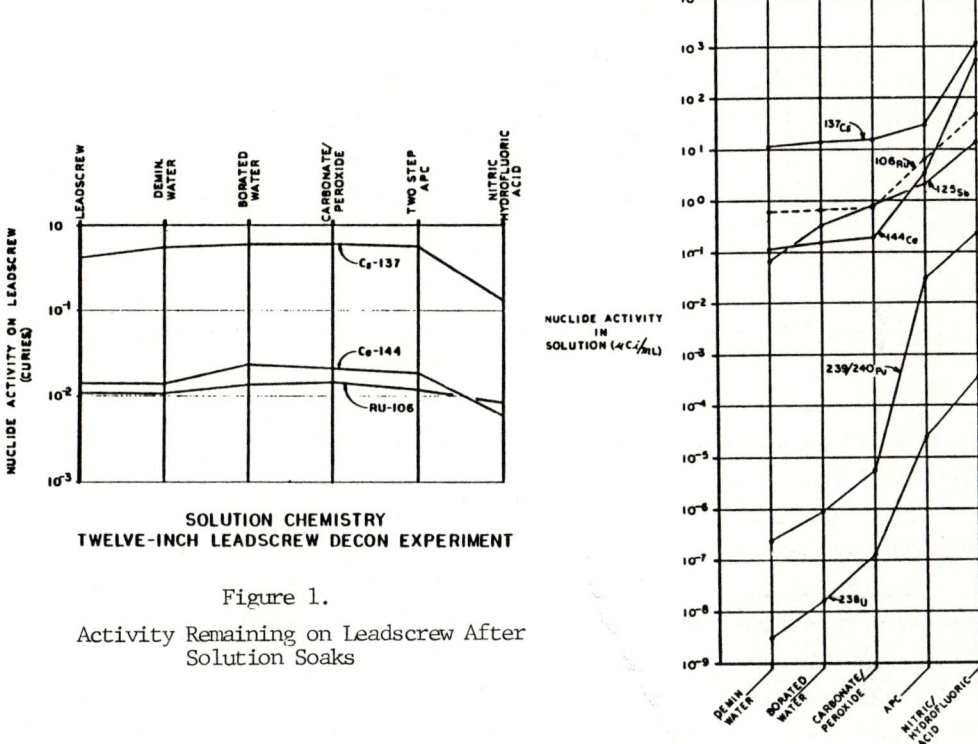

Figure 1.
Activity Remaining on Leadscrew After Solution Soaks

Figure 2.
Solution Concentrations

The chemical soak solutions were selected to evaluate the leadscrew surface chemistry. The carbonate/peroxide solution will dissolve fuel and associated debris. Examination of Fig. 2 shows a sharp increase in the fuel-related nuclides (i.e. ^{239}Pu, ^{240}Pu and ^{238}U) after soaking in this solution. The two step APC process will remove corrosion products and associated radionuclides. Increased quantities of radionuclides were removed by this solution as evidenced in Fig. 2. The nitric-hydrofluoric acid solution removes base metal and was the only solution which effectively removed the radionuclides on the leadscrew surface. Using the Fe concentrations in each solution as determined by XRF, the depth of base metal removed by all solutions was determined to be approximately 0.33 μm.

The general conclusions from the leadscrew solution soak tests are: 1) the ^{137}Cs and ^{134}Cs makeup the majority of gamma emitting radionuclides present on the surface of the leadscrew, 2) the cesium activities are located on the surface and have not penetrated the base metal more than 0.5 µm and 3) the mechanism of cesium absorption is unknown but is not correlated with any major elements associated with corrosion products, fuel or cladding.

More detailed analyses of the leadscrew surface on the other two short sections of the central leadscrew were performed. Approximately 4.5 grams of leadscrew debris was removed from one section by light brushing. Particle size analyses, scanning electron microscopy (SEM) and X-ray diffraction (XRD) analyses were performed on the debris. The results of these analyses showed that more than 90% of all the debris particles were less than 10 µm in diameter and the majority of the debris was Fe, U, Zr, Si, and Al (ref.2).

The analyses of the other sections revealed that the activity was contained in two distinct layers. The outer loosely adherent (LAD) layer contained 10% of the ^{137}Cs activity while the inner tightly adherent (AD) layer contained 90% of the ^{137}Cs activity. No cesium intergranular attack of the base metal was observed. Both the LAD and AD were analyzed for isotopic activity levels and elemental constituents. The analyses were performed on approximately 80 mg of deposits removed from about two square inches of leadscrew surface and included SEM examinations, microprobe and surface layer examinations (i.e. Auger, SEM/EDX), gamma spectroscopy and emission spectroscopy for each layer (ref.3).

DISCUSSION

The cesium activity data reported from the solution chemistry studies, the gamma spectroscopy of the threaded leadscrew, and the laboratory analyses of the LAD and AD surface deposits all indicate the cesium appears to be chemically bound to the leadscrew surface. A schematic description of the the leadscrew as determined by SEM examinations is illustrated on the following page. There is a corrosion oxide layer on the base metal, probably formed before the accident, approximately 1-3 µm thick. The AD layer is enriched in chromium oxides and is 20-40 µm thick. The principal cesium activity is associated with a layer less than 1µm thick. The LAD layer is very porous and is 70-90 µm thick. The presence of cesium tightly-bound to the surface was certainly unexpected.

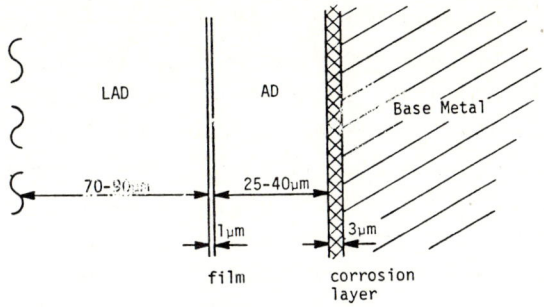

Figure 3.
Leadscrew Surface

Correlations of the solution data with microchemical data indicate that the cesium activity may be associated with a thin adherent "film" of less than 1 μm. In an attempt to understand the mechanisms responsible for the surface phenomena, emission spectrometry (ES) analysis of the surface deposits provided accurate elemental compositions in both regions (LAD and AD).

The ES results on the LAD and AD were compared to the XRF analysis results of the soak solutions after several assumptions. First, it was assumed that only the LAD was removed by the demineralized water and borated water soaks. Secondly, it was assumed that only the AD was removed by the HNO_3-HF soak. These data were used to calculate the relative quantity of elements removed by the demineralized and borated water soaks. A similar analysis was performed on the HNO_3-HF data. These results are summarized in Table 1.

TABLE 1

Comparison of Emission Spectrometry Analysis Results with The Solution Soak Analysis Results for The Loosely Adherent (LAD) and The Adherent (AD) Layers on The Leadscrew Surface

	LAD		AD	
	XRF	ES	XRF	ES
Element	Wt.%	Wt.%	Wt.%	Wt.%
Cr	3.9	3.2	2.6	15.7
Fe	34.1	31.5	26.1	23.1
Ni	1.7	1.6	0.9	1.0
Cu	2.2	2.4	2.1	0.6
Mo	5.0	0.5	0.8	0.2
U	23.0	15.0	9.2	0.9
Zr	25.6	8.1	52.6	1.6
Ag	----	14.8	----	2.0

Comparison of these data with the ES results suggest that the ES subsamples were representative. As can be seen from the data in Table 1, Fe, Cr, Zr, Cu, and U account for the majority of the LAD composition whereas Fe and Cr account for the majority of the AD composition. SEM/EDX and microprobe data indicate that Ag, Zr, U, and Sn are attached to the AD layer in discrete particulate phases. Using elemental ratios, the LAD debris can be shown to consist of core debris including fuel, zircaloy cladding, 304 stainless, Inconel, and Ag/In/Cd poison rods. The adherent layer is composed of corrosion products enriched in Cr and Mo relative to the base metal.

The mechanism of corrosion, transport of fuel debris, and tightly bound nature of cesium activities are not well understood. Chemical vapor deposition, quenching or brazing, and various chemical reaction mechanisms have been postulated. Additional microchemical evaluations are required to determine the mechanism of corrosion and of cesium adherence. These tests and analyses are planned for future artifacts from the TMI-2 core.

REFERENCES

1 J. A. Daniel, E. A. Schlomer, D. G. Keefer, T. L. McVey, K. J. Hofstetter, T. D. Lookabill, V. F. Baston and H. Lowenschuss, *Analysis of the H-8 Leadscrew from The TMI-2 Reactor Vessel*, Science Applications Inc. Report, SAI-139-83-04-RV (May 1983).
2 R. L. Clark, R. P. Allen, M. W. McCoy, *TMI-2 Leadscrew Debris Pyrophoricity Study*, Pacific Northwest Laboratory, GEND-INF-044 (August 1983).
3 G. M. Bain and G. O. Hayner, *Initial Examination of The Surface Layer of A 9-Inch Leadscrew Section Removed From TMI-2*, The Babcock & Wilcox Company Lynchburg Research Center, Research Project 2056-2 Task 1, Final Report (June 1983).

An n-TYPE HIGH-PURITY GERMANIUM DETECTOR-BASED ON-LINE RAPID COAL ANALYSIS SYSTEM

F. E. LeVert and W. W. Managan, Technology for Energy Corporation, One Energy Center, Pellissippi Parkway, Knoxville, Tennessee 37922

ABSTRACT

This paper describes an n-type high-purity germanium (HPGe) detector-based coal analysis system* that is capable of providing a real-time (200-second update time) analysis of sulfur and other elemental constituents of coal. The analysis system consists of a 234-μg californium-252 neutron source with suitable moderation and shielding, state-of-the-art pulse-shaping and processing electronics coupled to a 23.4% n-type HPGe detector, a 7000-series EG&G ORTEC multichannel analyzer, and a PDP/11 computer.

This system is capable of processing very high input pulse rates with low dead time. System input rates of 75 kHz to a gated amplifier were processed with a live-to-real-time ratio of greater than 0.55 for the pulses in the range of interest of this measurement (i.e., from 2 MeV to 10 MeV). Results are also presented which compare the measured and predicted throughput pulse rates for different spectroscopy systems.

When normalized to the ASTM weight percentages, the PGNAA results showed close agreement with the ASTM-measured values for sulfur and fixed carbon in the diverse coals. The comparative error in weight percent for the two techniques was less than 3.0%, and the net peak counts uncertainty was approximately 3.8%. In a typical 200-second run, the net peak counts uncertainty for the hydrogen, aluminum, silicon, iron, calcium, sodium, nitrogen, and chlorine in the coal varied between 0.2% (for hydrogen) and 17.5% (for calcium). These measurements demonstrated that a coal analysis system can be constructed that has an update time of 200 seconds with a single source and detection channel.

---

*Patent pending

Background

The increasing importance of coal for electric power generation as well as for gasification and liquefaction has led to a growing need for on-line analysis of coal composition and quality in large process streams. The need for real-time analysis of coal stems mainly from the large variability in the composition of coal which can affect the efficiency of these combustion or conversion processes. Several methods for performing on-line elemental analysis of coal have been reported. The method which has received the most study and development effort has been the Prompt Gamma Neutron Activation Analysis (PGNAA) technique. Systems based on the PGNAA method are designed to provide continuous monitoring of sulfur or full elemental analysis of the coal.

The U.S. Bureau of Mines initially developed and tested an on-line PGNAA sulfur meter in 1975 (refs. 1 and 2). This system, which used a NaI (Tl) detector as the gamma ray sensor, was installed at a coal preparation plant where it successfully monitored the sulfur content of coal flowing through a bin. Today there are three sulfur meter systems in operation in the USA. The three units are at Detroit Edison Monroe Power Plant, The Department of Energy site at Homer City, Pennsylvania, and the Paradise Steam Plant (ref. 3) near Drakesboro in Western Kentucky. These systems are used either for blending high- and low-sulfur coal for SO_2 compliance or to monitor coal preparation systems.

None of the deployed systems discussed above are designed to provide complete coal analysis. However, there has been an ongoing effort sponsored by the Electric Power Research Institute (ref. 4) and by DOE at Argonne National Laboratory (ref. 5) and by other researchers to develop a prototype continuous nuclear coal analyzer (ref. 6). These efforts have addressed the development of a continuous on-line coal analyzer with near real-time capabilities.

EXPERIMENTAL SETUP

Test Apparatus

The experimental setup for the TEC Coal Analyzer was designed to simulate, as closely as possible, a conveyor belt geometry (see Figure 1). That is, the system was designed to ensure that the geometrical

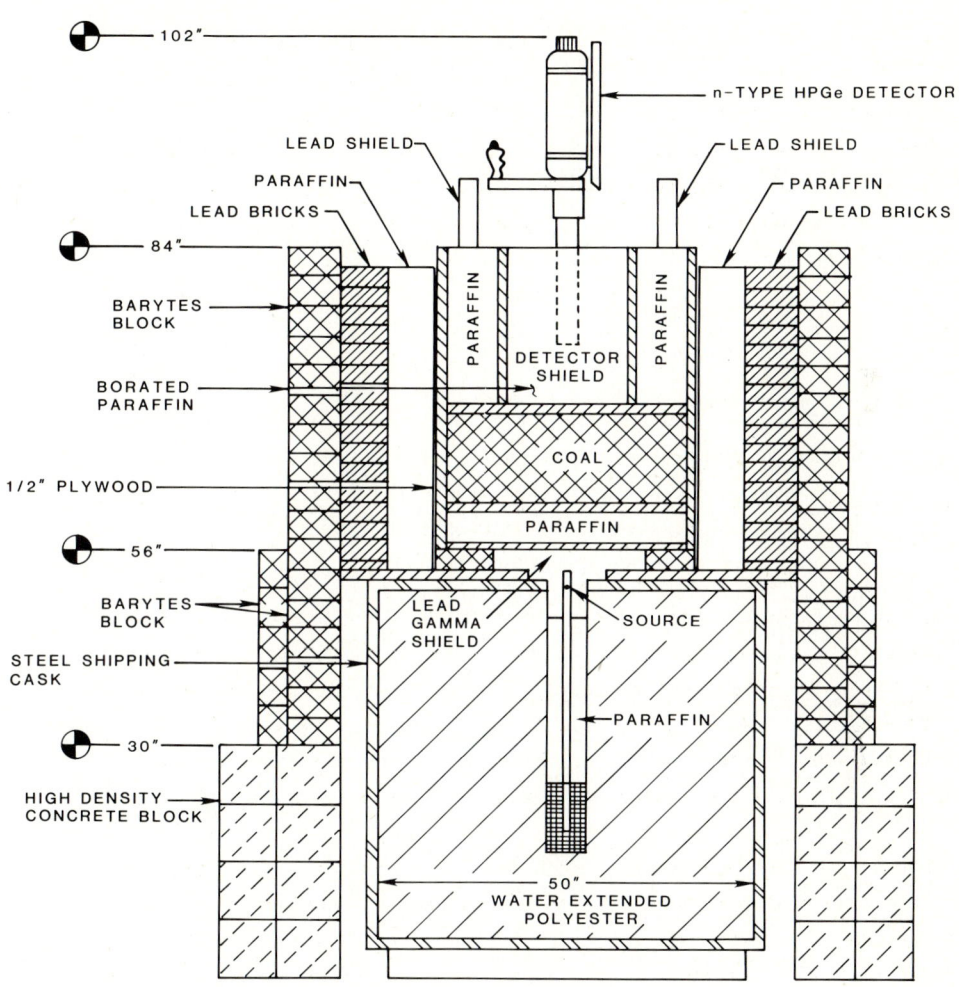

FIGURE 1. EXPERIMENTAL SETUP FOR COAL ANALYSIS EXPERIMENT

constraints most likely to be encountered in a fossil station would not seriously impact the results of this study. Therefore, the measured data could be used to predict design parameters in a prototype system. It is desirable to install the system without cutting the belt, i.e., to install it around an existing belt. However, since this is not practical in many existing plants, the system can easily be adapted to a chute or other coal-handling arrangements.

A n-type HPGe detector was used in these measurements. The detector had an efficiency of 23.4% compared to a 7.62 cm x 7.62 cm NaI(Tl) crystal at the 1.33 MeV Co-60 photon energy. The active volume of the 23.4% detector was 108 cm^3.

The neutron source used contained 234 µg of Cf-252 with an estimated neutron yield from spontaneous fission of ~5.8 x 10^8 neutrons/second. The source was double encapsulated, the outer stainless steel cylinder ~1.905 cm diameter and 3.81 cm long. The source also emits fission product gamma rays. During PGNAA operation, the source is surrounded by a lead shield (see Figure 1) which reduces the interference in the gamma spectra by fission product gamma rays from the source.

High Pulse Rate Throughput Circuitry

The electronic measurement system for a coal analyzer must be capable of processing high random rates with very good resolution. Many systems on the market will accept high input data rates, but the pulse processing time and pile-up rejection schemes often limit the maximum undistorted output data rates to 10,000 to 20,000 pulses per second. Typical pulse processing time for Gaussian filter amplifiers used with germanium coaxial detectors is 12 to 20 microseconds, while analog-to-digital converters (ADCs) typically require 20 to 40 microseconds to digitize and store the pulse information. Figure 2 is a schematic representation of the electronic system used during these measurements.

The system in Figure 2 was limited to an input count rate of approximately 80,000 cps because of the resistive feedback used in the preamplifier. Pulsed reset preamplifiers now available can extend the input count rate capability of this system to 350,000 cps. The input count rate should be limited to approximately 200,000 cps since this rate produces the maximum output data rate. Input count rate above

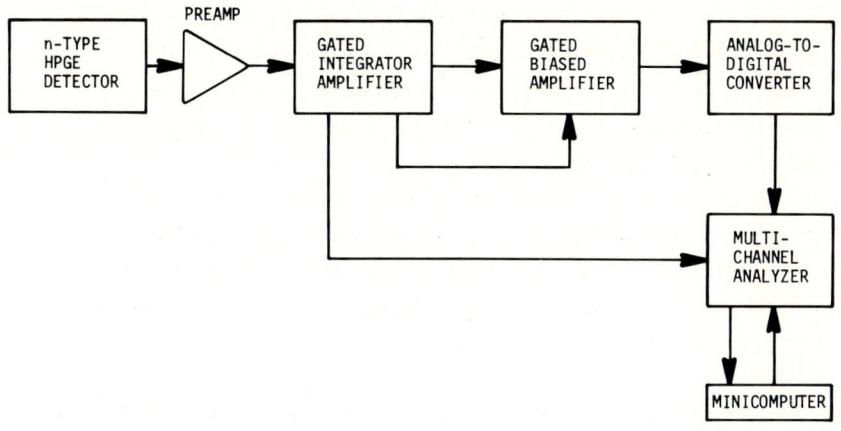

Figure 2. Detector, Signal Conditioning Electronics, and Data Processing Equipment Used in the Coal Analyzer System

200,000 cps decreases the output data rate and the detector lifetime without any corresponding benefit. The bias amplifier in Figure 2 was set to reject all pulses whose amplitudes corresponded to gamma rays with energies less than 2 MeV.

Discussion of Results

Two different coals were examined during these measurements. Coal A was ground to .635 cm x 1.27 cm size and packed in coal box 1 (see Figure 1) to a bulk density of 0.72 gm/cm^3. Coal B (a pyritic coal) was separated into two sizes with bulk densities of ~0.72 and ~0.80 gm/cm^3, respectively. Figure 3 shows the results of a 939-second live-time measurement on a 200 lb sample of Coal B.

Once the system was optimized in terms of electronics and source-detector distances, the GELIGAM-NAAC code (ref. 7) package was applied to Coal A and Coal B. Coal A was used as the known coal, and the two bulk-density coals from Coal B were considered as unknown coals. The measurements on Coal A (the known coal) were done in 5,000 seconds of live time. The sulfur and carbon weight percentages for Coals A and B for a 200-second real-time interval for a 23.4% HPGe detector are given in Table 1. Because of the sensitivity to the shape of the peak in known coal, this method was found to be extremely sensitive to random changes in the background, to the number of channels in the peak, and to gain shifts between runs. However, none of these potential problems were observed during the measurements of Table 1. The averaged results for Coals A and B show that the measurement is very precise, even with one detector/source channel.

CONCLUSIONS

Update analysis times of 200 seconds for sulfur and carbon weight percentages can be performed by this single source/detector system. This is achieved by use of (1) a strong Cf-252 neutron source; (2) a neutron-damage-resistant detector; and (3) state-of-the-art electronics.

The measured PGNAA results were in close agreement with those of classical analytical methods used by an independent laboratory to measure the sulfur and carbon weight percents.

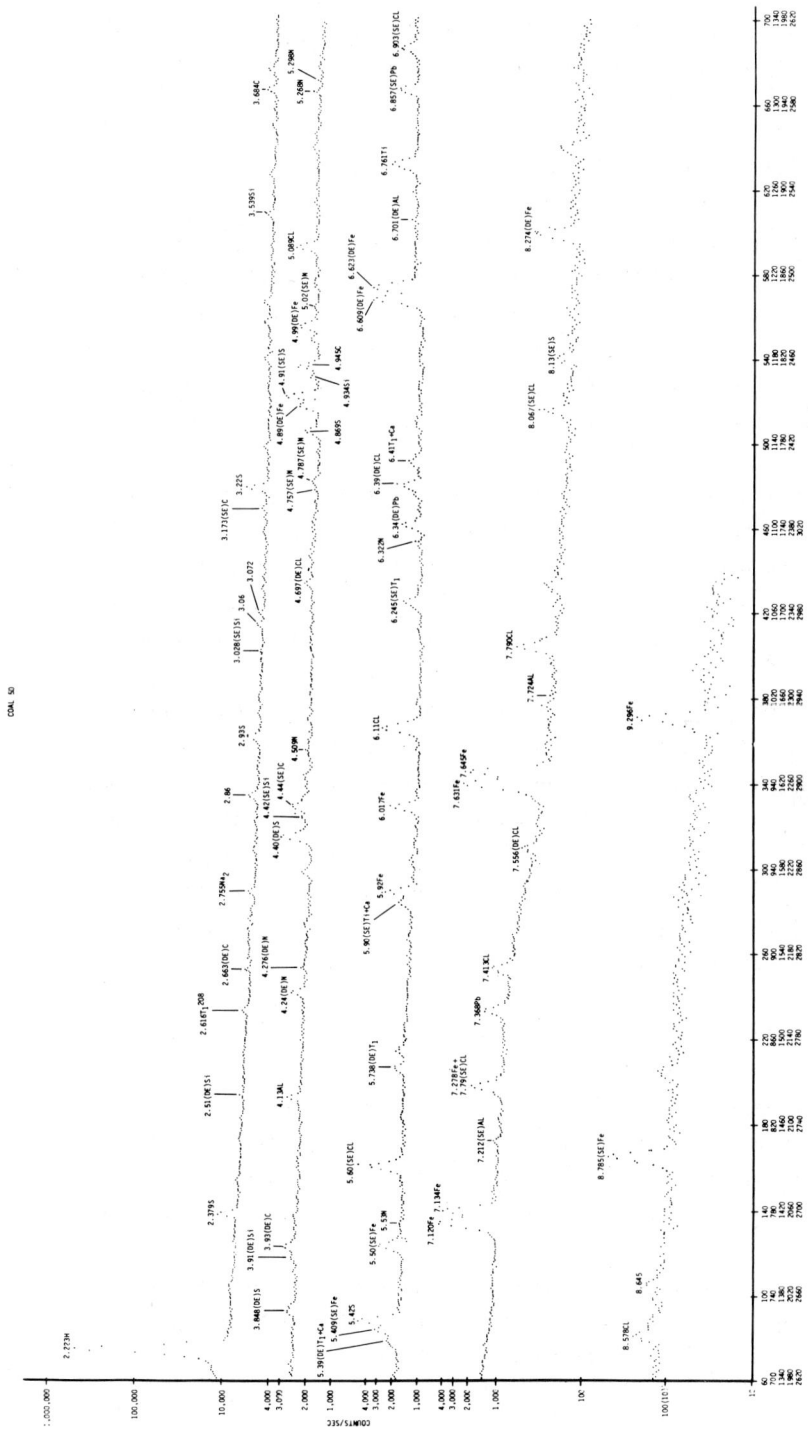

305

TABLE 1

On-line sulfur and carbon results obtained using GELIGAM with 5000 and 200 seconds of real-data acquisition time.

	Coal A		Coal B	
	Lab Analysis	PGNAA	Lab Analysis	PGNAA
	5000 seconds L.T.^{a}			
Fixed Carbon	53.65	53.6	51.03	51.00 ± 3.06
Sulfur	1.09	1.09	3.00	3.04 ± .18
		200 seconds R.T.^{b*}		200 sec. R.T.^{b*}
		52.95		51.2† ± 5.
		1.1		3.01† ± .1
				51.5†† ± 3.81
				3.05†† ± .24

†Bulk density estimated as 0.72 gm/cm^3

††Bulk density estimated as 0.80 gm/cm^3

*Values represent an average of ten 200-second real-time measurements

^{a}L.T. = Live Time.

^{b}R.T. = Real Time.

To function in a nonlaboratory setting where the constituents of the coal could vary widely, this system has to be integrated with systems capable of measuring moisture, density, and epithermal neutron leakage from the coal and coal handling to allow normalization for these secondary sources of variation.

REFERENCES

1. A. W. Hall, I. W. Martin, R. F. Stewart, and A. M. Poston, "Precision Tests of Neutron Sulphur Meter in Coal Preparation Plant," Bureau of Mines Report No. 8038, 1975.

2. R. F. Stewart, et. al, "Nuclear Meter for Monitoring the Sulfur Content of Coal Streams," TPR 74, Bureau of Mines, Jan. 1975.

3. D. A. Humpris, R. W. Lynch, and C. M. Huang, "On-Line Monitoring of Coal," presented at the Coal Testing Conference, Lexington, Kentucky, March 2-3, 1981.

4. G. Reynolds, H. Bozorgmanesh, E. Elias, T. Gozani, T. Maung, and V. Orphan, "Nuclear Assay of Coal, Volume 1, Coal Composition by Prompt Neutron Activation Analysis - Basic Experiments," EPRI IP-989, 1979.

5. C. L. Herzenberg, N. M. O'Fallon, B. S. Yarlagadda, R. W. Doering, C. E. Cohn, and K. G. Porges, "Neutron-Induced Gamma Spectrometry for On-Line Compositional Analysis in Coal Conversion and Fluidized-Bed Combustion Plants," Proceedings of the Third International Conference on Nuclear Methods in Environmental and Energy Research, Columbia, Missouri 1977.

6. H. R. Wilde and N. Herzog, "On-Line Analysis of Coal by Neutron Induced Gamma Spectrometry," Proceedings of the 1981 Symposium on Instrumentation and Control for Fossil Energy Processes, San Francisco, California 1981.

7. EG&G ORTEC, "Neutron Activation Analysis Preliminary Manual," 6529-CT, Vol. 1, 1982.

AUTOMATED GAMMA SPECTROMETRY AND DATA ANALYSIS ON RADIOMETRIC NEUTRON DOSIMETERS

W.Y. MATSUMOTO

Westinghouse Hanford Company, P.O. Box 1970, Richland, WA, 99352

ABSTRACT

An automated gamma-ray spectrometry system was designed and implemented by the Westinghouse Hanford Company at the Hanford Engineering Development Laboratory (HEDL) to analyze radiometric neutron dosimeters (ref.1). Unattended, automatic, 24 hour/day, 7 day/week operation with online data analysis and mainframe-computer compatible magnetic tape output are system features. The system was used to analyze most of the 4000-plus radiometric monitors (RM's) from extensive reactor characterization tests during startup and initial operation of the Fast Flux Test Facility (FFTF). The FFTF, operated by HEDL for the Department of Energy, incorporates a 400 MW(th) sodium-cooled fast reactor. Automated system hardware consists of a high purity germanium detector, a computerized multichannel analyzer data acquisition system (Nuclear Data, Inc. Model 6620) with two dual 2.5 Mbyte magnetic disk drives plus two 10.5 inch reel magnetic tape units for mass storage of programs/data and an automated Sample Changer-Positioner (ASC-P) run with a programmable controller. The ASC-P has a 200 sample capacity and 12 calibrated counting (analysis) positions ranging from 6 inches (15 cm) to more than 20 feet (6.1 m) from the detector. The system software was programmed in Fortran at HEDL (ref.2), except for the Nuclear Data, Inc. Peak Search and Analysis program and Disk Operating System (MIDAS+).

Operating Features

1. Automatic batch processing of 1 to 200 samples

2. Selectable count-rate controlled or operator selected counting positions

3. Selectable preset net count statistics defaulting to selectable preset maximum counting time limits

4. Automatic output of results to Master File, line printer and magnetic tape

5. Automatic peak search/analysis results and spectral data output to line printer and magnetic tape

6. Library of dosimeter types and sub-types for peak-of-interest selection and definition of analysis parameters

7. Bar-code label reader for automatic sample verification

8. Remote terminal dial-up access for checking system status, error correction and data entry

9. Manual operating option for the sample changer-positioner

10. Completely shielded operation for minimum personnel exposure

Support Systems

   A) Dosimeter sorting cell equipped with: two Central Research Laboratory (CRL) Model G master-slave manipulators, binocular telescopic magnifying optics, an analytical microbalance with remote digital readout, and a high resolution gamma spectrometry system used as a dosimeter identification aid.

   B) Dosimeter storage cell equipped with: two CRL Model G's and an indexed storage rack with a capacity for more than 4800 samples stored in bar-coded ASC-P vial trays (cassettes).

   C) Five additional high resolution gamma spectrometry systems: (ref.3), these systems range from a high geometry well-type Ge(Li) detector to a track mounted shielded sample carrier system where the sample can be positioned 30 ft (9.14 m) from the detector. Data handling and analysis capabilities of these systems range from simple net peak area calculations for operator selected regions of interest to computerized peak search and analysis with calculation of disintegrations per second per milligram at the end of irradiation time. All of these systems require operator attention for each sample.

   D) A Master Logbook, sample-follower-cards and card index comprise the basic data input and sample location records: all necessary sample data such as capsule number, sample identification, dosimeter type/subtype, weight, vial number and tray number in these records must be entered from this database into the computer system's Master File before the sample can be automatically processed under programmed batch control.

Summary and Conclusions

Data reduction by the automated gamma spectrometry and data analysis system stops with the calculation of the dosimeter specific activity, decay corrected to a reference (end of irradiation) time. The propagated uncertainty in this result at a nominal one standard deviation is also calculated. These data are transferred to a mainframe computer database and used for calculation of reaction rates and neutron flux, fluence and spectral results. Full use of the system allows efficient 24-hour operation with automatic data processing resulting in exceptional time and cost savings. Productivity and quality improvements through use of this system resulted in accomplishing high quality and efficient analysis of the majority of the FFTF reactor characterization test RM's with fewer than half of the personnel originally scheduled to work on this task. The ASC-P was designed for integral operation with other sub-systems to perform as a unit for efficient, high quality processing of RM's using a minimum number of operating personnel and keeping personnel radiation exposure to a practicable minimum. Others (refs.4,5) have reported on two different custom-built computer controlled sample changer systems for gamma-ray spectroscopy.

REFERENCES

1 R.L. Daubert and W.Y. Matsumoto, Automated Remote Positioning and Examination of FFTF Reactor Power Charaterization Dosimeters, Proceedings of the 29th Conference on Remote Systems Technology, Vol. 1 Summer 1981, Bal Harbour, Florida, ANS, Inc., La Grange Park, Illinois, 1982, pp. 30-33.
2 R.K. Lytz, Software for the HEDL Automated Neutron Dosimeter Analysis System, in preparation.
3 E.P. Lippincott and W.N. McElroy (Comp.), FTR Dosimetry Handbook, HEDL MG-166, Hanford Engineering Developmental Laboratory, Richland, Washington, March 1983, Appendix A2, p.. A2.1-1 - 2.1-14.
4 D.T. Jost, U. Krähenbühl and H-R. VonGunten, A Microcomputer Controlled Sample Changer System for Gamma-ray Spectroscopy, Nucl. Instr. and Meth. 197, 1982, pp. 365-369.
5 A.H. Andweg, A Large-capacity Sample-changer for Gamma-ray Spectroscopy NIM-2085, National Inst. for Metallurgy, Johannesburg, December 1980, 34 pp.

ENVIRONMENTAL APPLICATIONS FOR AN INTRINSIC GERMANIUM WELL DETECTOR

P. STEGNAR, J.S. ELDRIDGE, N.A. TEASLEY, AND T.W. OAKES
Oak Ridge National Laboratory, Oak Ridge, Tennessee 37831

ABSTRACT

The overall performance of an intrinsic germanium well detector for ^{125}I measurements was investigated in a program of environmental surveillance. Concentrations of ^{125}I and ^{131}I were determined in thyroids of road-killed deer showing the highest activities of ^{125}I in the animals from the near vicinity of Oak Ridge National Laboratory. This demonstrates the utility of road-killed deer as a bioindicator for radioiodine around nuclear facilities.

INTRODUCTION

Intrinsic Ge or Ge(Li) well type detectors greatly increase the efficiency of high resolution gamma spectroscopy measurements in the determination of many of the important radioactive contaminants in the environment. This is especially important in the case of radionuclides that emit low energy gamma-rays or x-rays.

In this study, the analysis of ^{125}I in deer thyroids was performed using the well geometry of an intrinsic germanium detector attached to a computerized multichannel analyzer.

The overall performance of this detector for ^{125}I measurements was investigated and the results for ^{125}I and ^{131}I concentrations in deer thyroids were determined.

EXPERIMENTAL

Thyroid glands were extracted from white-tail deer killed on roads within the Oak Ridge reservation from May 19 to October 21, 1983. After extraction of the gland, fat and supporting tissue were removed and the glands placed into standard Petri dishes for ^{131}I measurements. After counting, the thyroids were lyophilized and the dry tissues were placed into plastic counting vials (13 mm internal diameter) for ^{125}I measurements.

The samples and ^{125}I standard solution were counted in identical counting vials in the well (16 mm dia x 45 mm length) of an intrinsic type germanium detector (23% relative efficiency at 25 cm as compared to that of a 3 x 3 - in. NaI (Tl) detector at 1.332 MeV). The spectrometer was operated at a gain of $\sim$ 0.1 keV/channel yielding an energy range of 20 to 400 keV.

All thyroid specimens containing 0.2 to 2.0 g of freeze-dried tissue in counting vials were counted for 25,000 to 150,000 sec., while the counting time for the ^{125}I standard solution was 1000 sec.

Concentrations of ^{125}I in thyroids were calculated from net integrated counts in 3 different regions of interest including: a) K_α x-ray peak of 27.2 + 27.5 keV, K_β x-ray peak of 31 keV and γ-ray peak of 33.5 keV, b) coincidence sum peaks at 54.8, 58.3, 62.8 and 66.3 keV, and c) of regions a) and b) together.

The absolute activity of ^{125}I standard solution was calculated by Eldridge's calculation (ref.1). The activity concentration of the original ^{125}I solution obtained from the New England Nuclear Co. was 5400 disintegrations per second/mL in approximately 0.02 molar NaOH solution. The test samples were prepared by dilution, so that the activity of the final working standard was 540 dps in 0.02 M NaOH, on September 28, 1983.

RESULTS AND DISCUSSION

The decay scheme of ^{125}I is unique because the nuclide decays 100% by electron capture followed by a 35.5 keV gamma ray transition. The K_α x-rays of 27.2 + 27.5 keV and K_β x-rays of 31 keV emitted after electron capture events in the ^{125}I are in coincidence with the photons and x-rays following internal conversion events emitted in the decay of the 35.5 keV excited state of ^{125}Te (ref.1). The coincident photons give rise to the characteristic gamma-ray spectrum which is partly shown in Figure 1.

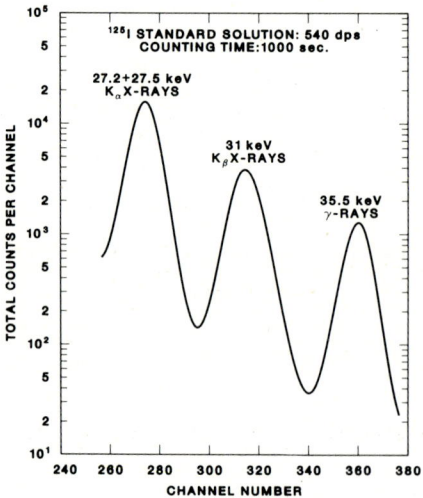

Fig. 1. Gamma-ray spectrum of 540 dps ^{125}I standard solution in the Ge well detector.

Counts per 1000 seconds vs. channel number are displayed. The prominent peak centered in channel 273 is due to the 114% K_α x-rays of 27.4 keV. The peak in channel 312 is due to the 31 keV K_β x-rays at 25.8% abundance and the third peak in channel 361 is due to the 35.5 keV gamma-rays at 6.7% abundance. By integration of these three peak areas and the areas of coincidence sum peaks of 54.8, 58.3, 62.8 and 66.3 keV (not shown in Fig. 1), an overall counting efficiency, defined as counts per second divided by disintegrations per second (cps/dps), was determined to be 0.72 for all peaks.

Figure 2 shows the gamma-ray spectrum of ^{125}I in the deer thyroid No. 80 counted for a period of 25,000 seconds. The points represent total counts per channel (sample + background), while the solid curve indicates the level of ^{125}I in the sample. The background counting rate in the total response region of ^{125}I in Figure 2 averaged 0.01 counts per second for 30 channels in the 27.4 keV K_α x-ray peak. Defining a detection limit as 3 standard deviations of background activity, a detection limit of 0.004 dps (0.11 pCi) for the K_α x-ray peak was calculated.

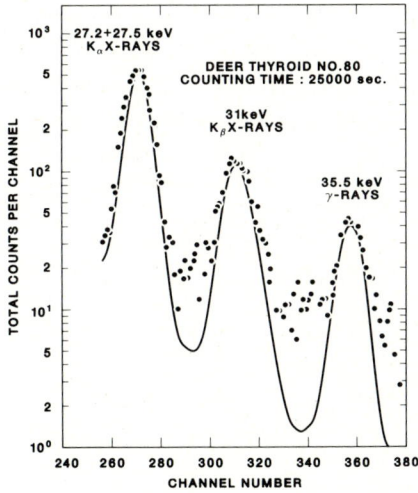

Fig. 2. Gamma-ray spectrum of ^{125}I in the deer thyroid No. 80.

Table 1 shows the relative activity of a ^{125}I standard solution as a function of position in the well of the Ge detector. A standard solution containing 0.5 mL of ^{125}I was counted for 300 seconds at 8 different levels (0.5 cm intervals) from the bottom towards the top of the well. Fairly constant activity in the 27.4 keV K_α x-ray peak was observed for 0.5 to 2.5 cm of distance from the bottom of the well; the activity started to decrease at 3.0 cm and at 4.0 cm was less than 40% of the activity in lower positions of the counting vial in the well.

In Table 2 the relative activity of a ^{125}I standard solution as a function of dilution is presented. The results clearly show relatively constant activity (249 ± 5.7 cps) for 0.1 to 3.0 mL of ^{125}I standard solutions which gives a useful volume of almost 4 mL for samples in the counting vial. This means that up to 3 g of freeze-dried sample can be easily placed into the counting vial and counted without losing counting efficiency due to the sample geometry in the well.

TABLE 1
Relative activity of ^{125}I standard solution as a function of position in the well of a Ge detector.

Distance from the bottom of the well (cm)	cps
0.5	256
1.0	261
1.5	254
2.0	253
2.5	243
3.0	220
3.5	159
4.0	100

TABLE 2
Relative activity of ^{125}I standard solution in the well as a function of dilution.

Volume (mL)	cps
0.1	257
0.5	256
1.0	242
1.5	250
2.0	248
2.5	247
3.0	241
3.5	228

In Table 3 the ^{125}I and ^{131}I contents of the 29 deer thyroids of this study are shown. The highest quantities of ^{125}I were found in those animals (Nos. 65, 77, 80, 96) from the near vicinity of Oak Ridge National Laboratory (ORNL). This indicates that ORNL was probably the origin of ^{125}I in the environment.

High amounts of ^{131}I were also determined in thyroids of some deer collected within the Oak Ridge reservation after August 22, when ^{131}I was released from ORNL to the environment.

TABLE 3
^{125}I and ^{131}I contents in thyroids of 1983 road killed deer in Oak Ridge, TN.

Deer (No. sex)	Collection date	Weight (kg)	Age (year)	^{125}I (pCi/g D.W.)	^{131}I (pCi/g D.W.)
59 m	May 19	40	1	20	
60 f	May 20	54	3	23	
61 m	May 23	75	2	10	
65 f	Jun 2	42	1	105	
67 f	Jun 22	51	3	5.1	
77 m	Jul 16	47	1.5	97	
79 f	Jul 20	38	1	15	
BL3 f	Jul 20	13	Fawn	33	
80 f	Jul 21	54	3	42	
82 m	Aug 1	62	1.5	9.5	
85 m	Aug 11	35	1.5	3.6	
93 m	Aug 19	55	1.5	4.4	
96 f	Aug 22	38	1	35	
98 f	Aug 27	50	4	20	64
99 m	Aug 29	15	9 wks.	12	
101 f	Aug 31	41	1.5	5.8	27
102 m	Sep 2	19	10 wks.	2.2	200
103 f	Sep 7	64	3	16	4.0
104 f	Sep 7	25	16 wks.	2.6	47
105 m	Sep 9	55	2		13
108 m	Sep 12	71	2	11	210
109 m	Sep 12	20	16 wks.		49
111 m	Sep 14	23	16 wks.		17
NR1 m	Sep 25	53	2		<8
116 f	Sep 27	55	2		46
118 f	Sep 30	67	5		54
127 m	Oct 10	51	2		5.0
128 f	Oct 11	46	3		9.6
136 f	Oct 21	30	20 wks.		11

m = male
f = female
D.W. = freeze-dried weight

High ^{131}I concentrations significantly interfered with ^{125}I determinations in the same thyroid. In particular, the K_α x-rays of 29.5 + 29.8 keV (4% abundance) and 0.9% K_β x-rays of 33.6 keV in the decay mode of ^{131}I contributed to the ^{125}I spectrum at this energy region (27.2 + 27.5 keV K_α and 31 keV K_β x-rays).

Our intrinsic germanium well-type detector was found to be an extremely suitable device for ^{125}I measurements. The limitation of the sample quantity was not a disadvantage in the case of deer thyroids for which the lyophilization was performed as a preparation technique prior to the final analysis. The lyophilization greatly reduced the original size of the fresh tissue (70 to 80% of water) and enabled the accommodation of the entire freeze-dried thyroid in the counting vial inside the well of the detector.

The high efficiency of this well type Ge detector at energies below 100 keV can have applications in the measurement of other important radionuclides in the environment such as ^{129}I, ^{241}Am, ^{210}Pb, etc. (ref.2,3).

Several previous low-level radioactivity measurement methods can also be improved by the use of Ge or Ge(Li) well type detectors (ref.4,5,6).

Another useful application is the measurement of higher energy emitters in larger volume samples placed in standard containers around ("Marinelli" beakers) or on the top (Petri dishes) of the detector (ref.3).

With respect to the environmental contamination by radionuclides, this study clearly demonstrates the utility of road-killed deer as a bioindicator for radioiodine around nuclear facilities.

Acknowledgement: The authors greatly appreciate the assistance provided by D.W. Parsons in collection of the deer thyroids.

This research was sponsored by the U. S. Department of Energy under contract W-7405-eng-26 with the Union Carbide Corporation.

REFERENCES

1 J.S. Eldridge and P.Crowther, Nucleonics, Vol. 22, No. 6, June 1964.
2 F.P. Brauer and W.A. Mitzlaff, IEEE Trans. on Nucl. Sci., Vol. NS-25, No. 1, (1978) 398.
3 J.S. Eldridge, T.W. Oakes and M.E. Pruitt, Am. Ind. Hyg. Assoc. J. 39, June 1978.
4 F.P. Brauer, J.E. Fager and W.A. Mitzlaff, IEEE Trans. on Nucl. Sci., Vol. NS-20, (1973) 57.
5 F.P. Brauer and J.H. Kaye, IEEE Trans. on Nucl. Sci., Vol. NS-21, (1974) 496.
6 F.P. Brauer, J.M. Kelley, R.W. Goles and J.E. Fager, IEEE Trans. on Nucl. Sci., Vol. NS-24, (1977) 591.

OTHER SPECTROSCOPIC TECHNIQUES

SAMPLING AND CHARACTERIZATION OF AEROSOLS FORMED IN THE ATMOSPHERIC HYDROLYSIS OF UF_6

W. D. Bostick,[1] W. H. McCulla,[1] P. W. Pickrell,[1], and D. A. Branam[2]
[1]Materials and Chemistry Technology Department
[2]Analytical Chemistry Department
Oak Ridge Gaseous Diffusion Plant, Oak Ridge Tennessee 37830

ABSTRACT

When gaseous UF_6 is released into the atmosphere, it rapidly reacts with ambient moisture to form an aerosol of uranyl fluoride and HF. As part of our Safety Analysis program, we have performed several experimental releases of UF_6 (from natural uranium) in contained volumes in order to investigate techniques for sampling and characterizing the aerosol materials. The aggregate particle morphology and size distribution have been found to be dependent upon several conditions, including the relative humidity at the time of the release and the elapsed time after the release. Aerosol composition and settling rate have been investigated using isokinetic samplers for the separate collection of UO_2F_2 and HF, and via laser spectroscopic remote sensing (Mie scatter and infrared spectroscopy).

INTRODUCTION

Uranium hexafluoride (UF_6) is the only uranium compound which is gaseous at moderate temperatures (boiling point $\sim 56°C$ at 760 torr); as such, it finds large scale application in both the gaseous diffusion and centrifugation uranium isotope separation processes. The United States Department of Energy has sponsored a Safety Analysis effort to determine the fate of UF_6 which might be released into the environment. This communication summarizes some of our recent investigations on the sampling and characterization of the reaction products formed under conditions of a controlled release within an enclosed volume. Emphasis is upon those characteristics of the product which affect their dispersion and detection.

When it is released into the atmosphere, gaseous UF_6 is rapidly hydrolyzed by ambient moisture to form uranyl fluoride and HF:

$$UF_6 \text{ (g)} + 2 H_2O \text{ (g)} \rightarrow UO_2F_2 \text{ (s)} + 4 HF \text{ (g)} \qquad (1)$$

"Anhydrous" UO_2F_2 is hydroscopic (ref. 1), as is HF (ref. 2), hence, both species tend to become hydrated to some extent. The particulate UO_2F_2 is readily visible as a white aerosol cloud or "smoke", which may rise and be dispersed into the environment.

PHYSICAL CHARACTERIZATION OF AIRBORNE PARTICULATE UO_2F_2

Particle size and morphology of airborne particulates are critical parameters in the dispersion of materials created in the atmospheric release of UF_6 and in the detection of these materials by physical measurements such as light scatter or ionized smoke detection. Pickrell (ref. 3) has summarized the results for the experimental release of UF_6 in a contained volume. Particle size and the degree of agglomeration were found to be dependent upon the temperature of the UF_6 at the time of its release, the relative humidity (R.H.) of the air into which it is released, and the time elapsed after the release (Fig. 1). Materials formed in high humidity (e.g., > 85% R.H.) environments usually form solutions and, therefore, tend to be spheroidized (Fig. 1). When formed under "normal" conditions of humidity (e.g., 20 to 80% R.H.), the materials collected from the UO_2F_2 "plume" tend to be chain-like agglomerates comprised of individual 0.1 to 0.2 μm spheroids. These agglomerates may reach 5 to 10 μm in greatest dimension, although most are smaller. Lux (ref. 4) has investigated the physical size distribution of fallout particulates formed under a variety of conditions, and found them to be concentrated in the lower end of the range 0.5 to 3.0 μm (electron microscopy).

Samples collected from UO_2F_2 plume and fallout have been examined by x-ray diffraction (XRD), (ref. 3). In general, these materials appear to be of variable composition, for which no standard data are available for comparison (ref. 1,3,5), although, on occasion, stable dihydrate or hydrogen fluoride hydrates of UO_2F_2 have been identified in the fallout material.

Fallout material from a typical UF_6-release experiment was determined to have a density of 4.13 g/cm^3, and a refractive index of ~1.54.

ISOKINETIC SAMPLING

Isokinetic samplers have application in aerosol measurements for health physics and in dispersion modeling due to their relative simplicity and low cost, and for the possible specific physical or chemical examination of the collected sample. Particulates, such as UO_2F_2, can be collected on a small pore size (e.g., 1 μm) membrane prefilter. Uranium may then be extracted and measured via spectroscopy or (if isotopically-enriched material is used) via alpha-radiation counting. HF vapor may be collected via chemisorption onto treated filters (ref. 6) or by impinging the vapor into a trapping solution (the former collection mode yields an integrated response over the sampling interval, whereas, the latter collection mode may be used for either integrated or continuous measurement). Fluoride ion is conveniently determined with use of an ion-selective electrode (ref. 6). Ishida and co-workers (ref. 7) have judged the sensitivity and response performance of the potentiometric measurement of fluoride ion (continuous mode) to be superior to that of an alpha dust

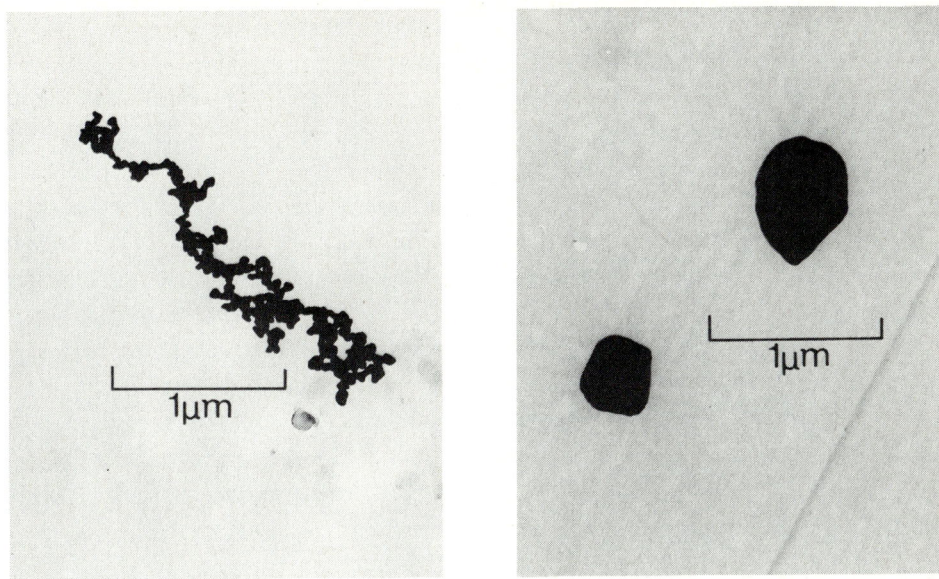

R.H. <2%   R.H. ~100%

Effect of relative humidity (R.H.) on particle size and shape.

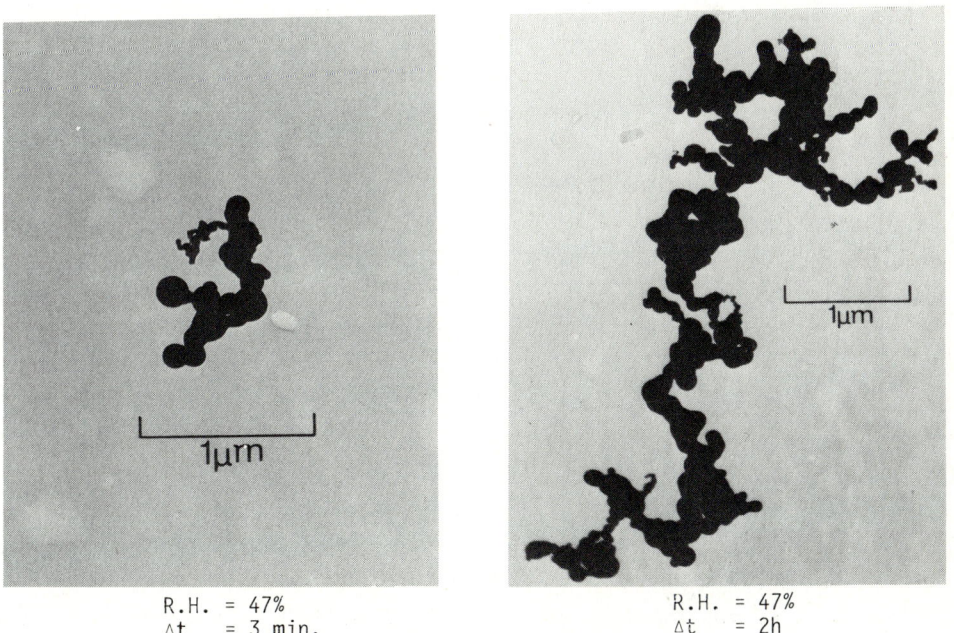

R.H. = 47%   R.H. = 47%
Δt = 3 min.   Δt = 2h

Effect of elapsed time (Δt) on particle agglomeration.

Figure 1

monitor or an ionized smoke detector for UF_6-release aerosols.

Our sampling system includes two membranes stacked together in a 47-mm filter holder made of hydrophobic polypropylene (Millipore Corp.); a 1-μm pore Teflon membrane is used as the prefilter, and a 5-μm pore formate-impregnated cellulose ester membrane (ref. 6) is used for HF recovery. In an UF_6-release experiment, 99.92% of the uranium drawn into the filter assembly was recovered on the prefilter. In a separate experiment, anhydrous HF (from a permeation tube apparatus, similar to that described in ref. 6) was metered into the filter assembly; recovery of HF was ~0.003% on the prefilter and >97% on the treated membrane.

The data presented in Fig. 2 represent analytical results obtained following the release of ~1 g UF_6 within an acrylic-walled test chamber with total volume ~0.2 m^3. A minature diaphram pump is used to pull an aerosol sample through the filter assembly and to return the remainder of the sample to the test chamber. Note in Figure 2A that both UO_2F_2 and hydrated HF are progressively lost from the gas-phase due to sedimentation.

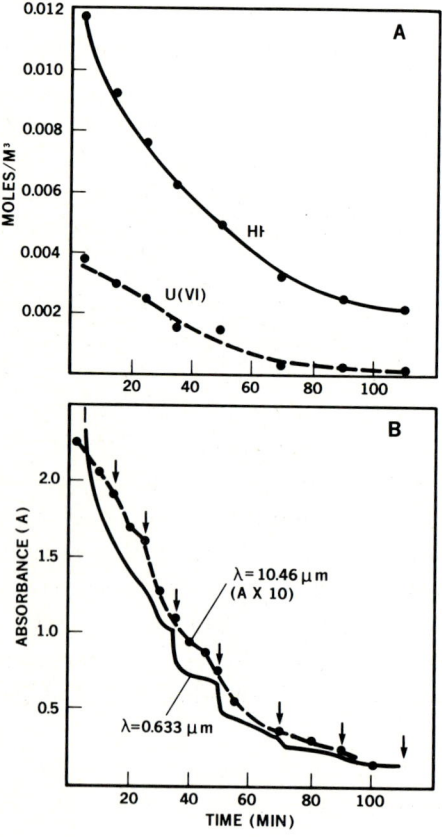

Figure 2. Aerosol measurement following the atmospheric release of UF_6 (~1 g in a contained volume of ~ 0.2 m^3). (A) Results from the chemical analysis of materials collected on membrane filters using an isokinetic sampling system. Particulate UO_2F_2 is collected on a 1-μm pore Teflon prefilter, and HF is chemisorbed on a formate-impregnated cellulose acetate membrane. (B) Absorbance measurements (~60 cm pathlength), as performed with use of (1) a HeNe laser (solid line; λ = 0.633 μm), representing Mie scatter, and (2) a CO_2 laser (broken line; λ = 10.46 μm), representing uranyl ion concentration. CsI windows were inserted into the test chamber to permit transmission of the IR radiation. Arrows indicate times at which samples were withdrawn for chemical analysis (see Figure 2A).

REMOTE SENSING OF PARTICULATE UO_2F_2 VIA SPECTROSCOPY

Remote sensing of aerosols would be advantageous in providing a real-time alarm to accidental releases of UF_6. Potentially, remote sensing devices could be employed to determine both the magnitude and the propagation of the aerosol.

Light-scattering is an obvious candidate for the remote sensing of particulates. Parameters which influence the intensity of light-scatter include the difference in refractive indices between the particulate material and the surrounding medium, the particle size distribution, and the radiation wavelength. The wavelength of visible radiation (0.4 to 0.7 µm) is comparable in magnitude to the particle sizes in the UO_2F_2 aerosol, and thus the scattered light is strongly concentrated in the forward direction (Mie scattering).

The change in absorbance (turbidity) following a release of UF_6 is shown in Fig. 2B, using a HeNe laser source ($\lambda = 0.633$ µm). Comparison with the data obtained with use of the isokinetic sampling system (Fig. 2A) indicates that light-scatter is a sensitive index to the aerosol composition. (Also note the perturbation in light-scatter signal that occurs when a volume of gas-phase [1 to 8% of total] is sampled.)

Although light-scatter is a sensitive means for the detection of the aerosol, it is inherently nonspecific. Techniques related to light-scatter, but which can provide specific chemical information, include fluorescence and Raman spectroscopy. Many uranyl salts are fluorescent; however, we noted little fluorescence response for solid-phase UO_2F_2 fallout material (i.e., only ∼0.3% of the response measured for $UO_2[NO_3]_2 \cdot 6H_2O$). The fluorescence of solid-phase uranyl salts appears to be strongly affected by crystal properties and the extent of hydration (ref. 8).

A sample of α-$UO_2F_2 \cdot 1.5H_2O$ (ref. 5, card 24-1151), with relatively large crystals, produced an intense Raman peak at 868 cm^{-1}, attributable to the symmetric stretching frequency of the uranyl ion. Unfortunately, the very small particles from UF_6-release fallout failed to give a Raman signal which was distinguishable from background (argon-ion laser excitation, $\lambda = 0.514$ µm).

Particulate material from the plume produced in the release of UF_6 was collected onto a polyvinylchloride membrane filter using the isokinetic sampling system. The filter was then scanned directly in an infrared (IR) spectrophotometer. A strong absorption was evidenced at ∼955 cm^{-1} (∼10.5 µm), due to the asymmetric stretching frequency of the uranyl ion. (This wavelength is long relative to the average particle size, minimizing scattering effects.) An additional absorption band was observed at ∼1625 cm^{-1}, attributed to hydrated water.

The 10P6 transition of a CO_2 laser ($\lambda = 10.46$ µm) is ideal for the specific detection of uranyl species in the aerosol. As shown in Fig. 2B, the absorption at this wavelength closely parallels the Mie scatter, but the sensitivity is reduced by an order of magnitude. The absorbance cross-section for uranium in the aerosol can be estimated using the expression: $A = 0.434 \sigma n l$, where A is the absorbance at 10.46 µm, n is the concentration of U(VI) in the aerosol (molecules/cm^3), and l is the absorption pathlength (cm); the cross-section σ is thus estimated to be $\sim 4 \times 10^{-18}$ cm^2 (a moderately strong IR absorption).

CONCLUSIONS

The most salient characteristic of the aerosol produced in the atmospheric hydrolysis of UF$_6$ is the small particle size of the UO$_2$F$_2$ which is formed. Uranyl fluoride and HF components of the aerosol may be separately collected with the use of membrane filters in an isokinetic sampling system. The collected samples are in a stable, concentrated form from which they may be subsequently assayed using sensitive and selective wet chemical procedures. Alternately, remote sensing of the aerosol is possible using Mie scatter (which is sensitive, but inherently nonspecific) or by selective (but modestly sensitive) IR detection.

ACKNOWLEDGEMENT

Oak Ridge Gaseous Diffusion Plant is operated by Union Carbide Corporation for the U.S. Department of Energy under contract W-7405-eng-26. By acceptance of this article, the publisher or recipient acknowledges the U.S. Government's right to retain a nonexclusive, royalty-free license to and for any copyright covering the article.

REFERENCES

1. L. H. Brooks, E. V. Garner, and E. Whitehead, Report IGR-TN/CA 277, Chemical and X-ray Crystallographic Studies on Uranyl Fluoride, February 1956.
2. W. H. McCulla, Report K/PS-155, Determination of the Rate of HF Hydration and the Effects of HF on Moisture Condensation, (in preparation).
3. P. W. Pickrell, Report K/PS-144, Characterization of the Solid, Airborne Materials Created by the Interaction of UF$_6$ with Atmospheric Moisture in a Contained Volume, April 1982.
4. C. J. Lux, Report GAT-T-3116, Evaluation of Techniques for Controlling UF$_6$ Release Clouds in the GAT Environmental Chamber, April 1982.
5. Powder Diffraction File, Joint Committee on Powder Diffraction Standards.
6. L. A. Elfers and C. E. Decker, Anal. Chem., 40(1968), 1658.
7. J. Ishida, G. Sakamoto, S. Takeda, and J. Kato, Report PNCT 831-79-02, Investigation of the UF$_6$ Aerosol Behavior in Air, IV, The Characteristics of Monitors Detecting UF$_6$ Release.
8. E. Rabinowitch and R. L. Belford, Spectroscopy and Photochemistry of Uranyl Compounds, Pergamon Press, Oxford, 1964, pp.55-56.

APPLICATIONS OF LASER SPECTROSCOPY FOR SPECIES DETERMINATION IN FISSION PRODUCT RELEASE EXPERIMENTS

W. H. McCulla, G. E. Nelson[1], and R. A. Lorenz[2]
[1]Oak Ridge Gaseous Diffusion Plant, Oak Ridge, Tennessee
[2]Oak Ridge National Laboratory, Oak Ridge, Tennessee

ABSTRACT

Identification of vapor-phase species and the kinetics of their reactions in simulated LWR nuclear accidents are important in predicting the ultimate fate of volatile fission products. Laser spectroscopic techniques such as laser Raman and laser-induced fluorescence (LIF) are valuable tools in determining molecular species and concentrations from these experiments. We have assembled equipment capable of vapor-phase measurements up to 1500K, and have applied these techniques to molecular iodine and CsI vapors. We describe the apparatus that we have used for stable isotope measurements and indicate the modifications necessary for operation with radioactive fission products.

The application of LIF to iodine vapor is reviewed and the problems associated with high-temperature concentration measurements (to 1200K) are discussed. The characterization of the LIF spectrum of CsI vapor at 1100K is presented.

INTRODUCTION

The accident to Unit 2 of the Three Mile Island Nuclear Power Plant on March 28, 1979, revealed a significant lack of understanding regarding the chemical forms of the radionuclides released from a damaged core. The very small release of radio-iodine has led to a new emphasis on the physical chemistry of fission product behavior in nuclear accidents. The current hypothesis is that the iodine was released as CsI which then condensed on cooler surfaces or dissolved in the cooling water (ref.1). To test this hypothesis a simulation of the accident is planned. Spectroscopic analytical techniques based on Raman and laser induced fluorescence (LIF) are being evaluated as monitors of the I_2 and CsI vapor species above damaged fuel elements up to 1100K and under simulated accident conditions.

EXPERIMENTAL APPARATUS

A spectrometer system originally designed for Raman spectra studies in vapors to 1500K was used in this study (ref. 2). An arrangement of the system is represented schematically in Figure 1. The exciting line source is a line tunable Control Laser Corporation Model 559A argon ion laser capable of 7 to 10 watts on either the 488.0 nm or 514.5 nm wavelength. The laser has a servo

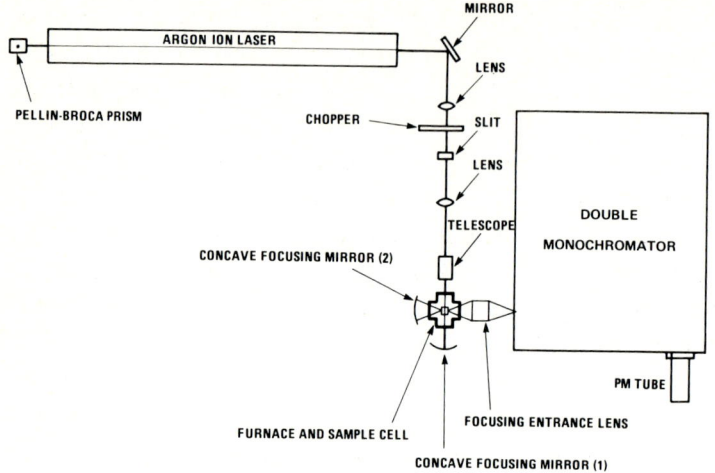

Figure 1. Schematic diagram of the optical system of the spectrometer.

feedback system to maintain constant output power. The output is monitored by a panel meter and the calibration is checked frequently with a Scientech, Inc., 36-0001 1-in. surface absorbing disc calorimeter. The output beam emerges vertically polarized from the laser and is passed through a filter monochromator consisting of two Pellin-Broca prisms mounted one above the other on a Rayleigh mount, and a telescope-slit assembly which removes the dispersed plasma lines, condenses the beam, and transmits the now monochromatic excitation beam. A second telescope and mirror assembly located in the sample chamber focuses and double passes the beam in the center of the sample cell.

The sample emission radiation is viewed at $90°$ and is focused onto the entrance slit of the monochromator by a Berthiat f/1.8, 80-mm focal length compound lens. The sensitivity of the instrument is improved by the reflection and refocusing of the "backscattered" radiation by the concave focusing mirror 2. The double monochromator is a 1 meter focal length f/8 Jobin-Yvon Ramanor HG.2S utilizing 2000 grooves/mm concave holographic gratings. Stray light rejection is 10^{-14} or better at 20 cm^{-1} from the Rayleigh line. The monochromator is stepping motor driven in 0.1 cm^{-1} steps and is capable of better than 0.5 cm^{-1} resolution. The slits are horizontal, 2 cm long and opening to 2 mm. At 514.5 nm, a 100 um slit opening represents 1.0 cm^{-1} resolution. A Hamamatsu R928 photomultiplier tube mounted in a Pacific Photometric Instrument 3461 thermoelectric cryostat maintained at $-30°C$ exhibits a dark count rate typically below 8 counts/sec.

The furnace cell holder is a solid Inconel piece machined to take 1 cm-quartz fluorimeter cells. It is fitted inside a stainless steel container with the window ports on the cell holder aligned with holes in the container wall. The region between the cell holder and container is filled with alumina for insulation. The furnace window ports are designed so that the spectrometer will see only the central quartz cell. This keeps the black body radiation entering the spectrometer to a minimum. The window ports are filled with quartz evacuated inserts. The furnace is heated resistively by making the Inconel cell holder part of an electrical circuit of the secondary of a low voltage power transformer. The transformer is controlled by a Variac in the primary circuit and can provide up to a 1.8 volt drop at 2000 amperes across the Inconel cell. The current is monitored by a 2000/5 ratio current transformer in the secondary lead. The lower electrode, container, and upper electrode are water cooled. The furnace is mounted on an XYZ translator for positioning. The busbar, connectors, and secondary of the transformer were silver-plated to maintain good electrical contact and reduce oxidation. Two stainless steel sheathed chromel-alumel thermocouples were used to monitor the temperature. One was inserted through a hole from the bottom of the cell holder to just beneath the quartz fluorimeter cell. The other was inserted from the top beside the quartz cell to just above a window. The temperature was monitored by two Omega 2160A digital thermometers.

The electronic arrangement for processing the data is diagrammed in Figure 2. The output pulses from the PM tube are amplified and discriminated by a Princeton Applied Research (PAR) 1121 amplifier-discriminator. The analog output from the PAR 1121 goes to one side of a PAR 116 differential preamplifier, and into the PAR 124A lock-in amplifier operated in the bandpass mode with a very high Q. The output of the PAR 1121 is also sent through an EGG LG105/N linear gate and stretcher to the ORTEC 449 log/lin ratemeter. A PAR 192 variable speed chopper with a two-position blade was used to chop the excitation laser beam and provide a reference signal to the PAR 124A lock-in amplifier. Using this phase sensitive technique, an approximately 1000-fold signal enhancement could be realized over a direct technique. The analog output from the lock-in amplifier is placed across a 1 megohm fixed load and an ORTEC 439 digital current integrator. This current to frequency conversion provides a digital signal to the ORTEC 550 single channel analyzer, which converts it to a standard 5 V pulse, and is then processed by a Nuclear Data 88-0574 multichannel scaling unit and a Nuclear Data ND-6600 data acquisition and processing system. The computer controls the Jobin-Yvon spectrometer and the multi-channel scaling unit, providing considerable versatility in the operation of the system through its wide range of scan rates, channel acquisition times, calculations of slit opening for required resolution, and adjustment of slits to the proper setting, plus its data handling and manipulation.

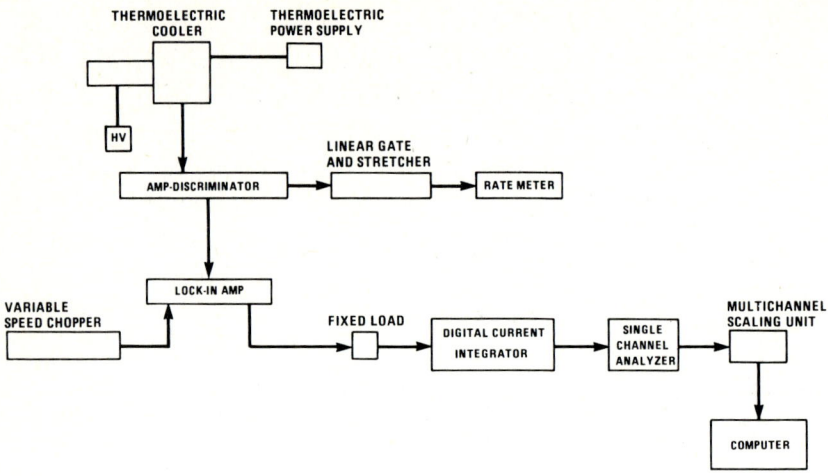

Figure 2. Schematic diagram of the electronic configuration.

RESULTS AND DISCUSSION

The Raman and LIF spectra of I_2 and CsI were determined for static vapor samples loaded in fused silica fluorimeter cells or sapphire tubes. The Raman spectrum of I_2 could easily be observed with 488 nm excitation at vapor densities of $10^{14}/cm^3$. It is a single band at 214 cm^{-1} from the Rayleigh line and $2\nu_1$ and $3\nu_1$ were also observed at this vapor density. The strength of the Raman transition is most probably due to a resonance or near resonance effect. With 514.5 nm excitation, a very strong LIF spectrum of the transition $X^1\Sigma$ state to the $B^3\Pi_o$ state of I_2 is observed.

This spectrum, shown in Figure 3, is a progression series of regularly spaced varying intensity peaks that are separated by the 214 cm^{-1} vibrational frequency of molecular iodine and extends below 15,000 cm^{-1}. Observation of the iodine LIF spectra intensity over the temperature range 350K to 1200K for 9 x $10^{14}/cm^3$ vapor density indicated very little temperature effect with only a slight broadening of the bands and almost no collisional quenching below 1100K. For a $10^{14}/cm^3$ vapor density of I_2 in 6 x $10^{19}/cm^3$ vapor density of water at 1100K an approximate order of magnitude reduction in the fluorescence intensity was observed.

In Figure 4 the spectrum of CsI vapor at 1100K for the 488 nm excitation is shown. The vapor density of the CsI was $10^{15}/cm^3$. The intensity of the transition is 10 to 100 times greater than expected for a Raman transition (even for a highly polarizable molecule). The band is composed of at least five distinct transitions separated by approximately 12 cm^{-1}, with the most intense transition occurring at 134 cm^{-1} from the Rayleigh line. This is not

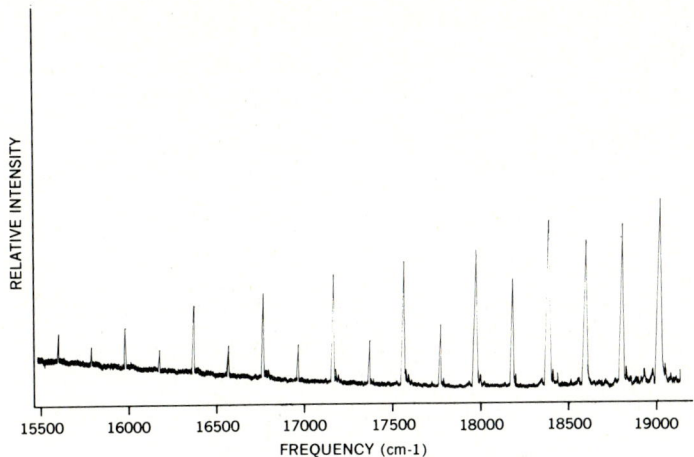

Figure 3. LIF spectrum of I_2 at 1125K (3 cm^{-1} resolution).

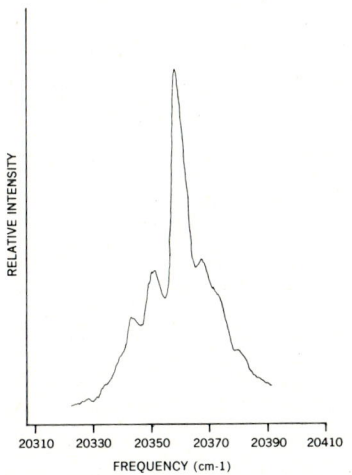

Figure 4. LIF spectrum of CsI at 1100K (2 cm^{-1} resolution).

what one would expect for the vibrational spectrum of a simple diatomic molecule. A similar progression series of transitions was observed for CsI. Thus, the CsI spectrum is attributable to fluorescence resulting from a resonant transition rather than a Raman effect. This progression was observed for three different exciting lines, i.e., 514.5 nm, 488.0 nm, and 457.9 nm. In each case the progression begins approximately 140 cm^{-1} from the exciting line and extends to about 18,000 cm^{-1}. This progression band in CsI has been observed before, first by Sommermeyer (ref. 3) in the absorption spectrum of CsI and again by Oldenborg, et al. (ref. 4) in the chemiluminescence of CsI from the reaction of a Cs beam in I_2 vapor. One difficulty that was experienced during the CsI experiment was associated with CsI's attack on quartz cells. CsI vapor is reactive toward quartz, therefore, sapphire tubes were used for optical cells.

CONCLUSION

A low noise spectrometer system designed to acquire emission type spectra (e.g., Raman, LIF) in vapors to 1500K has been described. An evaluation of this technique demonstrates that LIF is a potential monitor for I_2 and CsI at temperatures to at least 1100K. But, collisional quenching of the fluorescence will require, for quantitative determination of concentration, a calibration of fluorescence intensity as a function of vapor composition and pressure. Estimates of sensitivity based on preliminary results are at least $10^{12}/cm^3$ vapor density for I_2 and $10^{14}/cm^3$ for CsI. Additional work using a dye laser to vary the excitation and find the greatest response is needed. In experiments to simulate the TMI accident, a radioactive fuel element in a furnace and flowing gas system will be confined to a hot cell. The excitation source and spectrometer will be outside the hot cell but optically coupled to the experiment through mirrors and a telescope assembly.

ACKNOWLEDGEMENTS

This research was sponsored by the Nuclear Regulatory Commission through the U.S. Department of Energy under Contract No. W-7405-eng-26 with Union Carbide Corporation.

By acceptance of this article, the publisher and/or recipient acknowledges the U.S. Governments's right to retain a nonexclusive, royalty-free license in and to any copyright covering this article.

REFERENCES

1 F. R. Mynatt, Science, Vol. 216, No. 4542, 131, April 9, 1982.
2 W. H. McCulla, Report K/ET-833, The Polytopicity of Ionic Ternary Compounds, (1981).
3 K. Sommermeyer, Z. Phys., 56, 548(1929).
4 R. C. Oldenborg, J. L. Gole, and R. N. Zare, J. Chem. Phys., Vol. 60, No. 10, 4032 (1974).

SPECTROSCOPIC DIAGNOSTICS OF AN ARGON PLASMA JET SEEDED WITH URANIUM

GRAYDON K. ANDERSON

Chemistry Division, Los Alamos National Laboratory, Los Alamos, NM 87545

ABSTRACT

An 80-kW vortex-stabilized plasma torch is used to produce uranium vapor from submicron uranium powder. The torch operates at near atmospheric pressure with a mixture of argon and helium in the ratio of 4:1, at a flow rate of ~0.1 moles/s. The uranium powder (at flow rates up to 0.4 g/s) is entrained in the Ar:He mixture. This differs from the "normal" mode of injection for powder processing, in which the powder is fed into the arc jet downstream of the arc itself. The hot gases leaving the torch expand into a vacuum chamber, forming a supersonic free jet. The arc jet rapidly cools the gases, but also lowers the number density, so that condensation of the (now supersaturated) metal vapor is prevented.

A variety of spectroscopic methods have been used to characterize the plasma torch and the arc jet. Observation of Stark-broadened atomic emission from the torch allows a measurement of both temperature and degree of ionization of the arc gas. Visible emission spectra of the arc jet reveal which chemical species are present. Quantitative determinations of species concentrations in the jet are made by means of absorption spectra. The presence of incompletely vaporized metal powder in the jet is revealed by laser light scattering, and the particle velocities are measured by laser Doppler velocimetry (LDV). The results of these measurements will be discussed in the following paragraphs.

PLASMA TEMPERATURE AND ELECTRON DENSITY FROM STARK-BROADENED EMISSION

Atomic lines emitted from the core of the arc are broadened and shifted by the Stark effect. The arc gas mixture of argon and helium gives rise to a fortuitous excitation of a highly excited helium state. Presumably, argon metastables excite He by a resonant three-body process,

2 Ar(11.55 eV) + He → 2 Ar + He(23.09 eV)

The excited helium gives rise to a strong emission line at 587.6 nm. By measuring the wavelength shift of the line center from its normal position (typically 1Å) and the halfwidth of the broadened line (typically 3.5 Å) both electron temperature and number density may be determined, with no need to assume the existence of LTE. For typical operating conditions, this method gives the following results at the arc core:

$n_e = 2.0 \times 10^{17}$ cm^{-3}

$T_e = 12000$ K

The very high degree of ionization and relatively modest temperature indicate that the arc is in fact not in LTE. This is a commonly observed result for arcs with large superimposed gas flow.

EMISSION SPECTRA OF THE ARC JET

The arc jet is a bright source of atomic emission from both neutral and singly ionized uranium. When "strong" lines with approximately equal upper level energies are examined, the ion and neutral lines are of comparable intensities. The ion/neutral ratio cannot be determined this way since it is doubtful that LTE exists in the arc jet. However, it is clear that ions and neutrals each make significant contributions to the total uranium flux.

ABSORPTION SPECTRA IN THE ARC JET

Ground state neutral uranium number density is measured by laser absorption. The transition at 591.5 nm is used since it originates in the ground state and has a precisely known oscillator strength. A scanning single-frequency dye laser passes through the arc jet at right angles to the flow. The fully resolved absorption profiles are integrated and the uranium density number is calculated. An average uranium atom number density of 7.88×10^{11} cm^{-3} is observed in a typical case. This number is an average along the centerline of the jet at an axial distance of 3.0 cm from the torch exit plane. Spatially resolved concentration profiles could be obtained by absorption measurements at several vertical positions in the jet followed by Abel inversion of the height profiles.

Since U^+ has no suitably strong lines in the wavelength region of the single frequency dye laser, a white light absorption measurement was performed to determine ion number densities. The ion line at 385.9 nm was chosen because (1) it has a large oscillator strength, and (2) it originates from a low-lying (289 cm^{-1}) energy level. The light source is a 150 W Hg-Xe arc lamp. The light transmitted through the arc jet is dispersed by a spectrometer equipped with an echelle grating. The spectral bandpass of the apparatus is ~0.001 nm, which is very nearly equal to the expected Doppler width of the absorption feature. A problem arose in the measurement due to the presence of strong emission from the plasma at 385.9 nm. In fact, the plasma was several times brighter than the arc lamp at the peak of the line. To make the absorption measurement possible, the output of the arc lamp was chopped, and the transmitted signal was recovered by lock-in detection. The resultant ion number density (at the same location as the neutral number density measurement) is

$$n_{ion} = 1.78 \times 10^{12} \text{ cm}^{-3}.$$

This number is less precise than the neutral density because the oscillator strength is poorly known at present. The U^+/U ratio leaving the torch is therefore approximately 2.3/1, not surprising in view of the temperature and electron density measurements referred to earlier. Higher levels of ionization (particularly U^{++}) are also possible but line assignments and oscillator strengths are not available at present.

PARTICLE VELOCITY DETERMINATION BY LASER LIGHT SCATTERING

A laser Doppler velocimeter was constructed to measure the velocity of the particles which are not completely vaporized in the torch. The system consists of a 1-W argon-ion laser at 488 nm, split into two parallel beams 1.0 cm apart, focused to a single point in the center of the arc jet by an 80-cm f.l. lens. A portion of the forward-scattered light is collected and focused onto a pinhole, and detected by a fast photomultiplier tube. The Doppler-modulated scattering pulses are displayed on a storage oscilloscope. Particle velocities approaching 10^5 cm/s have been observed, whereas gas velocity in the arc jet is predicted to be approximately 3×10^5 cm/s. This confirms the expectation that the particles are in free molecular flow, i.e., the particle diameter is less than the gas mean free path in the torch, giving rise to velocity slip.

SUMMARY

Uranium particles are vaporized in a plasma torch with an efficiency of approximately 30%. More than two thirds of the uranium vapor is in the singly ionized species. The technique may be applied to produce vapors of virtually any metal or refractory compound.

ACKNOWLEDGMENTS

The author is grateful to William Maier for many useful discussions and to Thomas Gamble for his major role in constructing the apparatus and performing the experiments. This work was sponsored by the Denfense Nuclear Agency.

W.S. Lyon (Editor), *Analytical Spectroscopy*
Elsevier Science Publishers B.V., Amsterdam — Printed in The Netherlands

GENERAL-PURPOSE CHEMICAL ANALYZER FOR ONLINE ANALYSES OF RADIOACTIVE SOLUTIONS*

William A. Spencer and James W. Kronberg
E. I. du Pont de Nemours & Co., Savannah River Laboratory,
Aiken, South Carolina 29808

ABSTRACT

The Savannah River Laboratory is developing an automated analyzer to perform analytical measurements on radioactive solutions online in a hostile environment. This "General Purpose Chemical Analyzer" (GPCA) samples a process stream, adds reagents, measures solution absorbances or electrode potentials, and automatically calculates the results. The use of modular components, under microprocessor control, permits a single analyzer design to carry out many types of analyses.

This paper discusses the more important design criteria for the GPCA, and describes the equipment being tested in a prototype unit.

INTRODUCTION

Work now in progress will place the Savannah River Plant (SRP) separations areas under computer-based process control within ten years. This upgrade will reduce costs and energy requirements, improve safety, and decrease the exposure of personnel to radiation.

For maximum benefit, many analyses now carried out offline will be performed automatically online. Some of these analyses are unique to SRP; most involve radioactive product streams; proven online analyzers are available for none of them. Savannah River Laboratory (SRL) is developing analyzers to meet these requirements.

A single basic design has evolved, able to perform simple chemical analyses: the "General-Purpose Chemical Analyzer" (GPCA). Based on the design, SRL is developing a prototype unit which contains the computer system, software, reagent-dispensing system, and electronic interfaces to be used. That prototype is currently being assembled and testing of components is in progress.

APPLICATIONS

Table 1 lists online analytical tasks which the GPCA should be able to perform. The first three are being incorporated into the prototype unit; the final product, equipped with suitable enhancements, should perform any of them.

---

* The information contained in this article was developed during the course of work under Contract No. DE-AC09-76SR00001 with the U.S. Department of Energy.

The GPCA should also be adaptable to other tasks of about the same level of complexity.

TABLE 1
Initial applications for GPCA

Application	Parameter Measured
1. Free acid analysis	Temperature, pH
2. Reducing normality	Light absorption
3. Plutonium concentration	Light absorption
4. Uranium concentration (high)	Solution density
5. Uranium concentration (low)	Light absorption
6. Uranium concentration (trace)	Fluorescence
7. Redox potential	Electrode voltage
8. Radioactivity (gross gamma)	Scintillation

DESIGN CRITERIA

The GPCA must be capable of adding reagents to a sample, mixing, and measuring optical absorbance and electrode potentials. It must then calculate results, display them locally, and transmit them to process control equipment. To minimize radiation hazards, it must be divided into radioactive and nonradioactive modules or compartments.

The final design should be easily reproduced, and suitable for assembly by an outside contractor. Commercial subassemblies should be used as much as possible. Some design criteria for specific components are given in the system description.

Since up to 50 analyzers may be built, cost per unit should be kept to a minimum consistent with reliability and ease of maintenance.

PROTOTYPE SYSTEM DESCRIPTION

Figure 1 shows a block diagram of the prototype GPCA with all major components. These components are described in the following sections.

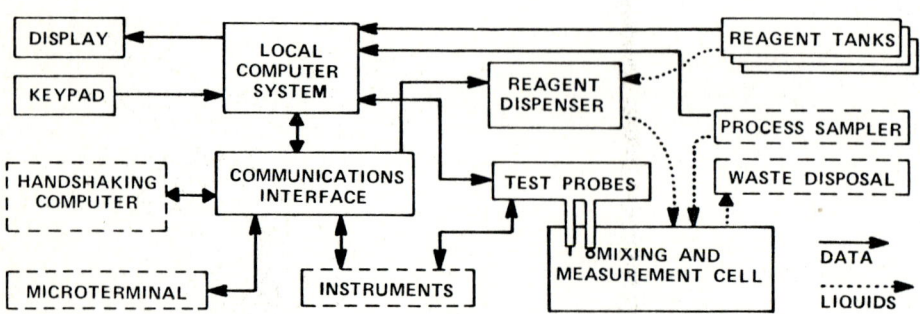

Fig. 1. Block diagram of General Purpose Chemical Analyzer (GPCA).

Computer control system

A reliable, rugged, and easily maintained computer is needed to control the GPCA. The use of interchangeable modules will simplify the changes needed for different applications. However, the computer need not be large; most applications require less than four kilobytes of memory.

The Rockwell AIM-65/40 single-board computer is to be used in the prototype. Additional serial, parallel, and analog I/O capacity is added by the compatible Rockwell RM-65 cards. These devices were chosen because of previous experience at SRP, versatility, low cost, and the availability of high-level languages in ROM. If contaminated, they may be simply replaced. Typical components are shown in Figure 2.

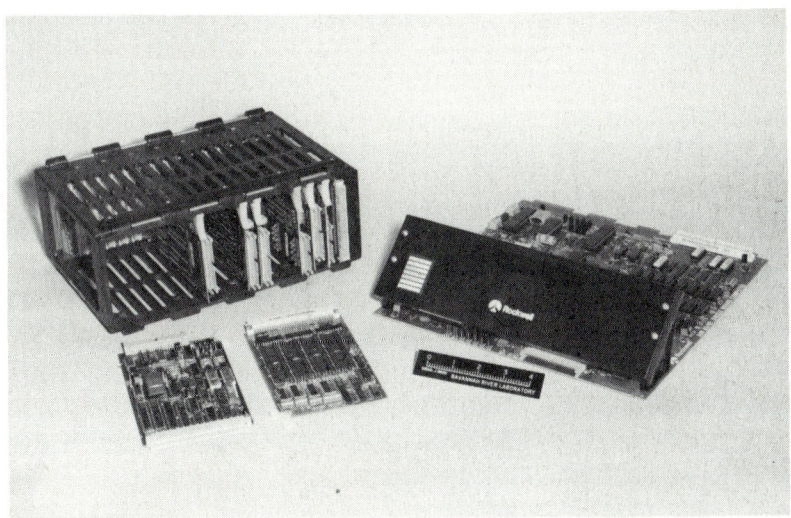

Fig. 2. Rockwell AIM-65/40 computer, RM-65 card cage, and typical cards.

Communications interface

The separations areas process control system will include a minicomputer for remote control of the analyzers. In remote mode, each analyzer will accept simple ASCII string commands and send out data through a two-way serial port. Other serial and parallel data ports are provided to control the reagent dispenser (see below), communicate with instruments, and allow field testing with an ASCII microterminal.

Operator interface

For local control, a visual display and keypad are provided.

The display is the single-line, 40-character vacuum-fluorescent module normally used with the AIM-65/40, and shown in Figure 2. Numerals and upper-case letters, 0.2 inch tall, are illuminated in a highly visible blue-green.

The keypad is a 20-key matrix type, of sealed membrane construction impervious to vapors and spills. It provides good tactile feedback, and is usable by an operator wearing gloves.

Reagent dispenser

Reagents are measured and transferred by motor-driven syringe pumps. The Zymark Z-510 Master Laboratory Station is shown in Figure 3; it contains three pumps, and accepts syringes in a wide range of sizes. Unfortunately, it is controlled through a nonstandard data link; modifications are needed to permit control through a standard serial port.

Reagents are made up in advance and stored in polyethylene bottles connected to the dispenser. A level sensor in each bottle signals the computer and causes a message to be sent to operators when the reagent needs to be replenished.

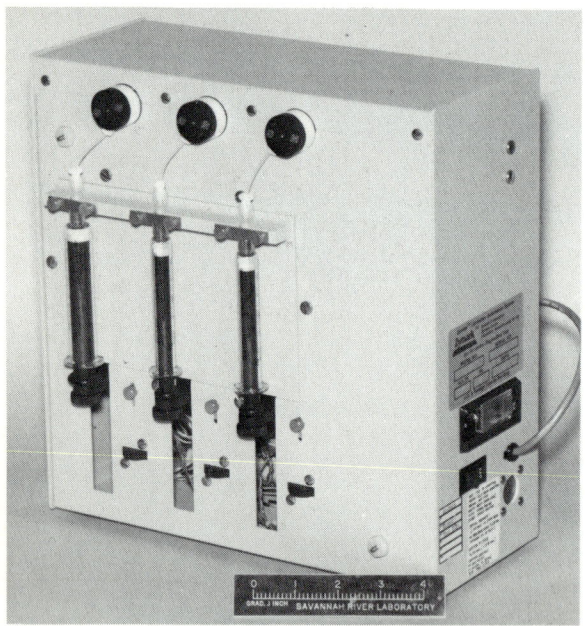

Fig. 3. Zymark Z-510 Master Laboratory Station.

Mixing and measurement cell

The syringe pumps transfer samples and reagents into a sealed cylindrical mixing cell made of stainless steel or Inconel® (Huntington Alloys, Inc.). Ports at the upper end accept test probes and a stirrer; the lower end is a shallow funnel leading to a drain valve. To simplify fabrication, the two sections are made separately, as shown in Figure 4.

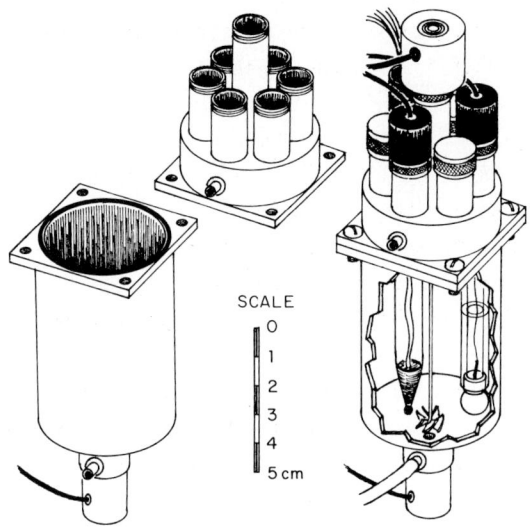

LEFT: Lower half of disassembled cell; exterior.

TOP: Upper half of assembled cell, showing probe ports.

RIGHT: Assembled cell, showing stirrer and typical probes in place.

Fig. 4. SRL design for two-part mixing and measurement cell.

Test probes

All measurements on the cell's contents are taken with interchangeable test probes the size of standard pH electrodes. Figure 5 shows probe designs being tested; most of these are commercially available.

Standard pH and redox electrodes ("A" and "B") are read and digitized by a high-impedance, differential-input digital panel meter.

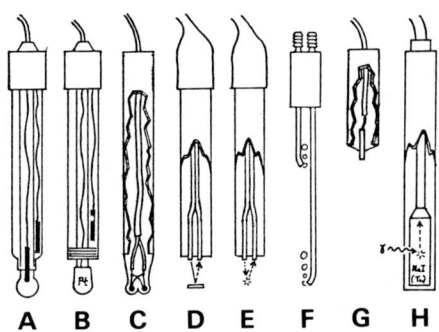

A - Combination pH electrode; Beckman Instruments.
B - Combination redox electrode; Innovative Sensors.
C - Temperature probe; Fenwal "Iso-Curve" thermistors.
D - Fiber-optic colorimeter probe; Brinkmann Instruments.
E - Brinkmann probe, modified to measure fluorescence.
F - Liquid-density probe; differential bubble tube.
G - Liquid-level probe; conductivity detector.
H - Fiber-optic scintillation probe; Harshaw Chemical Co.

Fig. 5. Test probes for mixing and measurement cell.

Temperature is sensed by precision thermistors ("C"). This method yields high-level output signals (zero to ten volts) which can be accepted by the RM-65 analog input cards without amplification or signal conditioning.

Light absorption is measured by a dual fiber-optic probe ("D"). Light passes down one fiber, penetrates a known thickness of liquid, is reflected by a small mirror, and returns through the liquid to a second fiber and to a photodiode or photomultiplier. _Fluorescence_ is excited by a small cadmium-vapor (blue) laser, and measured with a similar probe with no mirror ("E").

Solution density may be determined by a method already used in SRP process vessels: passing air through two stainless-steel tubes immersed to different depths ("F") and measuring the pressure difference between the tubes. A simple correlation exists between the pressure difference and the solution density.

The feasibility of using this method on the small scale of the analyzer cell is being investigated. If it fails to provide sufficient accuracy, a commercial "U-tube" densitometer may be added outside the measurement cell.

Liquid level is sensed by conductivity. The metal cell wall acts as one electrode; the probe holds the other ("G").

Gross gamma radiation level is measured with a miniature scintillation crystal ("H"), coupled to an optical fiber and detector as in "D" and "E."

All probes must be physically and chemically rugged, since they may be exposed to corrosive solutions and rough handling by technicians. Except for the glass pH and redox electrodes, all designs meet these criteria.

INFLUENCE OF URANYL DIBUTYLPHOSPHATE ON THE UV/VIS SPECTROPHOTOMETRIC ONLINE MONITORING OF URANIUM IN TRIBUTYLPHOSPHATE/HYDROCARBON SOLVENT

E. T. Creech, A. C. Rutenberg, R. W. Smithwick, and R. D. Seals, Union Carbide Corporation, Nuclear Division, P. O. Box Y, Oak Ridge, TN 37830

ABSTRACT

In the uranium recovery process at the Y-12 Plant uranium is recovered from aqueous uranyl solutions by extraction into a solvent consisting of 30% tributylphosphate (TBP) and 70% hydrocarbon solvent. Within this process the uranium is continuously monitored by a UV/VIS absorbance measurement of the uranyl/tributylphosphate complex in the organic phase. The uranium is then further extracted from the organic phase to a final water phase. Dibutylphosphate (DBP), which is a decomposition product of TBP, builds up in the organic solvent. A very strong complex of uranyl/dibutylphosphate is formed which cannot be extracted into the aqueous phase. Prior to this work the uranyl/dibutylphosphate complex absorbance was assumed to be the same as the uranyl tributylphosphate complex.

To determine the effect of the presence of uranyl/dibutylphosphate on the continuous UV/VIS monitor required (a) the purification of commercial dibutylphosphate, (b) the synthesis, and (c) the characterization of uranyl/dibutylphosphate.

Since no commercially available dibutylphosphate could be obtained in the purity desired, a method was developed to purify a commercial product to better than 99.5%. Briefly, the purification consists of extracting a single phase mixture of water, ethanol, and impure dibutylphosphate with petroleum ether with a continuous extraction apparatus which allows the solvent to percolate through the aqueous phase. Dibutylphosphate is recovered on evaporating the petroleum ether solvent.

Phosphorous-31 NMR spectra attest to the purity of a typical purification to be 99.6% DBP with the impurities of monobutylphosphate and tributylphosphate constituting the balance. Infrared analysis identified the synthesis product by computer search of the computerized Sadtler Infared Library. This could not be done prior to purification.

Uranyl chloride (UO_2Cl_2) prepared from hydrochloric acid and 99.9% uranium metal was dissolved in acetone. The 99.5% DBP was also dissolved in acetone. Nearly stoichiometric quantities of these two solutions were combined to obtain a clear solution of $UO_2(DBP)_2$. This clear solution was poured into demineralized water to form a yellow precipitate. The precipitate was filtered, washed with water, and dried overnight at 70°C in a vacuum oven. Chemical analyses, NMR, infrared, and UV/VIS spectroscopies were used to provide evidence for support of a structure.

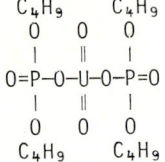

previously proposed by Tetrin [1].

After establishing a uranium concentration in the product by chemical analysis using the Davies-Gray [2] method, a portion was weighed and then dissolved in 30% TBP/70% hydrocarbon solvent and made to volume to provide a solution containing only the $UO_2(DBP)_2$ complex. The absorbance of the solution was measured at the wavelength used by the monitor. This reading was used to calculate a uranium concentration by a computerized calculation used for the online monitor.

uranium taken 0.00588 g/cc

uranium found by monitor 0.00878 g/cc

This result is of course a worse possible case but clearly demonstrates a large difference in absorbance between uranyl tributylphosphate and uranyl dibutylphosphate.

Approximately 0.15 g/cc of $UO_2(DBP)_2$ could be dissolved in TBP/hydrocarbon solvent with nitric acid present; however, if the nitric acid is removed by extraction with water the $UO_2(DBP)_2$ remains in solution for a short time but then starts to precipitate on the container walls. Over a period of several days the solubility dropped to 0.004 g/cc of $UO_2(DBP)_2$ in the absence of nitric acid.

By forming the $UO_2(DBP)_2$ complex in situ in TBP/hydrocarbon solvent using the 99.5% pure DBP a much higher solubility could be obtained; however, on removing the nitric acid the solubility drops and the $UO_2(DBP)_2$ precipitates.

Evidence has been presented based on chemical analysis, infrared, NMR, and UV/VIS spectroscopies to support a structure of $UO_2(DBP)_2$ for uranyl dibutylphosphate. Further, that absorbance measurements based on uranyl tributylphosphate standards can be in error where uranly/dibutylphosphate is present. Also, due to the limited solubility of $UO_2(DBP)_2$ at low pH, the complex can precipitate and contribute to emulsion formation.

ACKNOWLEDGMENT

The Oak Ridge Y-12 Plant is operated by Union Carbide Corporation, Nuclear Division for the Department of Energy under U.S. Government Contract W-7405-eng-26.

REFERENCES

1. Tetrin, E. G., Russian Journal of Inorganic Chemistry, 16, (3) 1971.
2. Tucker, H. L. and McElhaney, R. J., Determination of Uranium in Uranium Metal, Uranium Oxide, and Uranyl Nitrate Solutions, Union Carbide Corporation, Nuclear Division, Oak Ridge Y-12 Plant, Oak Ridge, Tennessee; August 1983 (Y/DK-345).

CONTROLLED-POTENTIAL-COULOMETRIC DETERMINATION OF URANIUM AT A PLATINUM ELECTRODE

N. M. Saponara and D. D. Jackson

Los Alamos National Laboratory, University of California, Los Alamos, New Mexico (USA) 87545

ABSTRACT

A controlled-potential-coulometric method has been developed for determining uranium at a solid electrode which features high specificity and a precision of 0.1% relative standard deviation at the 5 mg uranium level. Uranium and added iron, necessary for the electrolytic oxidation of uranium, are reduced to U(IV) and Fe(II) with excess Cr(II). At a sequence of controlled potentials, the excess Cr(II) is oxidized to Cr(III), Fe(II) and U(IV) are oxidized to Fe(III) and U(VI), then the Fe(III) is reduced to Fe(II). The difference in the measured number of coulombs for the oxidation of Fe(II) and U(IV) and for the reduction of Fe(III) to Fe(II) is proportional to the quantity of uranium.

INTRODUCTION

The current need for a controlled potential coulometric method for uranium is its use in an automated analyzer already constructed for plutonium determination (ref. 1). Initial objectives were high specificity, precision of 0.1% relative standard deviation at low-milligram uranium level and use of solid electrode with long-term stability. A mercury pool electrode is used almost exclusively for uranium because it provides adequate hydrogen overvoltage for reduction of U(VI) without reduction of hydrogen ion. No system with the desired features employing solid electrode has been described. Carbonaceous electrodes of glassy carbon, boron carbide and graphite have not provided reproducible surfaces necessary for precise measurements. Reticulated vitreous carbon (RVC), a honey-comb material with large surface area, had large, variable background currents and a limited cathodic range in acidic media (ref. 2). In a method developed by Davies et al. (ref. 3), Bi(III) and high chloride concentration provided the necessary overvoltage for hydrogen at platinum electrode. During our investigation, measurement of low milligram amounts of uranium was not sufficiently reliable (ref. 2). Phillips and Crossley (ref. 4) determined uranium at a solid electrode by reducing uranium with electrogenerated hydrogen. Removal of adsorbed hydrogen was slow and affected electrode stability. Because uranium is difficult to reduce electrolytically without hydrogen interference, a chemical reductant was sought. Cr(II) was selected because its potential is sufficiently low to reduce uranium and it is readily oxidized electrolytically without significant oxidation of U(IV).

In the method developed here, uranium and added iron are reduced with excess Cr(II). At a sequence of controlled potentials at a platinum electrode, the excess Cr(II) is oxidized to Cr(III), Fe(II) and U(IV) are oxidized to Fe(III) and U(VI), and Fe(III) is reduced to Fe(II). The difference in the measured number of coulombs for the oxidation of Fe(II) and U(IV) and for the reduction of Fe(III) to Fe(II) is proportional to the quantity of uranium. Added iron is necessary for the oxidation of uranium.

EXPERIMENTAL

Apparatus

The apparatus is centered about a Princeton Applied Research 173D potentiostat-galvanostat and a 179D digital coulometer. A strip chart recorder and a digital multimeter complete the system.

The electrolysis cell has been described in detail previously (ref. 5). The working electrode is a strip of platinum gauze formed into a cylinder; the platinum counter electrode and the saturated calomel reference electrode, isolated from the cell solution by separate compartments, are connected to the cell solution by porous Vycor disks. The counter electrode is located in the center of the cell to provide a uniform potential gradient. The reference electrode is located adjacent to the working electrode for best control. Stirring is provided by a glass paddle driven by a 1800 rpm synchronous motor.

Reagents

Deionized water and AR grade chemicals were used for all reagents.

Selection of Chemical Method

Sulfuric acid was selected as electrolyte. Because of its high boiling point, interfering volatile ions can be removed by fuming. Uranium determinations were independent of sulfuric acid concentration over the range of 0.5 to 2 M. To minimize the effect of acidity of a sample, 0.5 M H_2SO_4 was selected as electrolyte for the method.

Added iron was necessary for the oxidation of U(IV). For iron to uranium ratios of < 1, electrolyses times were very long. At molar ratios of 1 to 1 or greater, oxidation times were < 20 min.

A potentiometric titration of uranium and iron with Cr(II) to a -0.25 V endpoint assured quantitative reduction of uranium and iron but avoided addition of large excess of reductant. All potentials are vs saturated calomel electrode.

Electrolysis potentials for the method were established using the following conditions: 0.02 mmol (5 mg) uranium, 0.03 mmol Fe(III), 10 ml 0.5 M H_2SO_4, nitrogen sweep during electrolysis and titration with Cr(II) to a -0.25 V endpoint. A potential of 0.25 V was selected for the first oxidation; at this potential 99.99% of Cr(II) is oxidized to Cr(III) without oxidizing the U(IV) and Fe(II). At the second oxidation potential of 0.52 V, 99.95% of U(IV) is oxidized to U(VI). The Fe(III) is reduced to Fe(II) at a potential of 0.25 V. All electrolyses are continued to a 50 µA current endpoint.

RESULTS AND DISCUSSION
Diverse Ion Effects

The results of detailed investigation of the effect of 45 metal cations are presented in Table 1. Interference is defined as a change significant at the 95% confidence level relative to uranium alone. The level of uranium used for each test was 0.02 mmol (5 mg). The initial molar ratio tested was 1:1. If a result was significantly different, lower ratios were tested. Not all the elements in the alkali and alkaline earth groups were tested. The results of representative members are considered to apply to all members of the group. Most elements normally found in nuclear fuel cycle materials do not interfere. Several elements, osmium, selenium, and ruthenium did not interfere in the uranium determination but impaired electrode response so that sufficiently low current could not be attained in subsequent analyses. Bismuth, gold, iridium, molybdenum, rhenium and tellurium interfered even at 1:100 molar ratio.

TABLE I
TOLERANCE LEVEL OF METAL CATIONS

Mole Ratio With Respect to U	Cations
1:1	Al, As, B, Be, Ca, Cd, Ce, Cr, Co, Fe, Ga, Hf, In, La, Mn, Na, Nb, Ni, Os, Pb, Pu, Th, Tl, Zn, Zr
1:10	Sb, Se, Sn, Rh, Ta
1:100	Cu, Hg, Pd, Pt, Ru, W, V

Au, Bi, Ir, Mo, Re, Te interfered at 1:100

To ensure applicability of the method for determining uranium in uranium-plutonium mixtures the tolerance for plutonium was investigated in greater detail. There was no significant difference in results for five measurements of equal ratios of uranium and plutonium relative to five measurements of uranium alone.

Nonmetallic ions are yet to be studied. It is expected that many potentially interfering nonmetallic anions can be effectively removed by fuming with sulfuric acid.

Treatment to Restore Effective Electrode Behavior

The platinum-gauze working electrode was stable during the six months of method development. During the study of interfering ions, however, several elements impaired electrode response so that reaction endpoint of 50 μA could not be attained. The cleaning procedure to restore effective electrode behavior is as follows: after rinsing electrode in deionized water, immerse in concentrated nitric acid for 5 min, rinse with water and immerse for 1 h in 0.5 M H_2SO_4. A more severe cleaning was necessary after electrolysis of gold, iridium and rhenium: electrode was rinsed in water, immersed in aqua regia for 45 s, rinsed in water and immersed in 0.5 M H_2SO_4 for 1 h.

Measurement Precision

Within day precision was < 0.1% relative standard deviation at the 5 mg uranium level. The relative standard deviation for 32 analyses measured over a 2-week period was 0.17%. Even during investigation of diverse ion effects covering a 3-week period the precision of uranium standard determinations remained 0.17%.

REFERENCES

1 R. M. Hollen and D. D. Jackson, Automated Controlled-Potential Coulometer for Plutonium Determination, Los Alamos National Laboratory report LA-8653 (May 1981).
2 N. M. Saponara and D. D. Jackson, Controlled-Potential Coulometric Determination of Uranium, in Safeguards and Security Status Report, February-July 1982, J. P. Shipley, Compiler, Los Alamos National Laboratory report LA-9595-PR (February 1982).
3 W. Davies, W. Gray and K. C. McLeod, Coulometric Determination of Uranium with a Platinum Working Electrode, Talanta 17 (1970) 937-944.
4 D. Crossley and G. Phillips, A Secondary Controlled-Potential Coulometric Method for the Reversible Determination of Uranium at a Solid Electrode, UKAEA Report, R-8029 (1977).
5 D. D. Jackson, R. M. Hollen, F. R. Roench and J. E. Rein, Controlled-Potential Coulometric Determination of Plutonium with a Hydrochloric Acid-Sulfamic Acid Electrolyte and Phosphate Complexing, Anal. Chim. Acta 117 (1980) 205-215.

ATOMIC ABSORPTION SPECTROPHOTOMETRIC METHODS FOR THE DETERMINATION OF PHOSPHOROUS AND SILICON IN STEEL

J. P. McCARTHY, E. B. NUNN AND C. KINARD
Monsanto Research Corporation, Mound[*], Miamisburg, Ohio 45342

ABSTRACT

Several methods for the determination of phosphorus in steel by graphite furnace atomic absorption spectrophotometry (GFAA) have been put forth in the literature. The most promising of these procedures are evaluated and optimized parameters for this analysis are given. The methods evaluated are those involving zirconium carbide furnace tubes, chemical matrix modification with La, the use of the method of additions and calibration via NBS standards. Similarly, an overview of flame AA (FAA) methods for Si determination is given. An overall procedure is described which allows the GFAA determination of phosphorus (0.010 - 0.100%), with matrix modification using lanthanum and FAA determination of silicon (0.1 - 1%) and metal constituents in steel from a single sample dissolution using NBS steels as calibration standards.

INTRODUCTION

The composition of steels used in various energy related programs is of considerable interest, necessitating time efficient, yet accurate, analytical procedures for the determination of elemental constituents. In addition to metallic content, P and Si composition are often specified. P is present as a contaminant which adversely affects welding characteristics while Si is added to enhance wetting, flowing, and maching properties of the alloy.

Current methods utilized for P and Si determination involve labor intensive and time consuming wet chemical titrimetric or gravimetric procedures. A survey of the literature reveals several proposed graphite furnace atomic absorption (GFAA) procedures for the determination of P in steel. These methods include the use of Zr coated tubes, La matrix modifiers or Zeeman background correction (References 1-4). The most thorough evaluation of these techniques to date is that of Welz et al. (Reference 4). Their findings differed greatly from the originally reported results.

---

[*] Mound is operated by Monsanto Research Corporation for the U. S. Department of Energy under Contract No. DE-AC04-76DP00053.

Several methods for the flame atomic absorption (FAA) determination of Si in steels have also been proposed (References 5 and 6). These methods involve rather complex procedures to ensure Si dissolution except for the method of Price and Roos, in which hydrofluoric acid was employed. Their method also made some attempt at compensating the interferences produced by the steel matrix (Reference 7).

The intent of this work is threefold: 1) To investigate the discrepancies among the various methods proposed for GFAA determination of P in steel. 2) To establish an analysis scheme capable of the AA determination of P and Si along with metal constituents from the same sample solution. 3) To provide NBS traceability of the results. The procedure outlined provides a substantial time savings over present wet chemical methods for P and Si.

Experimental

Reagents: Baker analyzed reagent grade hydrogen peroxide (30%), nitric acid, hydrochloric acid, and hydrofluoric acid (48%) were used in the dissolution of the steels. Reagent grade $La(NO_3)_3$ (Baker) was used to make matrix modifier solutions 2500 µg/ml in La. 1000 µg/ml P standard solution was prepared from $(NH_4)_2HPO_4$ (Fisher) and standardized tritrimetrically (Reference 8). All dilutions were made with distilled-deionized water.

Instrumentation: A Perkin Elmer Model 403 atomic absorption spectrophotometer with deuterium background correction was used in conjunction with a model HGA 2000 (Perkin Elmer) graphite furnace for phosphorus determinations. A phosphorus electrodeless discharge lamp operated at 7.5W was used as the source. All measurements for P determination were done at 213.6 nm using a 0.2 mm bandpass.

Silicon determinations were carried out on a Perkin Elmer Model 5000 atomic absorption spectrophotometer using a nitrous oxide/acetylene flame. The silicon hollow cathode lamp used as the line source was operated according to manufacturer's specifications. All measurements were done at 215.6 nm using a 0.2 nm band-pass.

Procedure: 2.5 grams of sample are weighed into a 400 ml teflon beaker and 30 ml 6M HCl and 30 drops H_2O_2 are added. The beaker is covered and heated gently on a hot plate until the initial reaction subsides (ca. 1 hr). 1.5 ml nitric acid is added and the beaker returned to the hot plate until the vigorous reaction ends. Nitric acid is then added dropwise until no reaction is observed. The solution is allowed to cool and 15 drops of hydrofluoric acid are added with swirling. After standing 15 minutes the solution is transferred to 50 ml polyethylene volumetric flasks and diluted to volume. Five milliliters of this solution is transferred to a 10 ml volumetric flask and diluted to volume. Ten microliter aliquots of this solution are pipetted into the furnace and ten microliters of a 2500 µg/ml La solution is pipetted directly on top of the sample. The furnace temperature program is given in Table 1.

TABLE 1

Furnace conditions for GFAA determination of phosphorus.

Step:	Dry	Ash	Atomize	Clean
Temperature (°C)	125	1350	2750	2750
Time (sec.)	40	55	6	10

Calibration curves were constructed using NBS SRM steels. The NBS steels used for standardization were SRM #344, SRM #129C and SRM #339 containing 0.018%, 0.076%, and 0.129% P respectively. Quantification was achieved using peak height measurement.

Silicon determinations were made by diluting a 10 ml aliquot of the initial sample solution to volume in a 50 ml polyethylene volumetric flask. This solution was read against a calibration curve constructed with NBS SRM steels dissolved as above. Calibration standards were NBS SRM 19G, SRM 345, SRM 101f containing 0.186%, 0.610%, and 0.88% Si respectively. Si absorption measurements were made using deuterium background correction.

Results and Discussion

Whiteside and Price (Reference 3) reported the successful determination of P in steels utilizing the method of standard additions. While we were able to achieve adequate sensitivity from phosphorus solubilized from the steel, we observed substantially less sensitivity from the P used for additions which was in the form of $(NH_4)_2HPO_4$. This was probably due to the difference in chemical form of the phosphorus and led to erroneous results.

Ediger (Reference 2) was the first to report the use of lanthanum as a matrix modifier to enhance the GFAA signal of phosphorus. In our evaluation of lanthanum matrix modification, 10 µl of La at 2500 ppm was pipetted directly on top of 10 µl of phosphorus containing sample since lanthanum fluoride would precipitate if the lanthanum was added directly to a dissolved steel sample. The addition of lanthanum substantially increased the signals obtained from phosphorus standards (Figure 1), but the signal produced from phosphorus solubilized from steel remained relatively unaffected. The amount of phosphorus necessary to produce a 1% absorption was 7 ng. Ten to twenty sample runs were required before the graphite tube was conditioned to the point that reproducible results at full sensitivity were obtained.

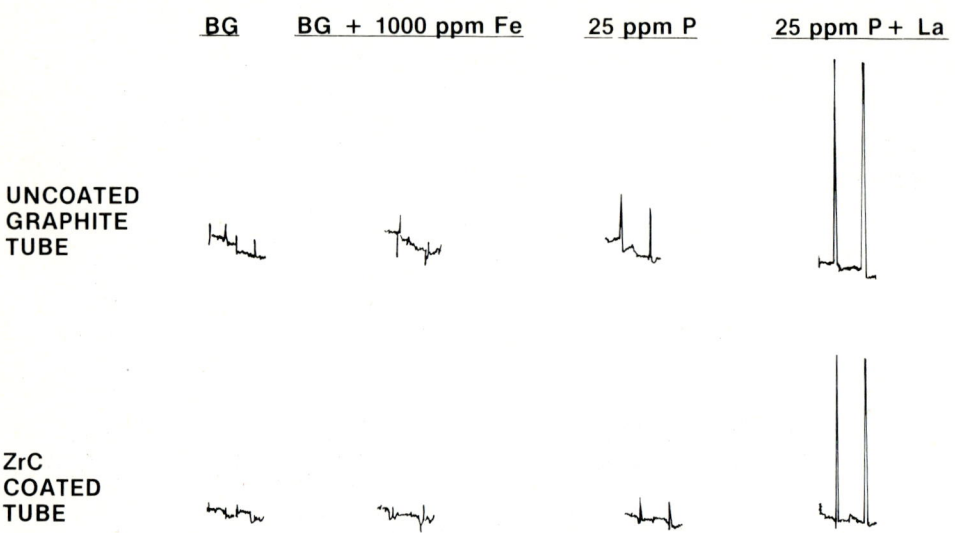

Figure 1. Enhancement effect of matrix modification with La and results of ZrC furnace tubes. (BG = Background Signal)

The standard additions experiment was repeated using the addition of lanthanum. These results were non-reproducible for several reasons. The somewhat outdated furnace design and manual pipetting of the sample led to poor reproducibility of peak heights. Also, the negative dip preceding the absorption peak in steel samples made location of the baseline difficult. The background overcompensation producing the negative dip is due to the inability of deuterium background correction to correctly compensate iron spectral interference within the 0.2 nm bandpass.

Graphite tubes treated with zirconium oxychloride to form zirconium carbide surfaces have been reported (Reference 1) to enhance the phosphorus signal without lanthanum addition. Our results for these tubes is shown in the lower portion of Figure 1. Essentially the same sensitivity was observed with a slight decrease in background signal; otherwise no significant advantages were observed.

To compensate for the iron interference, matrix matching was mandatory. NBS SRM steel samples with phosphorus content covering the range of interest were used to construct a calibration curve from which sample concentrations were determined (Figure 2).

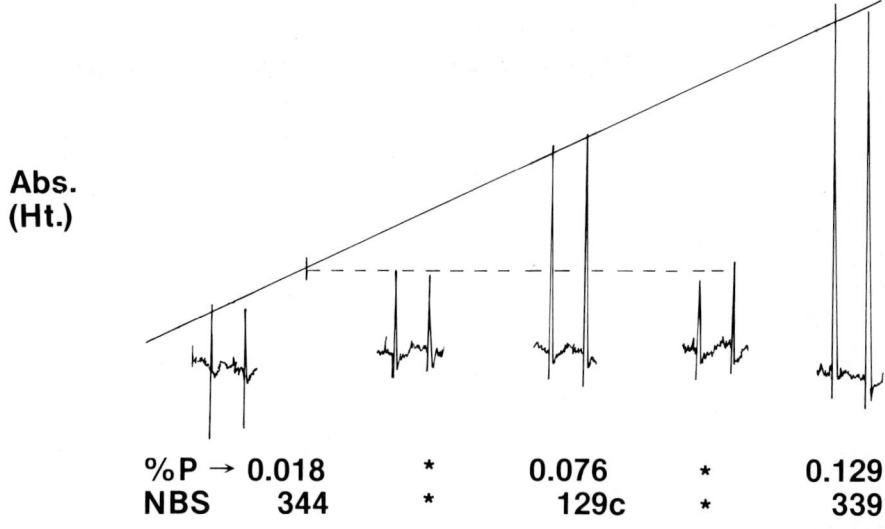

Figure 2. GFAA determination of P using NBS calibration scheme.
* NBS SRM 160A found 0.025% versus 0.027% reference value.

Since all standards and samples experienced the same background offset, background shift was not a problem. Results are shown for NBS steels in Table 2.

TABLE 2
Analysis of NBS SRM Steels.

NBS-SRM	%P		%Si	
	Found	Reference	Found	Reference
19g	0.046	0.046	*	0.186
121c	0.030	0.028	0.636	0.64
160a	0.025	0.027	0.602	0.600
344	*	0.018	0.379	0.395
8i	0.054	0.080		

* Used in Standardization

Good correlation was obtained except for SRM 8i, Bessemer steel. This discrepancy is most likely due to a matrix constituent not compensated for in the above procedure. Peak area measurements would most likely give more accurate results, however, this option was not available to us.

Hydrofluoric acid is included in the sample preparation scheme to dissolve silicon. Hydrofluoric acid was added after the solutions had cooled to room temperature, since loss of silicon as the volatile tetrafluoride occurs at elevated temperatures. Extensive interferences with many metals in silicon determination has been reported (Reference 7). In order to circumvent these interferences, calibration was done with NBS SRM steels as the phosphorus determinations. Deuterium background correction provided the most stable absorbance signal. The linear range of the method extends up to 2.5% silicon content in steel. Results for reference materials are given in Table 2 and show excellent correlation.

CONCLUSIONS

The results of the current study demonstrate a method capable of atomic absorption determination of P, Si and metal constituents from a single sample solution. This method circumvents the time consuming wet chemical techniques for P and Si determinations. The majority of interferences are eliminated in these determinations by incorporating matrix matching techniques using NBS SRM's as calibration standards.

REFERENCES

1. I. Havezov, E. Russeva, N. Jordanov, Fresenius Z. Anal. Chem 296, pp 125-127 (1979).
2. R. D. Ediger, Atomic Absorption Newsletter, 15, pp 145 (1976).
3. P. J. Whiteside, W. J. Price, Analyst 102, pp 618-620 (1977).
4. B. Welz, U. Voellkopf, Z. Grobenski, Anal. Chem. Acta, 136, pp 201-214 (1982).
5. W. J. Price, J. T. H. Roos, Analyst, 93, pp 709-714 (1968).
6. G. E. Peterson, J. D. Kerber, Atomic Absorption Newsletter, 15, pp 134-143 (1976).
7. J. Musil, M. Nehasilová, Talanta, 23, pp 729-731 (1976).
8. "Phosphorus by the Alkalimetric Method", 1960 Book of ASTM Methods for Chemical Analysis, Parts 37-40, 42, 43, ASTM, Phila., 1961, pp 113-115.

ANALYSIS OF MOLECULAR HYDROGEN USING SPONTANEOUS RAMAN SPECTROSCOPY

Kirk Veirs and Gerd M. Rosenblatt
Chemistry Division, Los Alamos National Laboratory, Los Alamos, NM 87545

ABSTRACT

Relative number densities of all molecular isotopic species of hydrogen (H_2, HD, HT, D_2, DT, T_2) are determined by Raman spectroscopy. Unambigous identification is due to the unique Raman shift of each species. Modern laser methods lower the detection limit to 6 ppm in atmospheric samples.

INTRODUCTION

Sensitive analytical methods for all molecular isotopic species of hydrogen are required for system diagnostics and dynamic accounting methods in fusion energy installations. In the next decade five experimental facilities will come on-line with a total inventory in excess of a million Curies. Traditional analytical methods, mass spectrometry and gas chromatography, are unable to quantitatively or qualitatively analyze mixtures of these gases. For example, in a sample of pure protium ($^1H_2$) mass spectrometry, with a peak at $m/e^-=3$ due to H_3^+, will incorrectly indicate the presence of HD. Spontaneous laser Raman spectroscopy offers advantages for both system diagnostics and dynamic accounting methods including:

1. Convenience--in-situ, non-destructive analytical method;
2. Specificity--isotope shifts for hydrogen are large and species identification is unambiguous;
3. Sensitivity--modern laser methods have reduced the detection limit to a partial pressure of 0.01 Torr;
4. Range--signal depends only upon partial pressure; accurate determinations may be made over at least 6 orders of magnitude in total pressure.

EXPERIMENTAL METHOD

A highly sensitive laser Raman apparatus has been assembled from commercial components. A Spectra-Physics 171 Ar^+ laser capable of 8W cw power at 488 nm is used as the scattering light source. An external resonator cavity consisting of the output coupler of the laser and a second dielectric mirror increases the power to 160W cw in the scattering region. Two matched 55 mm focal length anti-reflection coated lenses are used within the external cavity to form a beam waist and to shape the beam for maximum power. Careful attention has been paid to the collection optics which magnify the beam waist to fill both the slits and the grating of the spectrometer.

A Spex 1403 double monochromator with high dispersion (10 cm^{-1}/mm) holographic gratings achieves a spectral purity of 10^{14} at ± 20 cm^{-1}. Detection is by photon counting using a RCA C31034A photomultiplier tube selected for low dark count (10 counts per second at -50°C) and high sensitivity. A computer interface stores and manipulates the inherently digital data.

The intensity of a Raman line is given by

$$I = I_o\, C\, R(\nu)\, \nu^3\, \frac{N_{v,J}}{(2J+1)}\, \langle\alpha\rangle^2$$

where I_o is the incident laser intensity, C is a collection of constants, $R(\nu)$ is the frequency response of the collection optics, monochromator, and photomultiplier tube, ν is the frequency of the scattered radiation, $N_{v,J}$ is the number density of the initial level with vibrational quanta v and rotational quantum number J, and $\langle\alpha\rangle$ is the matrix element for the transition, containing the line strength factor and corrections due to centrifugal distortion to the rigid rotator, harmonic oscillator line intensities.

Relative number densities are given by

$$\frac{N_1}{N_2} = \frac{I_1}{I_2} \frac{(2J+1)_1}{(2J+1)_2} \frac{R(\nu)_2}{R(\nu)_1} \frac{\nu_2^3}{\nu_1^3} \frac{\langle\alpha\rangle_2^2}{\langle\alpha\rangle_1^2}$$

where I_o has been held constant.

RESULTS

The Isotope Separation System at the Tritium Systems Test Assembly at Los Alamos was used to prepare a sample anticipated to be pure protium from bottled hydrogen (0.015% HD). Gas chromatography and mass spectrometry were unable to provide an accurate analysis of this sample. Analysis of a 0.7 atmosphere aliquot by Raman spectroscopy yielded 2×10^5 counts per second for the Q(1) line of H_2 and a count rate for the Q(3) line of HD indistinguishable from the background count rate of 0.7 per second. Using $R(\nu)_{HD}=7$, $R(\nu)_{H_2}=6$, $\nu_{HD}=16872$, and $\nu_{H_2}=16340$ then $N_{HD}/N_{H_2} = 5 \times 10^{-6}$. Since, at room temperature, J=3 contains 39% of the HD population and J=1 contains 65% of the H_2 population, then the maximum relative partial pressure of HD would be 9 ppm.

Spontaneous Raman scattering is useful for isotopic analysis of all hydrogen species down to the 10 ppm range. Improvement can be made by using a cooled silicon photodiode array detector. An array detector coupled with a computer could be capable of a complete analysis in less than 30 seconds.

W.S. Lyon (Editor), *Analytical Spectroscopy*
Elsevier Science Publishers B.V., Amsterdam — Printed in The Netherlands

DIFFUSE REFLECTANCE INFRARED FOURIER TRANSFORM SPECTROSCOPY: A VERSATILE AND PROMISING ANALYTICAL TECHNIQUE FOR SOLIDS

N. R. Smyrl, E. L. Fuller, Jr., and G. L. Powell, Union Carbide Corporation, Nuclear Division, P. O. Box Y, Oak Ridge, TN 37830

ABSTRACT

Diffuse reflectance (DR) is a solid sampling technique which has seen a resurgence of activity in the past several years, particularly in the infrared spectral region. This resurgence is due in large part to the widespread use of Fourier transform infrared spectrometers which with their very high optical throughput and signal averaging capabilities require only very simple, less efficient optical accessories for collection of the diffusely reflected radiation. We have applied the diffuse reflectance infrared Fourier transform (DRIFT) technique to a variety of organic and inorganic solids both as powders and in the bulk forms in an effort to obtain molecular information that is often difficult to obtain by other means.

The experimental equipment needed for these types of studies includes a commercial set of transfer otpics (Harrick Model DRA-SID Diffuse Reflectance Accessory) which is utilized in conjunction with our Digilab FTS-15 Fourier transform infrared (FTIR) spectrometer. Also, in a great many of our studies it is desirable to control very accurately the environment of the sample in terms of both temperature and gas over-pressure. For this purpose we utilize an evacuable cell constructed from a 2.75 in. Varian Conflat flange [1]. This cell possesses two ZeSe windows through which the radiation can enter, strike the sample, and then be reflected back out of the cell. Valves are available which permit evacuation and exposure of the sample to various gas environments in the pressure range of 10^{-7} torr to 1 atm. The temperature of the sample can be varied from room temperature to 700°C.

One of our first and more fruitful applications of DRIFT spectroscopy involved corrosion studies related to LiH, which is a very reactive inorganic material produced at the Oak Ridge Y-12 Plant. The material is highly susceptible to corrosion by water vapor even at very low levels of exposure with the principal corrosion product being LiOH. The O-H stretch of LiOH observed at 3675 cm^{-1} can be used to monitor residual amounts of this corrosion product on the surface of both powder particles and warm-pressed disks of LiH and LiD. A detailed <u>in situ</u> study has been made by DRIFT spectroscopy for the kinetics of the moisture corrosion process on the surface of warm-pressed ^{6}LiD exposed to 35 ppm of water vapor at ambient temperature.

Another application of DRIFT spectroscopy involved the <u>in situ</u> oxidation of uranium metal powders to UO_2 by reaction with CO_2 at 300°C and also by reaction with H_2O vapor at ambient temperature. The spectra of the resulting oxides indicate that the surfaces differ structurally due to adsorption of H_2O in the case of H_2O formed UO_2. Two different surface structures are observed for the H_2O formed UO_2 resulting from a two step irreversible dehydration process. The first step, which involves evacuation, produces a metastable structure which then converts to the final stable dehydrated state on heating to 400°C.

The study of coal and its related chemistry is another area in which we are utilizing DRIFT spectroscopy very successfully. One segment of our investigations has been directed toward studies of coal oxidation carried out in the intermediate temperature range of 300° to 500°C [2]. A basic understanding of the oxidation processes occurring at these lower temperatures where reactions are slow enough to follow spectroscopically may be useful in gaining insight into the more rapid processes occurring in the combustion phase.

Powdered and chunk forms of coal of different ranks have been oxidized in the DRIFT cell in the intermediate temperature range (400°C). Two major chemical changes occur which are reflected in the spectra as the oxidation proceeds. One of these changes involves the loss of aliphatic C-H stretching intensity with increasing oxidation indicating that the initial oxidative attack occurs primarily at aliphatic sites in the coal structure. The other change involves the formation of carbonyl type species noted by the appearance of bands in the 1700 cm^{-1} region. A very complex band structure was generated which was observed to change continually throughout the oxidation indicating the formation of a variety of carbonyl type species. These bands appear to represent a logical sequence of carbonyl products that might be expected from increasing reaction. Fourier self-deconvolution and band fitting techniques are being utilized to analyze the complex band structures in both the carbonyl and C-H stretching regions as a function of oxidation. Application of these techniques should result in a better understanding of the relative types and quantities of the carbonyl species formed and the aliphatic hydrocarbon sites that are attacked during oxidation.

The DRIFT technique can be used as a quantitative tool, although its application is not as straight forward as in normal transmission spectroscopy. We have utilized the DRIFT technique to quantitatively determine certain types of mineral matter in coal and have been able to establish fairly large analytical working ranges for inorganic materials such as $CaCO_3$ dispersed in nonabsorbing media using weaker combination bands and overtones as the analytical bands.

The DRIFT technique is proving to be a very versatile and promising analytical technique for solids. The most important attributes of the technique can be summarized as follows. The technique ia applicable to both powder and bulk solid samples and can be utilized for the analysis of neat materials undiluted in any supporting medium. It is perhaps the most easily adapted infrared sampling technique for _in situ_ measurements where both temperature and pressure need to be accurately controlled and is applicable to opaque samples such as coal, graphite, and UO_2.

ACKNOWLEDGMENT

The Oak Ridge Y-12 Plant is operated by the Union Carbide Corporation, Nuclear Division for the Department of Energy under U.S. Government Contract W-7405-eng-26.

REFERENCES

1. N. R. Smyrl, E. L. Fuller, Jr. and G. L. Powell, J. Appl. Spectrosc., 37 (1983) pp 38-44.
2. N. R. Smyrl, E. L. Fuller, Jr. in E. L. Fuller, Jr. (Ed.), Coal and Coal Products: Analytical Characterization Techniques, American Chemical Society, Washington, D.C., 1982, pp 133-145.

DIODE LASER SPECTROMETRY FOR MONITORING ACETYLENE ON-LINE IN A LIQUEFACTION PLANT

W. L. Maddox and C. M. Turner, Process Support Division, Oak Ridge Gaseous Diffusion Plant,* Oak Ridge, Tennessee 37830

ABSTRACT

A diode laser source spectrometer has been acquired for use as an on-line monitor for acetylene in the rich (oxygen-enriched) liquid stream of the Oak Ridge Gaseous Diffusion Plant (ORGDP) liquefaction plant. The liquefaction plant produces high purity liquid nitrogen for use in the gaseous diffusion plant and for development activities on the ORGDP site. The facility is of a design that runs continuously with considerable automation, but in addition to the built-in hydrocarbon analyzer in the final stage, a special instrument is required to monitor acetylene to ensure that its concentration does not exceed the specification limit--0.5 ppm in the product reflux condenser shell.

The instrument chosen for this application, a research-grade infrared spectrometer, requires somewhat special handling in order to be useful in this context but affords selectivity and sensitivity substantially in excess of that available with conventional "on-stream" monitors. Its long path absorption cell allows monitoring a trace constituent, and its superior spectral resolution permits determination of the sought substance in the presence of other constituents of the stream. We were fortunate in having an existing air-conditioned laboratory close by the new liquefaction facility, where the spectrometer could be housed and from which reasonably short lines could be installed to conduct samples to it from the plant. Samples may be withdrawn continuously from several locations in the process piping. At present, the three sampling locations selected are in the silica-gel "guard beds" and in the condenser shell where the rich liquid provides the final cooling for the product nitrogen. The latter location is the critical one. If the C_2H_2 concentration reaches 0.5 ppm there, the condenser must be flushed and drained, since there is no way to clear this contamination by boil-off. The sample lines lead to a manifold from which control valves admit one stream at a time to the spectrometer's White cell. A vacuum pump continuously evacuates the cell to maintain the pressure therein at ca. 30 torr. (A pressure in this vicinity is necessary to prevent excessive line broadening.) The spectrometer provides control of the diode laser current and temperature at values chosen by the operator and maintains frequency by

*The Oak Ridge Gaseous Diffusion Plant is operated by Union Carbide Corporation Nuclear Division for the U.S. Department of Energy under contract W-7405-eng-26.

By acceptance of this article, the publisher and/or recipient acknowledges the U.S. Government's right to retain a nonexclusive royalty-free license in and to any copyright covering this paper.

means of feedback through the current source. A software tracking function provides computer-controlled adjustment of the current as necessary to compensate for long-term drift. Software supplied by the manufacturer controls the action of the spectrometer as it measures the second derivative of laser intensity at the absorption line of acetylene and at a nearby line due to HCN, the "reference" gas. The ratio of these is fitted to a previously established calibration curve stored in memory to compute the concentration. Local and remote readouts of the concentration are provided, and alarms may be actuated as desired.

Experience to date has been off-line: Grab samples have been analyzed by standard additions. Adequate sensitivity has been demonstrated and unattended operation of the instrument for periods up to 3 days has been achieved. At the time of writing, no concentration of acetylene above the 20-ppb limit of detection has been found. The sample lines to the plant have recently been installed, and work is now underway to complete the remote display and alarm system.

Optical Measurement of SO_2 in Combustion Environments

Donald Lucas and Nancy J. Brown
Applied Science Division
Lawrence Berkeley Laboratory
University of California
Berkeley, CA 94720

Sulfur dioxide has been measured in a high temperature combustion environment by ultraviolet absorption spectroscopy and by Tunable Atomic Line Molecular Spectroscopy (TALMS). Measurements are made in the post-flame region of a flat flame burner using stoichiometric and lean methane/air mixtures doped with SO_2.

TALMS is an <u>in situ</u>, line of sight technique that uses a Zeeman-tuned atomic emission lamp as the light source. In these experiments the Zn line at 213.8 nm ($^1S_0 \rightarrow {}^1P_1$ transition) is used with the light parallel to the magnetic field. Detection of SO_2 is accomplished by magnetically tuning one of the Zeeman components into resonance with a specific absorption in SO_2 while using the other component as the reference signal. The two components are very close in frequency (~20 GHz), but differ in polarization, with the $\Delta M = +1$ transition right circularly polarized, and the $\Delta M = -1$ transition left circularly polarized. The differences in frequency and polarization form the basis for molecular detection with TALMS. A variable phase retardation plate then converts these beams into two perpendicular linear polarized beams, which are subsequently separated by a linear polarizer prior to detection by a photomultiplier tube. Both beams traverse an identical optical path, thus permitting good correction of the signal in highly scattering or absorbing systems, such as those encountered in sooting flames. The high selectivity of the technique greatly reduces interference by other molecules that absorb in the same spectral region.

Because of the complexity of the ultraviolet absorption spectrum of SO_2 in the 200.0 to 220.0 nm region it is not possible to identify the particular transition that causes the TALMS signal. The dependence of signal strength on magnetic field obtained at 1500K is equivalent to that determined at 300K, confirming that the same transition is measured in the flame system as under ambient conditions. The maximum TALMS signal occurs at a magnetic field strength of 14.5 kgauss.

The sensitivity of the TALMS technique varies with the temperature of the sample. For a 7cm path length at 1 atmosphere, the mininum detectable concentration is ~40ppm at 300K, and ~2000ppm at 1500K. Density differences due to combustion account for a factor of 5, and broadening of the absorption features at the higher temperatures reduces the sensitivity by an additional order of magnitude. The ultraviolet absorption spectrum of SO_2, measured as a function of temperature in the 300 to 1500K range, indicates the discrete absorption features gradually broaden as the temperature increases, and the underlying continuum becomes even more absorbing. This broadening is largely due to redistribution of the rotational level populations at the elevated temperatures, with the result that fewer molecules are in the particular state probed by the Zn line.

The TALMS signal is linear with SO_2 concentrations below 20,000ppm, while the absorbance at aproximately the same wavelength is linear at concentrations as high as 30,000ppm. This indicates that the fall-off in the TALMS signal is not attributable to self absorption or deviations from Beer's law, but is inherent to the TALMS technique. The non-linear region can be avoided by the proper choice of path length.

ACKNOWLEGEMENT

This work was supported by the Director, Office of Energy Research, Office of Health and Environmental Research, Physical and Technological Research of the U.S. Department of Energy under Contract No. DE-AC-03-76SF00098.

QUANTITATIVE DETERMINATION OF HYDROGEN BY PULSED NUCLEAR MAGNETIC RESONANCE SPECTROSCOPY

ALBERT ATTALLA and ROBERT C. BOWMAN, JR.
Monsanto Research Corporation, Mound*, Miamisburg, Ohio 45342

ABSTRACT

Hydrogen can be determined routinely in solids and liquids by pulsed nuclear magnetic resonance (NMR) when the proton spin-spin relaxation time is relatively long (>20 μsec) and the proton spin-lattice relaxation time is relatively short (<100 sec). Short spin-spin relaxation times result in intolerable loss in signal amplitude due to short-lived free-induction-decay signals. Long spin-lattice relaxation times lengthen the analysis time by lengthening the delay time between successive pulses during data accumulation. The technique is non-destructive and requires only a few milligrams of sample for solutions, and approximately 100-500 mg for direct analyses of solid samples. The precision (relative standard deviation) of repeated determinations is approximately ±2% relative to the average value, and the accuracy varies from ±2 to ±5% depending on the relaxation time parameters and size of the sample. The accuracy can be improved by the method of standard additions of hydrogen and by providing standard samples which reflect the physical and chemical properties of the analytical samples. The technique has been applied to the determination of moisture in solids and liquids, to the determination of hydrogen stoichiometry in solid metal hydrides, impurity analyses, compound identification, and molecular structure confirmation.

INTRODUCTION

Any atomic nucleus having a magnetic moment (odd number of protons or neutrons or both) will exhibit the NMR phenomenon (ref.1), that is, the absorption of radiofrequency energy in the megahertz range when placed in a large external stationary magnetic field. With the proper selection of available NMR instrumentation, any magnetically active nucleus can be detected and measured in any type of material or substance, whether it be solid, liquid, or gas; or element, compound, or mixture. The versatility of NMR is reflected in its applications in every field of science: chemical analyses, molecular structure determination, kinetics, and rates of reaction; determination of magnetic moments in physics; and whole-body holography in biology and medicine.

---

*Mound is operated by Monsanto Research Corporation for the U.S. Department of Energy under Contract No. DE-AC04-76DP00053.

THEORY

At constant magnetic field each nuclear species will absorb rf energy at a unique frequency. Thus, each magnetically active nuclear species can be detected by scanning either the rf frequency range or the external magnetic field according to the following relationship,

$$\nu_0 = \frac{\gamma H_0}{2\pi}, \tag{1}$$

where ν_0 is the frequency of energy absorbed, γ is the magnetogyric ratio of the nuclear species, and H_0 is the strength of the external magnetic field.

Factors influencing detection of NMR signal

The ability to detect an NMR signal from any nuclear species depends on the magnitude of the rf frequency (or magnetic field strength), the number (N) of nuclei being observed, and its magnetic moment (μ) and nuclear spin (I). The signal amplitude is directly proportional to the abundance of the nuclear species, its magnetic moment, and the stationary magnetic field strength, and is inversely proportional to the nuclear spin. Other factors affecting the magnitude of the NMR signal are the natural abundance of the nuclear species, duration of the rf pulse, its temperature, and spin-lattice and spin-spin relaxation times (ref.2). The signal amplitude is inversely proportional to the absolute temperature of the sample, since the distribution of the nuclear spins among the various nuclear magnetic energy levels is more equalized at higher temperatures, and thus reduces the number of nuclei that can absorb rf energy.

Spin-lattice relaxation time. The spin-lattice relaxation time is a measure of the time necessary for the nucleus to lose the rf energy it absorbed during the NMR experiment. If the energy is not lost to the lattice, the energy states (see Fig. 1) of the nuclei will become equally populated, and no net absorption of rf energy will take place.

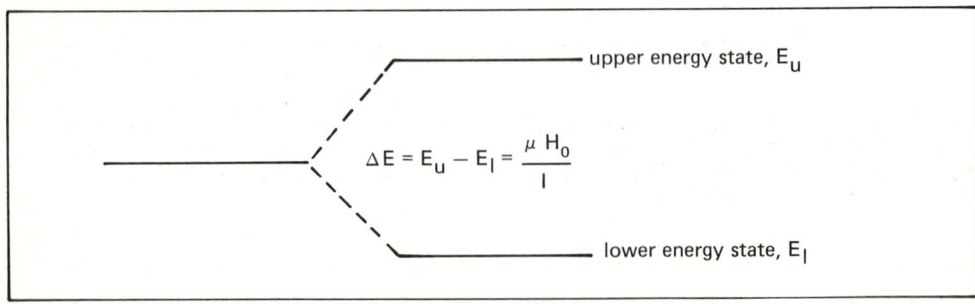

Fig. 1. Diagram illustrating the splitting of the nuclear magnetic energy level in the presence of an external magnetic field, H_0.

However, if sufficient time (at least five times the spin-lattice relaxation time) is allowed between successive rf pulses for the nuclei to return to their original state, then the ability to detect a signal is regained.

rf pulse duration. To attain the maximum instrument response during the NMR experiment, the nuclear magnetic moment vector must be rotated 90° with respect to the direction of the external magnetic field, H_0; for it is the interaction between these two fields which generates the free induction decay (FID) signal seen after the application of an rf pulse. The time necessary to accomplish the 90° rotation is a function of the nuclear spin (I), the magnetic moment (μ), and the rf field strength (H_1),

$$(time)_{90° \text{ pulse}} = \frac{Ih}{4\mu H_1} \tag{2}$$

where h is Planck's constant.

Spin-spin relaxation time. The spin-spin relaxation time, T_2, is a measure of the time for the decay (see Fig. 2) of the NMR signal after application of the rf pulse. The effect of a short T_2 is

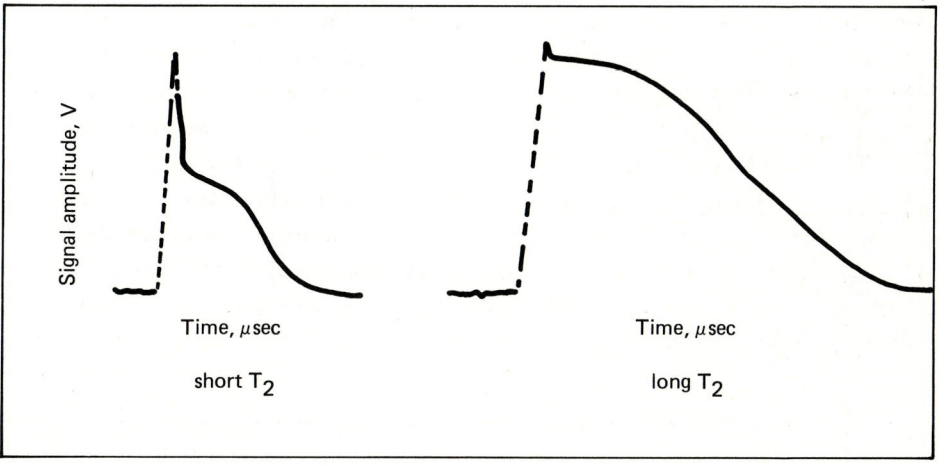

Fig. 2. Illustration of the concept of spin-spin relaxation time. These two free induction decay signals were obtained from the same sample having different T_2 relaxation times. The signal amplitude and duration are greater in the sample having the longer T_2.

the loss in signal amplitude and the introduction of an error in the quantitative determination of the given nuclear species. This error is minimized by preparing standard samples that are similar in chemical and physical properties to the analytical samples.

EXPERIMENTAL

Preparation of standards

For samples having long T_2, solutions are prepared containing known concentrations of the substance or element being analyzed or determined. Otherwise, samples identical to the unknowns are first analyzed by other techniques, such as volumetric and gravimetric quantitative methods, P-V-T analyses, and x-ray fluorescence spectroscopy, and then are used as standards.

Determination of sample size

The optimum (maximum instrument response) sample size is determined by plotting the signal amplitude as a function of sample size (see Fig. 3). This is necessary because the rf coil which encloses the sample is finite in length, resulting in a non-uniform rf field towards the ends of the coil. Samples are placed in the center of the rf coil.

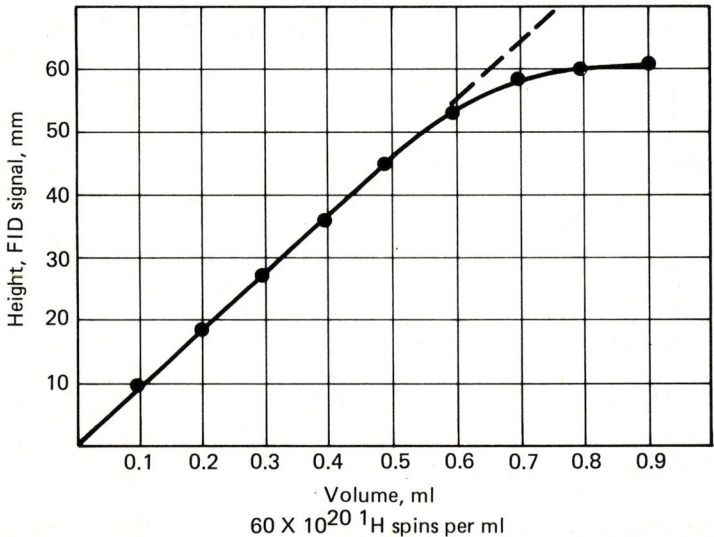

Fig. 3. Plot of signal amplitude as a function of sample size for determining optimum instrument response.

Free induction decay signals

Figures 4-7 illustrate the various types of FID signals obtained from different kinds of samples. Two main factors influence the appearance of the line-shape of the signals. One is the spin-spin relaxation time which determines the length of the signal (compare Figs. 4 and 5). The second is the physical or chemical state of the nuclear species being observed. Figure 6 illustrates the appearance of the signal when the sample contains a nuclear species existing in two different chemical environments, that is, hydrogen bonded to the metal titanium in $TiH_{0.15}$ (fast-decay, rigid-lattice hydrogen) and hydrogen bonded to oxygen in H_2O (slow-decay, mobile hydrogen). Figure 4 is a photograph of a series of FID signals from a set of 1/2-ml water standards used for the determination of hydrogen in solutions and solids possessing long spin-spin relaxation times. The first four amplitudes are plotted in Fig. 8 as a function of hydrogen concentration in units of 10^{20} nuclear spins or hydrogen atoms. Figures 5 and 6 illustrate the technique for measuring the signal amplitudes of various FID signals.

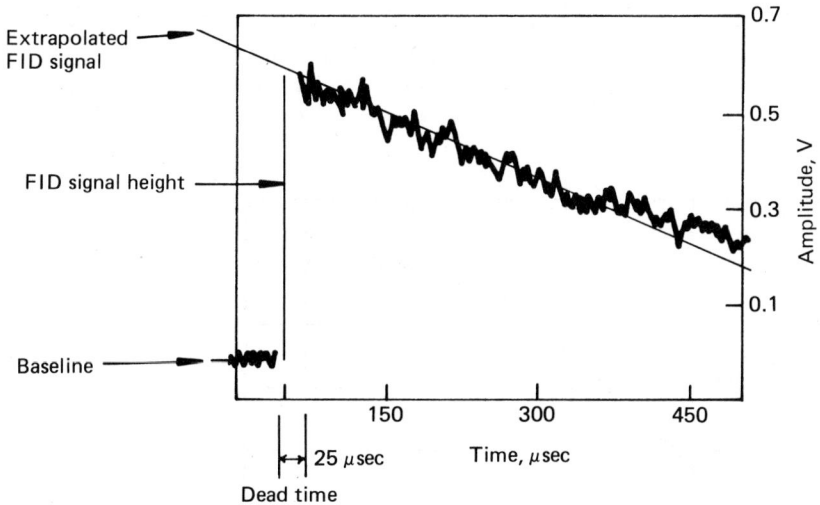

Fig. 4. Photograph of oscilloscope trace of six H_2O standard samples.

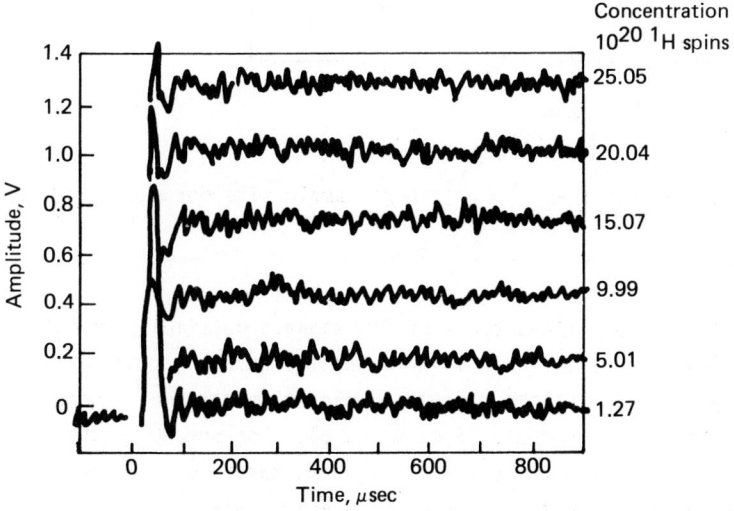

Fig. 5. Free induction decay signal of tritium molecules ($^3H_2$) in solid lithium tritide (LiT).

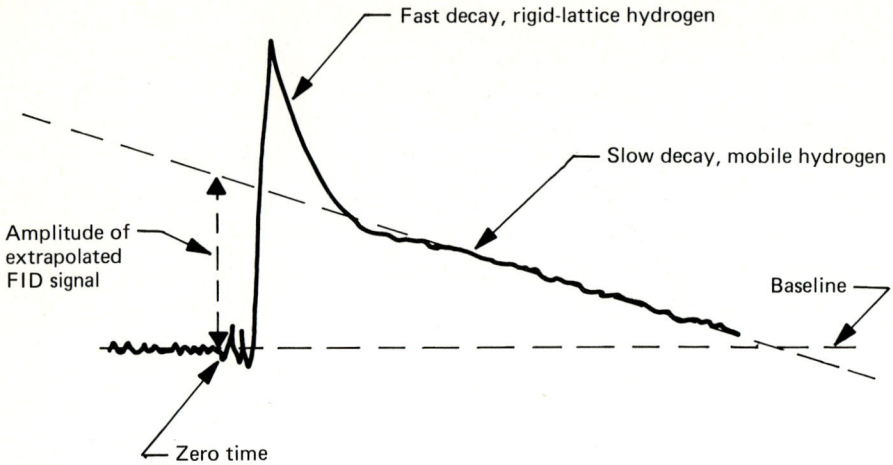

Fig. 6. Free induction decay signals of hydrogen in $TiH_{0.15}$ (fast decay) and water (slow decay).

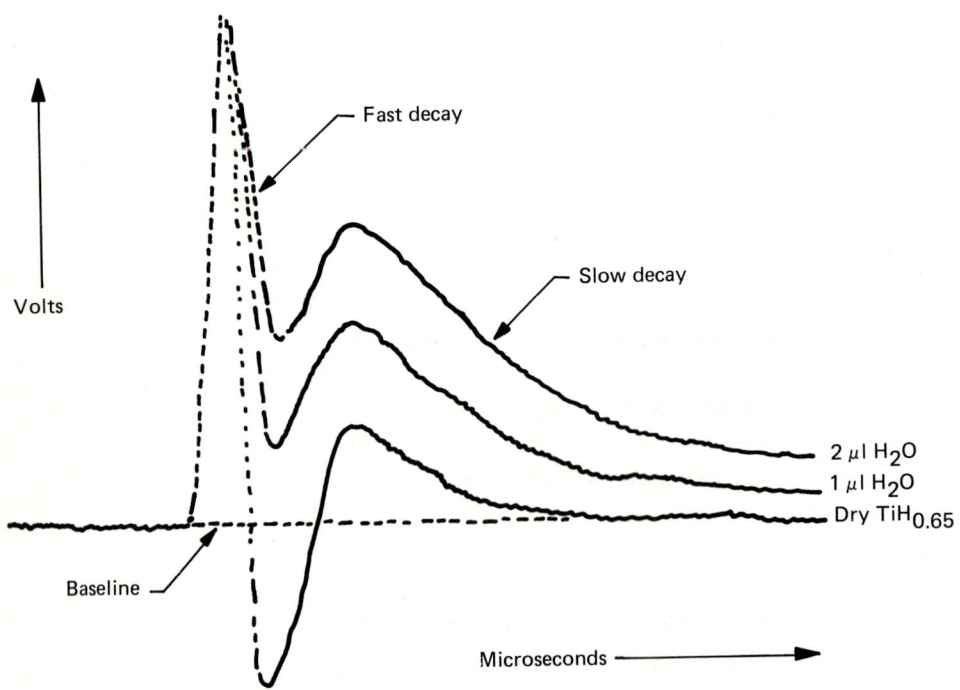

Fig. 7. Free induction decay signals of dry $TiH_{0.65}$ and $TiH_{0.65}$ with successive additions of microgram (1 μl H_2O) quantities of water.

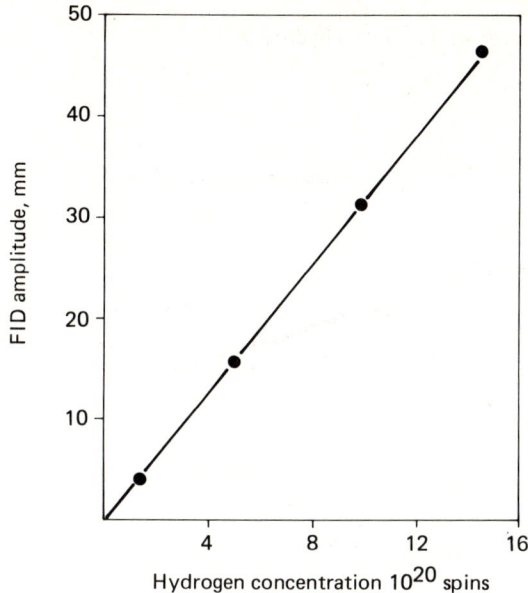

Fig. 8. Calibration curve for hydrogen NMR response in standard water samples.

Method of standard addition

Figure 7 is a photograph of the FID signals obtained after the successive additions of 1 μl quantities of H_2O to solid $TiH_{0.65}$ for the determination of residual moisture in dry $TiH_{0.65}$. The amplitudes of the peaks of the water FID signals are plotted as a function of added H_2O (see Fig. 9), and the plot is extrapolated to the concentration coordinate to obtain the amount of residual moisture in the sample.

Precision and accuracy

The precision (relative standard deviation) of repeated determinations of hydrogen by pulsed nuclear magnetic resonance, as found here, is approximately ±2% relative to the average value. The accuracy varies from ±2 to ±5% depending on the relaxation time parameters and size of the sample. Accuracy is estimated from multiple runs on reference samples similar in physical and chemical properties to the analytical samples.

CONCLUSIONS

Nuclear magnetic resonance spectroscopy is a specific technique for quantitative analysis since a specific nucleus can be uniquely selected in a sample, and either the nucleus or the compound in which the nucleus resides can be measured. The technique is nondestructive of the sample, requires a small sample size, and is fairly rapid and precise. The technique can be applied to any material containing a magnetically active nucleus in any state of matter.

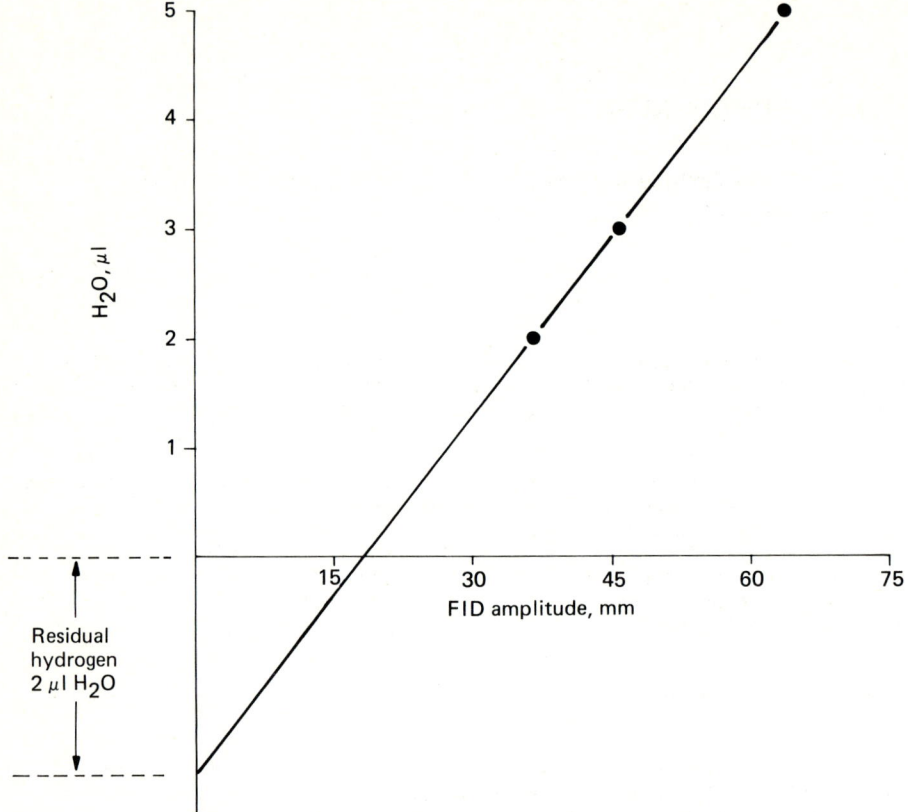

Fig. 9. Plot of free induction decay signal amplitudes of solid TiH$_{0.65}$ as a function of successive additions of water.

REFERENCES

1 F. A. Bovey, Nuclear Magnetic Resonance Spectroscopy, Academic Press, New York, 1969, pp. 3-5.
2 N. Bloembergen, Nuclear Magnetic Relaxation, W. A. Benjamin (Ed.), New York, 1961, 178 pp.

APPLICATIONS OF FOURIER SELF-DECONVOLUTION

R. L. Howell, Union Carbide Corporation, Nuclear Division, P. O. Box Y, Oak Ridge, TN 37830

ABSTRACT

Spectroscopists have sought to increase the information contained in and extractable from infrared spectra from the earliest days. This has been accomplished by improving the instrumentation with the use of brighter sources, more efficient monochromators, and more sensitive detectors. The advent of the minicomputer led to practical Fourier transform systems with the corresponding advantages which increased resolution, sensitivity, and accuracy. At the same time, mathematical enhancement of the spectra was developing. For example, digitized data permitted the spectrum of a pure component to be subtracted from that of a mixture, if sufficient spectral accuracy was obtained.

The inverse relationship between the instrument slit width and spectral resolution has long been recognized. This loss of resolution is best described by the convolution integral,

$$E(\nu) = \int_{-\infty}^{\infty} E'(\nu) \, S(\nu-\nu') \, d\nu' \tag{1}$$

where $E(\nu)$ is the observed spectrum; $E'(\nu)$ is the real spectrum; and $S(\nu)$ is the slit function. This operation is generally regarded as directly irreversible. Attempts to compute an approximation of $E'(\nu)$ are known as deconvolution and have included spectral subtraction techniques [1], a pseudo-deconvolution approach [2], factor analysis [3], band fitting [4], and an iterative method [5] which is based on Van Cittert's work [6].

When the convolution integral is expressed as the Fourier transformations of the functions in Equation 1, the result is greatly simplified.

$$E(\nu) = E'(\nu) \cdot S(\nu) \tag{2}$$

Now, in the Fourier, or time, domain, convolution is reduced to multiplication. Recently, a general theory of Fourier self-deconvolution was presented [7] in which the slit function, $S(\nu)$, is replaced by the intrinsic lineshape function, $G(\nu)$. In an infrared Fourier transform (FT) system, which has no slit, the shape of the intrinsic lineshape function is a result of the truncation of the necessarily finite interferogram. Fourier self-deconvolution, then, involves taking the inverse Fourier transform of the spectrum to be deconvoluted. It is then multiplied by $D(x)/F^{-1}\{G(\nu)\}$ where F^{-1} implies an inverse Fourier transformation and $D(x)$ is an apodization function [8]. In practice, a Lorentzian line is a good approximation for the intrinsic lineshape function. The critical parameters are the width of the Lorentzian line, which controls the overall quality, the degree of width reduction, and the apodization function, which determines the amplitude of the side lobes.

Fourier self-deconvolution seems to be applicable to spectra where the instrumental resolution is higher than the width of the intrinsic lineshape.

This generally includes infrared, fluorescence, UV-visible, Raman, and NMR spectroscopies. A portion of this paper presents the deconvolution of spark source mass spectrometry and x-ray diffraction data. In these cases some of the spectral degradation caused (at least partially) by the use of an instrumental slit is being removed. In the Fourier transform infrared spectral data, the degradation is a consequence of sampling a finite portion of the interferogram. The chromatographic system, however, is basically different in that no slit or interferometer is used. A chromatographic column can be considered as a system which converts a single input signal to a series of broader peaks [9]. If the chromatogram is adequately sampled, deconvolution can yield narrower peaks with conservation of area. This could enable components to be identified and quantified accurately despite rather poor chromatographic resolution.

Fourier self-deconvolution can be a powerful tool when used to obtain peak parameter data on bands comprising contours and shoulders. These are often not easily discernable even to experienced spectroscopists. This information may be optimized by a band fitting program to yield the positions, widths, and heights of the component bands.

ACKNOWLEDGMENT

The Oak Ridge Y-12 Plant is operated by the Union Carbide Corporation, Nuclear Division for the Department of Energy under U.S. Government Contract W-7405-eng-26.

REFERENCES

1. J. W. Frazer, L. R. Carlson, A. M. Kray, M. R. Bertoglio, and S. P. Perone, Anal. Chem., 43, (1972) 1479.
2. N. R. Jones, R. Venkataragharan, and J. W. Hopkins, Spectrochim. Acta., 23A (1967) 925.
3. M. K. Antoon, L. D'Esposito, J. L. Koenig, Appl. Spec., 33 (1979) 351.
4. P. C. Gillette, J. B. Lando, J. L. Koenig, Appl. Spec., 36 (1982) 401.
5. P. E. Siska, J. Chem. Phy., 59 (1973) 6052.
6. P. H. Van Cittert, Z. Phys., 69 (1931) 298.
7. J. K. Kauppinen, D. J. Moffatt, H. H. Mantsch, and D. G. Cameron, Appl. Spec., 35 (1981) 272.
8. J. K. Kauppinen, D. J. Moffatt, H. H. Mantsch, and D. G. Cameron, Anal. Chem., 53 (1981) 1454.
9. D. W. Kirmse and A. W. Westerberg, Anal. Chem., 43 (1971) 1035.

CAPILLARY SUPERCRITICAL FLUID CHROMATOGRAPHY - FOURIER TRANSFORM INFRARED SPECTROMETRY

S.V. OLESIK, S.B. FRENCH AND M. NOVOTNY
Chemistry Dept., Indiana University, Bloomington, Indiana 47405

One of the most demanding tasks asked of an analytical chemist today is to separate and identify the components of a nonvolatile complex mixture. An efficient separation technique combined with a universal detector that provides structural information, therefore, would be a great asset to analytical chemists. Capillary supercritical fluid chromatography (SFC) - Fourier transform infrared spectrometry (FTIR) shows great potential for being such a technique.

The most important properties of a mobile phase in chromatography are solvent viscosities, solute diffusivities, and solvating power. Supercritical fluid chromatography takes full advantage of the unique properties of compressed gases near their critical temperature. Supercritical fluids have solvating powers similar to those of liquids, viscosities comparable to those of gases, and binary diffusion coefficients intermediate between those of gases and liquids. In comparison to liquid chromatography, this unique combination of properties provides efficient separations of nonvolatile compounds but with shorter analysis times. With the low critical temperatures of fluids, such as carbon dioxide (T_c = 31 $^\circ$C), the efficient separation of thermally labile compounds can be achieved. The use of capillary columns with SFC gives the further advantages of low volumetric flow rates (microliters/minute) and increased peak concentrations relative to those from analytical scale HPLC columns. With the total volume of mobile phase needed for a chromatographic run substantially reduced, the use of exotic mobile phases, such as xenon, is quite feasible.

The attempts at interfacing high performance liquid chromatography (HPLC) with FTIR illustrate the extreme interest in infrared spectroscopic detection for separations of nonvolatile compounds. Presently there are two approaches to HPLC-FTIR and both have substantial drawbacks. One approach is to use an on-line flow cell interface [ref. 1]. The major drawback of this technique is poor detection limits due to mobile phase absorption of mid-IR radiation. Also, gradient elution techniques are incompatible with this approach. The second approach [ref. 2] involves solvent evaporation followed by solute detection.

This approach is not compatible with thermally labile compounds or gradient elution techniques involving aqueous mobile phases. In comparison, SFC-FTIR should provide spectra of solutes with low detection limits since SFC mobile phases with low absorbance in the mid-IR can be used. Also, because solvent selectivity is varied in SFC by varying the density rather than changing the composition of the mobile phases as in HPLC, the possibility of changing the selectivity over the course of a chromatographic run using FTIR detection exists.

To successfully couple SFC with FTIR, a high pressure (0 - 3500 psi), low volume (microliter) IR flow cell was developed. Important parameters which were optimized include: volume (to be compatible with capillary SFC), pathlength, cross sectional area of the detection region, and construction. The present pathlength, flow cell volume and cross sectional area of the cell are: 1 mm., 2 microliters and 20 mm. The flow cell has been designed for use with chromatographic columns with an inner diameter of 250 microns. Construction problems of maintaining a high pressure fluid to a low volume will be discussed with various tradeoffs listed. One important decision was the choice of window material. We chose zinc sulfide as a window material because of its high resistance to rupture and its wavelength cutoff (700 cm^{-1}).

The spectral characteristics of several SFC mobile phases were studied. Solvents considered were dichlorofluoromethane, hydrocarbons, such as propane, and carbon dioxide. In these initial studies, carbon dioxide has proven to be an excellent choice as a mobile phase for this new hypenated technique. The infrared spectrum of supercritical carbon dioxide contains large transparent regions in the mid-IR. Regions in the mid-IR that are opaque when carbon dioxide is used as a mobile phase are caused by absorption bands centered at 3716 cm^{-1}, 3609 cm^{-1}, and 2349 cm^{-1}. Initial studies into varying the solvent strength by density programming will be discussed. Preliminary results are very promising.

Chromatograms and spectra of model compounds using carbon dioxide as a mobile phase will be shown. Detection limits of the present configuration of flow cell are in the low microgram range with many variables yet to be optimized. Planned improvements to lower the detection limits and provide more efficient separations will also be discussed.

In conclusion, we feel that SFC-FTIR shows great potential as a very powerful technique for separation and identification of thermally labile and nonvolatile compounds. Research is continuing in these labs to further optimize the technique.

REFERENCES
1 R.S. Brown and L.T. Taylor, Anal. Chem., 55 (1983) 1492-1497.
2 D. Kuehl and P.R. Griffiths, J. Chromatogr. Sci., 17 (1979) 471-476.

SUPERCRITICAL FLUID METHODS IN ANALYTICAL CHEMISTRY

R.D. SMITH, B.W. WRIGHT and H.R. UDSETH
Chemical Methods and Kinetics Section, Pacific Northwest Laboratory, (Operated by Battelle Memorial Institute), P.O. Box 999, Richland, WA 99352

ABSTRACT

Increasing the pressure of a supercritical fluid allows its properties to be varied continuously from those of a gas to those of a dense fluid with the solvating power of a liquid. The potential for improved chromatographic methods was recently realized using capillary columns suited for supercritical fluid chromatography. Supercritical fluids also offer distinct advantages for transport of thermally labile compounds to the mass spectrometer ion source.

INTRODUCTION

The application of supercritical fluids is providing an array of new analytical techniques for the chemist. These techniques include both methods of separation and those where supercritical fluids are used for improved transfer and detection. These techniques are exemplified by supercritical fluid chromatography (SFC)[1,2] and direct fluid injection (DFI) mass spectrometry[3,4]. In SFC the advantageous combination of gaseous and liquid properties can be utilized to allow the separation of many compounds impossible by gas chromatography, with greatly improved efficiencies compared to liquid chromatography. The recent introduction of open tubular capillary columns for SFC has resulted in additional advantages which include significantly extended resolution, improved pressure (density) programming, and flow rates compatible with mass spectrometry[1]. The DFI technique offers significant advantages for characterization of supercritical fluids, provides a near optimum detector for capillary column SFC, and in itself provides a powerful new direct introduction method for mass spectrometry[4]. The instrumentation used in our laboratory for capillary column SFC, DFI-mass spectrometry, and SFC-MS has been described in detail elsewhere[3-5]. While these methods are still in an early stage of development their promise seems to assure them a future role in the analytical laboratory.

RESULTS AND DISCUSSION

The direct fluid introduction (DFI) mass spectrometry method allows any compound soluble in a supercritical fluid to be transferred to the gas phase for ionization using conventional methods[2-5]. The DFI method has the following advantages: quantitation is straightforward; the chemical ionization (CI) process is applicable to essentially any compound; various CI reagents may be selected to vary the degree of spectral complexity (fragmentation); and the technique is inherently sensitive and rapid.

DFI-CI mass spectra have been obtained using both n-pentane, ammonia, water, isobutane, and carbon dioxide as the supercritical fluid often with small amounts of solvent modifiers. Methanol, 2-propanol, water and methylene chloride solvent modifiers, at 0.1-10% by (liquid) volume, have been used as solvent modifiers. The operating temperature is selected so that it is typically 5-30°C above the estimated critical temperature of the solvent system. (The critical temperatures for n-pentane, ammonia, and carbon dioxide are 197°C, 132°C, and 31°C, respectively.)

Subtle effects related to solubility enhancement by the sample solvent and dependent upon flow rate may be observed in the DFI-MS analysis for compounds of marginal solubility. For 0.2 uL sample injections, the sample solvent may have a significant effect upon solubility in the absence of any chromatographic separation due to the 3 uL total volume available for mixing during transfer. The enhanced solvating power of the resultant fluid mixture can result in transfer and detection of otherwise insoluble compounds. Thus, when the sample solvent is observed to affect the supercritical fluid solvating characteristics, all sample injections are best preceded and followed by blank solvent injections to remove any such material prior to analysis and to confirm complete transfer of material.

Compounds of primary interest for DFI analysis include those which are thermally labile and relatively nonvolatile. An example of a relatively labile compound is given in Figure 1 which shows the DFI mass spectrum obtained for a 20 ng injection of 2-deoxyadenosine. The mass spectrum was obtained by injection of 0.2 uL of an aqueous solution into a pressurized stream (400 atm) of subcritical ammonia followed by a rapid increase in temperature to supercritical conditions. As shown in Figure 1, the mass spectrum obtained for 2-deoxyadenosine is extremely simple and is dominated by the protonated molecular ion, $(M+1)^+$.

The DFI-mass spectra are typically quite simple and, hence, further techniques are useful for obtaining additional information on the structure of unknown materials or to distinguish compounds having similar molecular weights. A promise of the direct supercritical fluid injection method is the potential to utilize the selective solvating power of the supercritical fluid to develop rapid methods for direct extraction from complex substrates into the mass

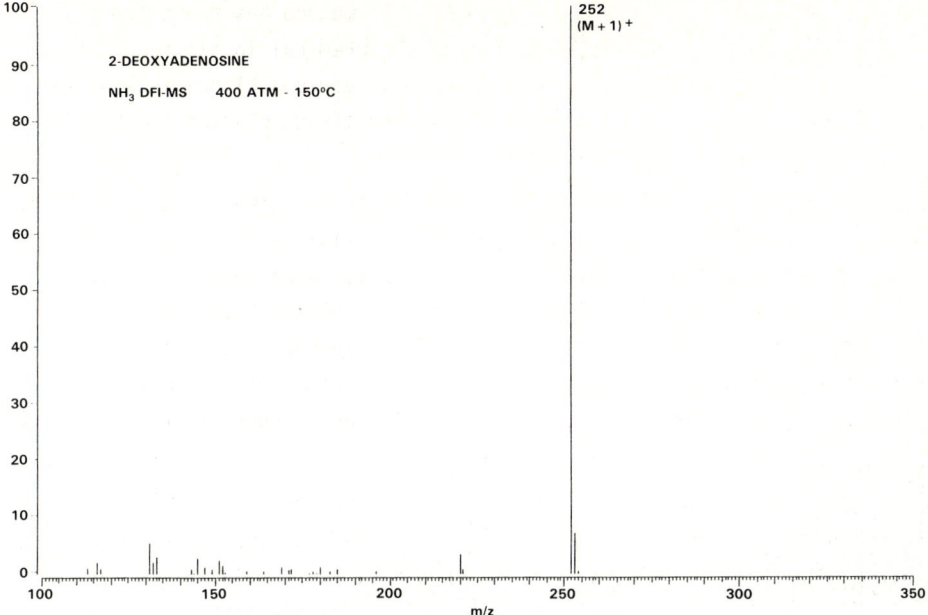

Fig. 1. DFI Mass Spectrum of 2-Deoxyadenosine.

spectrometer for analysis. The combination of the selective extraction from substrate materials and detection with more selective collision induced dissociation tandem mass spectrometric methods (MS/MS) may provide a rapid analytical method for complex mixtures. An added benefit of this approach is that the chemical noise sometimes associated with high DFI flow rates can be avoided, and in principle improved detection limits should be possible in many applications.

Analysis of humic and fulvic acid materials has provided a striking example of the application of DFI methods. In these studies the HPLC injection valve was replaced by a small volume (∼100 uL) cell containing the sample. The sample cell is elevated to supercritical temperatures for the chosen fluid. Details of the experimental arrangement have been described previously[3]. Figure 2 gives an example of a mass spectrum obtained for a humic acid sample obtained for supercritical carbon dioxide modified with ∼0.4% water at 90°C and 230 atm. Under these conditions humic acid solubility amounts to only a few percent, however, substantial material of moderate molecular weight is evident in the mass spectra. Solubility was found to be significantly greater for ammonia and much less for pure carbon dioxide. Similar studies have been reported for coal, where the feasibility of substantial sample fractionation has been noted[6].

The potential advantages of SFC accrue from the nature of the supercritical fluid. The lower viscosities and higher diffusion coefficients relative to liquids result in the potential for significantly enhanced chromatographic

HUMIC ACID EXTRACTION WITH SUPERCRITICAL CO$_2$-H$_2$O
90°C - 230 atm

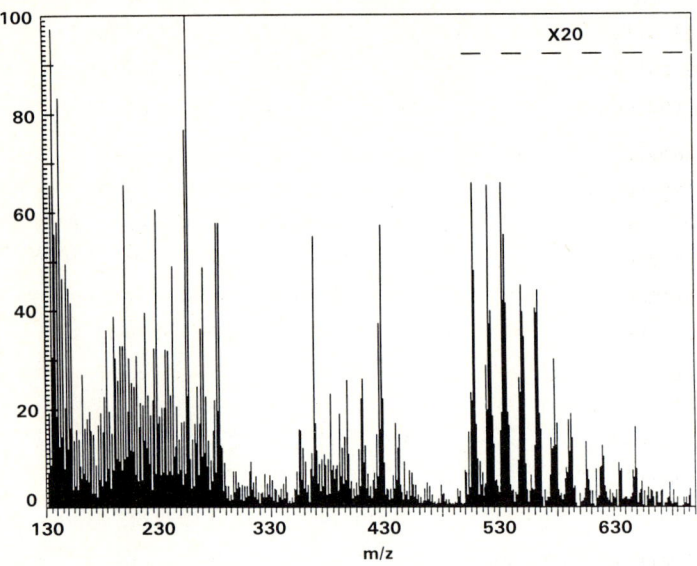

Fig. 2. DFI Mass Spectrum During Extraction of a Humic Acid.

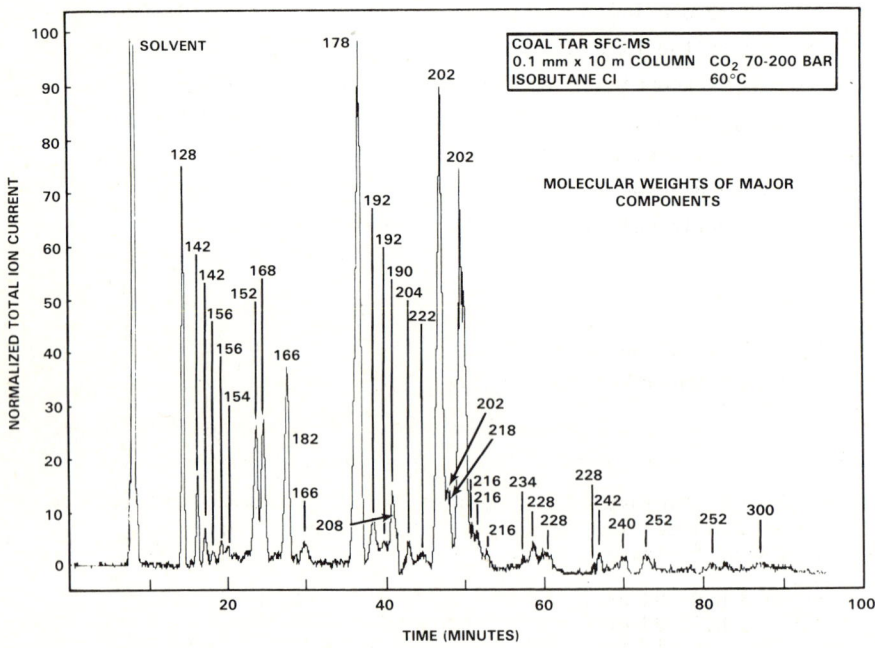

Fig. 3. SFC-MS of Coal Tar Using Pressure Programming.

efficiency relative to liquid chromatography. In SFC the mobile phase is maintained at a temperature somewhat above its critical point, typically at reduced temperatures of 1.02 to 1.1. The mild thermal conditions (determined by the SFC mobile phase) allow the application to thermally labile compounds. Additionally, since the solvating capability of the mobile phase is dependent upon pressure, one has the capability for accurate control of this parameter.

Figure 3 gives an example illustrating results typical of current open tubular SFC-MS. This figure shows a total ion chromatogram obtained for coal tar separated using a 0.1 mm ID x 10 m fused silica column coated with cross-linked SE-54. The separation used supercritical carbon dioxide at 60°C as the mobile phase and a programmed linear pressure ramp from 70 atm to 200 atm starting at 10 minutes. The mass spectrometer utilized isobutane as the chemical ionization reagent gas. Figure 3 gives the molecular weight of major components in the coal tar sample, assuming a dominant $(M+17)^+$ molecular ion for most compounds. Figure 4 suggests the actual complexity of the sample and utility of mass spectrometric detection evident from eight selected ion chromatograms obtained for this analysis.

These results illustrate the present state of the SFC-MS techniques. Since both the SFC and DFI-mass spectrometry methods are in their infancy further substantial progress may be anticipated. For example, the development of an interface for high resolution and high molecular weight mass spectrometry will allow exploitation of the full potential of the DFI method. In addition, capillary column SFC technology continues to progress. The development of improved stationary phases, improved methods for preparation of smaller bore (50 to 70 μm I.D.) columns, and materials appropriate to more polar compounds is required. These advances will ultimately determine the role and define the limitations of supercritical fluid methods in analytical chemistry.

ACKNOWLEDGEMENTS

We thank the Office of Basic Energy Sciences, U.S. Department of Energy for support of this work under contract DE-AC06-76RLO 1830.

REFERENCES

1. P.A. Peaden, J.C. Fjeldsted, M.L. Lee, M. Novotny and S.R. Springston, Anal. Chem., 54 (1982) 1090-1093.
2. R.D. Smith, W.D. Felix, J.C. Fjeldsted and M.L. Lee, Anal. Chem., 54 (1982) 1883-1885.
3. R.D. Smith and H.R. Udseth, Sep. Sci. Tech., 18 (1983) 245-252.
4. R.D. Smith and H.R. Udseth, Anal. Chem., to be published.
5. R.D. Smith, J.C. Fjeldsted and M.L. Lee, J. Chromatog., 247 (1982) 231-243.
6. R.D. Smith and H.R. Udseth, Fuel, 62 (1983) 466-468.

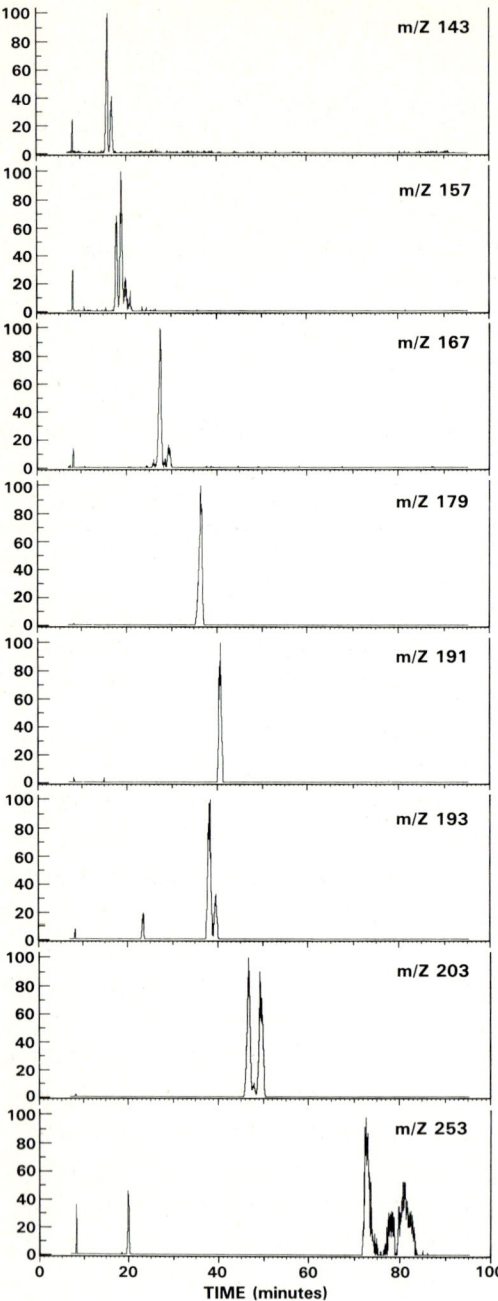

Fig. 4. Selected Ion Chromatograms Obtained During the SFC-MS of a Coal Tar (See Figure 3).

AIR-STABLE REFERENCE MATERIAL FOR MEASUREMENT OF THE OXYGEN-TO-METAL (O/M) RATIO OF NUCLEAR FUEL

C. F. Hammond, R. E. Honnell and J. E. Rein
Los Alamos National Laboratory, University of California, Los Alamos, New Mexico (USA) 87545

ABSTRACT

As a reference material for the oxygen-to-metal (O/M) ratio of nuclear oxide fuel, pellets prepared by calcining powder blends of titanium suboxide and thorium oxide in hydrogen atmosphere have desirable features of controllable O/M-ratio values, long-term stability in air, absence of plutonium, and high resistance to fracture. Under the conditions of the O/M-ratio measurement method of heating at 800 to 900°C in an atmosphere of 6% hydrogen - 94% helium containing water vapor, the titanium suboxide oxidizes stoichiometrically to titanium dioxide. Pellets have been prepared having O/M-ratio values equivalent to values of 1.94 to 1.98 for 3/1 uranium/plutonium mixed-oxide fuel, and have been used effectively for calibration and quality-control surveillance of an automated, predictive-thermogravimetric analyzer being developed by the Hanford Engineering Development Laboratory.

INTRODUCTION

The Hanford Engineering Development Laboratory is constructing a Secure Automated Fabrication facility for the production of mixed uranium-plutonium oxide fuel pellets (ref. 1). This facility will have a close-coupled chemical analysis system in which various specified components of the fuel will be determined at-line. The oxygen-to-metal (O/M) ratio is to be determined by a predictive thermogravimetric apparatus (ref. 2) having an objective elapsed time of 15 min. The analysis basis is dynamic measurement of the increasing weight of a fuel pellet having a hypostoichiometric O/M ratio as it is heated at 800 to 900°C in a 6% H_2 - 94% He atmosphere saturated with water vapor at 0°C to stoichiometric $(U,Pu)O_2$ (ref. 3,4). The total weight gain is predicted from the increasing weight gain.

The development of the apparatus required a reference material having features of (1) controllable O/M ratios equivalent to values of 1.94 to 1.98 for 3/1 uranium/plutonium mixed-oxide fuel, (2) long-term stability in air, (3) absence of plutonium, and (4) pellet form having high resistance to fracture. Pellets prepared by calcining powder blends of titanium suboxide and thorium dioxide meet all these features. This reference material, either in pellet or powder form, is considered applicable to O/M-ratio measurement methods in which hypostoichiometric oxide fuels are adjusted directly to the stoichiometric oxide by heating in a controlled atmosphere.

EXPERIMENTS AND DISCUSSION

The required properties of a chemical compound for use as the reference material are a low-oxidation form that is stable in air and a high-oxidation form that is produced under the conditions of the O/M-ratio measurement method of heating at 800 to 900°C in the water-containing 6% H_2 - 94% He atmosphere. The selected mode of producing fracture-resistant pellets of the O/M reference material was (1) preparing a powder blend of a low-oxidation metal oxide, an inert metal oxide matrix, and an organic binder, (2) pressing pellets, and (3) calcining the pellets at 1600°C in a hydrogen atmosphere.

Evaluation of Metal Oxides with Variable Oxidation States

Seven metal oxides, selected on the basis of the redox potential of the metallic element for reduction by hydrogen to a stable low-oxidation state, were investigated. The other required property of oxidation of the low-oxidation form to a high-oxidation form under the O/M-ratio measurement conditions was not considered predictable based on redox potentials.

Table 1 lists the experimental products of hydrogen reaction at 1600°C with the seven metal oxides followed by heating at 900°C in the water-containing 6% H_2 - 94% He atmosphere. Also listed are the products of subsequent heating at 1000°C in air.

TABLE I

Products of seven metal oxides heated sequentially under three conditions.

Initial Compound	Reaction with H_2 at 1600°C	Reaction with H_2O-containing 6% H_2- 94% He at 900°C	Reaction with air at 1000°C
CeO_2	Ce_2O_3	Not tested	Not tested
Cr_2O_3	None	None	None
Fe_2O_3	Fe metal	Partial oxidation	Not tested
MnO_2	MnO	None	MnO_3
Nb_2O_5	NbO_2	Partial oxidation	Nb_2O_5
TiO_2	Ti_3O_5	TiO_2	None
V_2O_5	V_2O_3	None	V_2O_5

Only titanium oxide had the properties required for the reference material of reduction to a low-oxidation state upon heating with hydrogen and oxidation to a stable high-oxidation state under the O/M-method conditions. It, therefore, was selected as the active component for the O/M-ratio reference material.

Although CeO_2 reduced to Ce_2O_3 with hydrogen at 1600°C, it is not practically useful as the component for an O/M-ratio reference material. The Ce_2O_3 powder is unstable, in fact is pyrophoric, so that O/M-ratio reference material prepared with it would not be stable in air. Iron oxide is not suitable because the hydrogen reaction product of iron metal is not stable in air nor is it oxidized to a stable high-oxidation form under the O/M-ratio method conditions. Chromium, manganese, and vanadium oxides are not useful because the hydrogen-reaction low-oxidation forms do not oxidize to stable high-oxidation forms under the O/M-ratio method conditions.

Evaluation of Metal Oxides as the Inert Component

Three metal oxides, Al_2O_3, ThO_2, and UO_2, selected on the basis that the products produced upon reaction with hydrogen at 1600°C would not oxidize under the O/M-ratio method conditions, were evaluated. Powder blends of each metal oxide, Ti_3O_5, and organic binder were prepared to have O/M ratios of 1.94 to 1.98 equivalent to mixed uranium-plutonium oxide, and pellets were pressed and calcined in hydrogen at 1600°C. These pellets were analyzed for their O/M ratios and their resistance to fracture was tested as described in a later section.

The pellets prepared using the ThO_2 as the inert component had the highest long-term O/M-ratio stability in air and had excellent fracture resistance. Also their density, and hence weight for equal-size pellets, was close to that of uranium-plutonium mixed oxide fuel to serve as an effective reference material for the calibration of apparatus and methodology used for determining the O/M ratio of this fuel.

The pellets prepared using UO_2, which had an O/U ratio slightly greater than 2 had variations in their O/M-ratio values between and within batches that were unacceptably large. We observed porosity and density differences which could have affected the rates of hydrogen reduction of the hyperstoichiometric UO_2 and the rates of oxidation of the hydrogen-reduced sintered pellets in the O/M-ratio measurement. The fracture resistance of these pellets was satisfactory.

The pellets prepared using Al_2O_3 had long-term O/M-ratio stability in air; however, they were fragile so as not to withstand handling with the type of equipment being fabricated for the Secure Automated Fabrication facility. Also their density and weight for equal size pellets was low relative to that of uranium-plutonium mixed oxide pellets.

Fabrication of Ti_3O_5 - ThO_2 pellets

The Ti_3O_5 was prepared by heating high-purity TiO_2 powder in a tungsten crucible in a dry-hydrogen atmosphere at 1625°C for 3 h. The calcined product was pulverized and passed through a U. S. Standard No. 325 Sieve (44-µm opening).

The ThO_2 used was reactor-grade quality having an agglomerate structure, 5.2 - m^2/g BET surface area, 1.57 g/cm^3 bulk density, and 7.7-μm average particle size. It was heated in air to remove volatile components, then passed through a U. S. Standard No. 325 Sieve.

Weighed quantities of the Ti_3O_5 and ThO_2 to give O/M-ratio values of 1.94, 1.96, and 1.98, equivalent to 3/1 uranium/plutonium mixed oxide fuel, were roll-blended for 63 h in a glass bottle containing aluminum breaker bars and having an aluminum-foil liner in the lid. The masses of Ti_3O_5 per gram of blend that give O/M-ratios of 1.94, 1.96, and 1.98 are 52, 35, and 18 mg.

The powder blends were pressed at 20 kpsi in a 6.09-mm steel die with double punches. A solution of stearic acid in acetone served as the die-wall lubricant.

The pellets, in a molybdenum V-block-boat assembly, were calcined in a dry-hydrogen atmosphere at 1625°C for 3 h. Their shrinkage was 15%.

Stability of Pellets and Reliability of the O/M-Ratios

The produced Ti_3O_5 was evaluated for its stability, especially relative to moisture. The ultimate test was submersion of the powder for 24 h in water heated at 80°C. No change in appearance indicative of oxidation was observed. Afterwards the powder was dried in air to constant weight and weighed portions were ignited in air to produce TiO_2. The computed composition of the powder after its submersion in water still was Ti_3O_5.

The pellets prepared with O/M ratios of 1.94 to 1.98 have been stored in air having about 50% relative humidity for over 1 yr. Their O/M ratios have not changed.

The calcining of pellets has been done in batches of 50 to 100 and the O/M ratio was measured on four pellets from each batch. Table II summarizes the obtained O/M ratio results. The method measurement uncertainty, based on analysis of homogeneous powders, is 0.0015 standard deviation in absolute O/M-ratio values. The pooled standard deviation of the O/M-ratios, computed from the

TABLE II

Reliability of O/M-ratio in fabrication pellet batches.

O/M ratios, average ± standard deviation, in pellet batches having three Ti_3O_5-ThO_2 compositions.

18 mg Ti_3O_5/g	35 mg Ti_3O_5/g	52 mg Ti_3O_5/g
1.9798 ± 0.0021	1.9588 ± 0.0030	1.9390 ± 0.0008
1.9845 ± 0.0010	1.9595 ± 0.0006	1.9395 ± 0.0013
		1.9392 ± 0.0010
		1.9408 ± 0.0005

data given in Table II, is 0.0015, indicating that pellets are homogeneous within calcined batches.

Fracture Resistance of Pellets

For application to the automated, predictive thermogravimetric apparatus, the pellets must resist fracture during operations with manipulators exerting a force of 28 g. Tests assessed resistance to compressive force and to fragmentation free-fall.

Compressive force was measured by placing a pellet on its curved surface between parallel, flat metal plates, then placing weights on top. Weight losses of pellets consistently were less than 0.02% at a maximum applied force of 1500 g. The pellets, therefore, resisted fracture at a force more than 50 times that exerted by manipulators having 28-g force and V-groove contacts.

In the free-fall test, pellets were dropped onto metal plates from a height of 25 cm, both vertically and at 45°, through glass tubes that had a diameter slightly larger than the pellets. Weight losses consistently were less than 0.02% for both the vertical fall where the pellets landed on their flat ends and for the 45° fall where they essentially landed with point contact.

SUMMARY

A reference material for the measurement of the hypostoichiometric O/M ratio of nuclear fuel consists of pellets prepared from powder blends of Ti_3O_5 and ThO_2. The powder also can be used. Features include controllable O/M ratios, greater than 1-yr stability in air, homogeneity, and high fracture resistance for the pellets.

REFERENCES

1 W. W. Bowen, D. L. Sherrell and M. J. Wieners, Trans. Am. Nucl. Soc. 39 (1981), 982.
2 G. C. Swanson, in W. S. Lyon (Ed.), Analytical Chemistry in Nuclear Fuel Reprocessing, Science Press, Princeton, 1978, pp. 361-368.
3 C. E. McNeilly and T. D. Chikalla, National Ceramic Society Meeting, Washington, D. C., May 3-8, 1969.
4 1982 Annual Book of ASTM Standards, Part 45, Nuclear Standards, Designation C-698-80, Sections 150-158, ASTM, Philadelphia, 1982.

ACKNOWLEDGMENT

This work was supported by Division of Reactor Research and Technology, U. S. Department of Energy.

INDEX

absorption cell
  long path 81-86, 359
acetylene
  monitoring 359-360
accelerator
  (see intense pulsed neutron source)
actinides
  migration in basalt 255-262
  (see also specific isotopes)
aerosols
  remote sensing 325-326
  in UF_6 hydrolysis 321-326
Al
  in stainless steel 153-154
alloys
  studied by HIXSE 253
alpha containment
  facility 228-229
aluminum
  energy levels 181
^{241}Am
  by IDMS 111-118
  ^{241}Pu interference in 114-115, 144-148
^{243}Am
  Preparation 112-114
  (see also ^{241}Am)
Anderson, G. K. 333
Apel, C.T. 213
Argonne National Lab
  neutron activation at 243-248
atomic absorption (AA)
  flame 349-354, 204-211
  graphite furnace 349-354
  of impurities in Na 289-291
  of radioactive samples 219-225
Attalla, R. 363
automated analyzer
  SRL description 336-342

B
  determination by ICP 187-193
basalt repository
  analog 255-256
Bear, B. R. 187
Bostick, W. D. 321
Bowers, D. L. 255
burnup
  methods 273-278
Bushaw, B. A. 57

CANDU Reactor
  (see thorium fuel)
capillary supercritical fluid
  chromatography 373-374, 375-380
Cd
  determination by ICP 187-193
^{140}Ce
  in burnup determination 273-278
^{142}Ce
  in burnup determination 273-278
^{252}Cf
  for coal analysis 299-307
  in neutron monitor 264-266
Chastagner, P. 119
Christie, W. H. 167
coal
  DRIFT applications 358
  on line analysis of 299-307
cobalt
  energy levels 181
Coleman, C. J. 195
conformations
  by PAC 76-78
controlled potential coulometric
  determination of U 345-348
    interferences 349
Crawford, D. W. 111
Creech, E. T. 343

diffuse reflectance
  with Fourier transform   357-358
    applied to coal   358
    applied to U   357
diode laser spectroscopy   81-86
direct fluid injection
  in mass spectrometry   375-380
DNA
  analysis by FLN   34-36
Donohue, D. L.   143
Doppler widths
  in gas measurements   84
dosimeters
  (see neutron)
DRIFT
  (see diffuse reflectance)

Edelson, M. C.   227
excited states
  in sputtered atoms   180-184
Fang, H. L.   75
Fassett, J. D.   137
Fast Flux Test Facility (FFTF)
  monitoring leaks in   127-131
fiber optics
  in remote sensing   13-18
  for temperature, pressure   19-23
    lasers in   19-23
  in U determination   25-30
fission product
  vapor release   327-332
flow cytometry   53-55
fluorescence
  laser induced   7-12
    matrix isolation   31-32, 43-47
    for single molecule determination
      53-55
  by Raman   49-52
  of uranium   26,29
fluorescence line narrowing (FLN)
      spectroscopy   31-36, 45-46

  for DNA analysis   34-36
  for PAH determination   32-34
fluorimetry
  laser   13-18
Fourier
  self-deconvolution   371-372
Fourier transform infrared (FTIR)
      spectrometer
  with chromatography   373-374
  for solids   357-358

gadolinium
  poison measurement   267-268
gallium
  isotope ratios   151-152
gamma ray spectrometry
  at ANL   247
  automated   309-311
  in coal analysis   232-307
  with Ge well   313-318
  of ^{129}I 283
  in Na coolant   287
  of ^{237}Np 259-260
  of TMI screws   294-298
Gantner, E.   263
gas analysis
  (see tritium)
  (see Kr, Xe)
gas dynamic focusing   158
Ge
  instrinsic well   313-318
glove boxes
  for ICP work   214-217
glow discharge
  with dye laser   174-177
grain boundaries
  analysis by SIMS   169-172
Green, L. W.   273
groundwater
  actinides in   255-262

Halouma, A. A. 201
Hammond, C. F. 381
Hanford Nuclear Reactor
  for ^{129}I determination 280-284
Harris, T. D. 49
Harrison, W. F. 173
heavy-ion induced x-ray
        satellite emission (HIXSE)
  chemical sensitivity 250-254
Heinrich, R. R. 243
High Flux Isotope Reactor (HFIR)
  for ^{129}I determination 279-284
Hiller, J. M. 219
Hirschfeld, T. 13, 19
Hoffstetter, K. J. 293
hollow cathode discharge
  aqueous inorganics 99-101
hollow cathode source
  for MPRI 180
Howell, R. L. 371
HPLC
  for burnup analysis 274-278
  with FTIR 373-374
hydrogen
  determination by NMR 363-370
  isotopic analysis 355-356

^{125}I
  in deer 313-318
^{129}I
  comparison of methods 284
  determination by NAA 279-284
^{131}I
  in deer 313-318
indium
  energy levels 181
  in Si 154
ICP
  with quadrupole MS 95-99
    for aqueous inorganics 96-99

ICP spectroscopy
  determination of B, Cd by
        187-193
  for nuclear fuel 195-199,
        235-239
  for nuclear waste 195-199,
        203-211
  for Pd 203-211
  for radioactive samples 213-217
        223-225, 227-233
    glove boxes for 214-217
    hoods for 223-225, 238
  for Rh 203-211
  spectra 215
  for U-Pu solutions 271-272
infrared spectra
  (see Fourier)
infrared laser spectroscopy 81-86
intense pulsed neutron source (IPNS)
  description 244-246
iridium
  (see grain boundaries)
iron
  by RIMS 137-142
isokinetic sampling
  of particulate UO_2F_2 322-324
isotope dilution mass spec
  U determination 161-165
isotope ratios
  (see specific elements)

Karlsruhe
  research at 263-271
Kaye, J. H. 279
Kim, K. C. 81
Kimock, F. M. 179
kinetic phosphorimetry
  (see time resolved
        phosphorescence)
Kr
  tag for FFTF tests 127-131

^{139}La
    in burnup determination   273-278
laser
    Doppler velocimeter   335
    enhanced ionization   63-68
        for Ni   63-68
laser enhanced ionization   63-68
    for Ni   63-68
laser mass spectrometry
    for C, N, O   94-95
    history   90-91
    for Ta,Th, Mo   92-93
lasers
    limits of detection   4
    line narrowing   4
    ruby, temperature effects   21-23
    (see also fluorescence)
    (see also fiber optics)
    (see also laser spectroscopy)
laser spectroscopy
    for acetylene   359-360
    for aerosols   325-326
    for I_2, Cs I   328-332
    Raman   51-52, 355-356
laser types
    argon   51-52, 54-55, 327-332,
        355-356
    CW   155
    diode   81, 359
    dye   32-36, 38-42, 44-47, 75-80,
        144-148, 173-177, 333, 335
    nitrogen   4, 7-12, 58-62
    rare gas   4, 54
    ruby   20-23
    YAG   64-68, 91-101, 138-142, 150
    (see also resonance ionization mass
        spectrometry)
LeVert, F. E.   299
Lucas, D.   361
Lytle, F.   3

McCarthy, J. P.   349
McCulla, W. H.   327
McGlade, M. J.   31
Maddox, W. L.   359
Malstrom, R. A.   25
mass spectrometer type
    double processing   89
    quadrupole   90, 97-99, 119-126,
        174-177
    thermal ionization   103-110,
        111-118, 274
    (see also secondary ion mass
        spectrometry)
    (see also resonance ionization
        mass spectrometry)
Matsumoto, W. Y.   309
matrix isolation spectroscopy
        31-32, 37-42, 43-47
    reasons for   43-44
May, M. P.   161
metal hydrides
    hydrogen in   366-370
metal oxides
    oxidation for O/M   383
    reaction with H_2 382
Michel, C. E.   235
molybdenum compounds
    differences in HIXSE   252-253
multicollector   133-136
multiphoton ionization/mass
        spectrometry   5
multiphoton resonance ionization
        (MPRI)
    of Al, In, Co   180-184
    description of   179-180

^{145,146}Nd
    in burnup determination   273-274
^{148}Nd
    in burnup determination   273-278

neutron
  dosimeters analysis of 309-311
  flux at ANL 246
  monitoring 263
  spectrum at ANL 245-246

neutron activation analysis
  at ANL 245-248
    SRM's 246-248
  instrumental (INA) 274
  prompt γ 230-307

nitrogen
  liquefaction plant 359-360

Nogar, N. S. 155

^{237}Np
  determination in ground water 257-259
  determination in rock core 259-262

nuclear fuel
  O/M ratio 381-385

nuclear magnetic resonance (NMR)
  hydrogen by 366-370
  theory 364-365

nuclear reactors
  (see specific reactors)

nuclear waste
  Rh in 203-211
  Pd in 203-211

oil
  determination of Ni in 64-68

Olesik, S. V. 373

on line
  analysis of coal 299-307
  analyses at SRL 335-342
    types of 336
  U monitoring 343-344

optrodes 16

organics
  by mass spec 375-380

overtones
  by PAC 77-80

oxygen-to-metal (O/M) ratio
  in fuel 381-385

palladium
  (see nuclear waste)

Parks, J. E. 149

phosphorous
  determination in steel 350-354

photoacoustic spectroscopy 37-42, 75-80
  of butadiene 79-80
  of gas phase ethanol 76-77
  of PAC's 37-42
  of propane 78

plasma jet
  U in 334-335

platinum
  determination 268-269

Plucinski, C. E. 127

plutonium
  analysis for Am, Pu 111-118
  in ICP 214, 231-233, 235-239
  isotope ratio 146-147
  metal analysis companion 118
  monitoring by neutrons 266-272
  (see also specific isotopes)

polycyclic aromatic compounds (PAC)
  determination by photoacoustic spectroscopy 37-42
  (see also polycyclic aromatic hydrocarbons)

polycyclic aromatic hydrocarbons (PAH)
  determination 32-34, 43-47
  (see also polycyclic aromatic compounds)

Polson, C. A. 103

protons
　for neutron production　244
Pt electrode
　regenerating　348
　in U determination　346-348
^{238}Pu
　as hot source　167
　　analysis of cladding alloy
　　　168-172
　　impurities in 169-172
^{239}Pu
　determination in ground water
　　258-259
　determination in rock core
　　259-262
^{242}Pu
　emission spectrum　232

quenching
　kinetic analysis correction
　　59-61
　reduction of　51-52
　of U　8-9, 26-27, 57-58

radioactive solutions
　remote analysis　335-342
radionuclides
　in Na　287-289
Raman
　molecular hydrogen by　355-356
　scattering from metals　49-52
　spectra I_2, CsI　330-332
remote sensing　13-18, 25-30
resin bead
　in RIMS　144-148
　in U determination　161-165
RIMS
　of Fe　137-142
　intracavity　156-160
　of K　175-177

　at Los Alamos　155-160
　method discussion　144-145
　of Pu　143-148
　sputter induced　149-154
　　of Ga　151-152
　　of indium　153-154
　　of stainless steels　153-154
　of U　143-148
rhenium
　determination　270-271
rhodium
　(see nuclear waste)
Rockwell Hanford
　containment facilities　219-225
Rocky Flats
　analysis of Pu at　235-239
Rosseel, T. M.　249

Saponara, N. M.　345
Savannah River Lab
　on line analyzer　335-342
　waste analysis at　198-199
scrubber systems
　at Rockwell Hanford　219-225
secondary ion mass spectrometry
　　(SIMS)　167-172
Shaw, R. W.　37
Shpol'skii
　spectroscopy　31-32
　solutions for anal.　41-42, 45
silicon
　determination in steel　350-354
Smith, R. D.　375
sodium cooled reactor
　monitoring for impurities
　　285-292
　sampling　285-287
solvation　69-74
solvent extraction
　of Np　258

of Pu   235-238
  of U   8, 196-198
Smyrl, N. R.   357
SO_2
  determination   361-362
Spencer, W. A.   335
spectrophotometric
  U on line determination   343-344
sputtering
  in MPRI   181-183
Stamm, H. H.   285
steel
  definition of Si, P in   249-354
Stegnar, P.   313
supercritical fluid methods
  capillary   373-374, 375-380
    in mass spectrometry   375-380
supersonic expansion
  laser application   5
surface studies
  of TMI screws   294-298
  using DRIFT   357-358
Svec, H. V.   89

^{99}Tc
  electrochemical determination
    269-270
  thermal ionization MS
    for $^{241}Am/^{243}Am$ 111-118
    for $^{241}Am/^{239}Pu$ 115-116
    description MAT-261   103-106
  Three Mile Island (TMI)
    lead screws from   293
      analysis   294-298
      soaking tests   294
ThO_2
  for O/M reference   383-385
thorium fuel
  burnup methods   273-278

thyroid
  I in deer   313-318
time resolved fluorimetry   4-5
time resolved phosphorescence
    57-62
titanium
  hydrides   365-367
  satellite spectrum   250
TOPO
  for U separation   187-189
tritium
  impurity monitoring   119-126
  by NMR   367
Trkula, M.   53
tunable atomic line molecular
      spectroscopy (TALMS)
  for SO_2 361-362
Turk, G. C.   63
Turner, P. J.   133
two photon spectroscopy
  theory and solvation   69-73

^{234}U
  by multicollection   134-136
  ratio to ^{236}U 109-110
^{235}U
  emission spectrum   232
  by multicollector   134-136
  ratio to ^{258}U 107-110, 146
  in ^{236}U 85
UF_6
  hydrolysis of   321
UV absorption spectroscopy
  for SO_2 361-362
uranium
  coulometry of   345-348
  determination in H_2O 61
  determination in Pu   7-12
  determination in urine   62,
    161-165

dibutylphosphate structure
      343-344
  DRIFT application 357
  fluorescence 29
  by ICP 195-199
  on line determination 25-30,
      343-344
  by RIMS 157
  separation in exchange 161-162
  spectra from ICP 232
  by thermal ionization 104-110,
      133-136
  trace impurities in 187-193,
      196
  vapor 333-335
uranyl
  phosphate determination 57-62
UO_2F_2
  sampling and measurement
      322-326

Van de Graaff
  tandem for HIXSE 250
vapor
  CsI, I, analysis of 327-332
Veirs, K. 355

Wehry, E. L. 43-47
West, M. H. 203
Wirth, M. J. 69
W. S. U. Reactor
  for ^{129}I determination 280

Xe
  tag for FFTF tests 127-131
x-ray fluorescence
  heavy ion excited 249-254

Young, J. E. 7